Qualitative Research in Technical Communication

T0199672

This collection will inform academics and practitioners about the breadth and strength of qualitative inquiry. Qualitative techniques such as interviews and participant observation have become a natural part of the work of technical communicators, and this work will help practitioners bring more rigor to their use of these techniques.

This book will be of interest to students and academics seeking up-to-date information on current industry practices in technical communication, as well as practitioners in the field of technical and professional communication. It will be suitable for use as a text for undergraduate seminars and courses at the master's level.

James Conklin is assistant professor of applied human sciences at Concordia University, where he teaches group dynamics and human systems intervention in the graduate and undergraduate programs. He is also an associate scientist at the Élisabeth-Bruyère Research Institute in Ottawa. For five years he has led the evaluation team that is using a mixed-methods design to investigate processes of knowledge exchange within the Seniors Health Research Transfer Network. He has more than 25 years experience as a technical communication and organization development professional.

George F. Hayhoe is professor of Technical Communication and director of the MS program in technical communication management at Mercer University's School of Engineering. A fellow of the Society for Technical Communication, he was editor of its journal, *Technical Communication*, from 1996 to 2008. He has co-authored two books, and published book chapters, articles, and conference papers on a variety of topics in technical and professional communication.

Qualitative Research in Technical Communication

Edited by

James Conklin
Concordia University

and

George F. Hayhoe
Mercer University School of Engineering

Routledge
Taylor & Francis Group

NEW YORK AND LONDON

First published 2011
by Routledge
711 Third Avenue, New York, NY 10017, USA

Simultaneously published in the UK
by Routledge
2 Park Square, Milton Park, Abingdon, Oxon OX14 4RN

Routledge is an imprint of the Taylor & Francis Group, an informa business

Typeset in Goudy
by Keystroke, Tettenhall, Wolverhampton

Library of Congress Cataloging-in-Publication Data
Qualitative research in technical communication / edited by James
Conklin and George F. Hayhoe.
p. cm.
1. Communication of technical information. I. Conklin, James. II. Hayhoe,
George F.
T10.5.Q54 2010
601'.4—dc22
2010010839

ISBN 13: 978–0–415–87635–3 (hbk)
ISBN 13: 978–0–415–87636–0 (pbk)
ISBN 13: 978–0–203–84663–6 (ebk)

Contents

Preface

James Conklin and George F. Hayhoe

Qualitative research in the field of technical communication has experienced explosive growth in recent years. This fact is evident in the results of a recent bibliographical exercise (see Chapter 15 in this book) that revealed that between 2003 and 2007, 225 articles in five journals reported the results of qualitative research studies in the field of technical communication. These publications represent nearly 39 percent of the total number of articles published in these journals over that five-year period.

This exceptional number of qualitative research studies in technical communication builds upon a foundation set in place in the 1980s, with the publication of the watershed collection, *Writing in Nonacademic Settings*. In their introduction to the collection, Odell and Goswami (1985) acknowledged that in the mid-1980s we knew relatively little about the writing that took place in workplaces. Several articles attempted to address this gap in our knowledge by looking at writing and reading as interactive social processes that occurred within a web of workplace relationships. In that volume, for example, Faigley (1985) urged us to take note of how communication in the workplace serves to constitute and sustain social groups. In another article in this early collection, Doheny-Farina and Odell speculate that "writing does more than reflect the social context in which it exists; it may be that writing helps shape that context" (1985, 503). Faigley suggested prophetically that "Because qualitative research offers the potential for describing the complex social situation that any act of writing involves, empirical researchers are likely to use qualitative approaches with increasing frequency" (1985, 243).

These early qualitative studies exploring the nature and impact of workplace writing seem especially prescient in the context of the recent work on adaptive organizations, and the way that learning, knowledge, and interaction bring moments of stability and instability to organizational life (Brown and Duguid 2000; Taylor and van Every 2000; Weick 2009). This work, much of which focuses on organizational behavior and change, suggests that workplace writing takes place within an endless dynamic of "texting and talking" (to borrow the evocative language of Taylor and van Every), which helps to stabilize organizational life during periods of rapid change and dissonance, and which also contributes to a ready-to-hand repertoire of ideas, solutions, and processes that can be accessed and implemented when needed.

Eight years after Odell and Goswami published their collection, Rachel Spilka (1993) produced a new anthology of 18 research articles. In her preface she looked back to the Odell and Goswami anthology, and suggested that the earlier volume's focus on social perspective could enliven our interest in research on workplace writing. The result was a sudden surge of descriptive, qualitative studies of writing in real-world contexts that revealed the web of social relations and interactions that simultaneously gave rise to writing endeavors and made sense of these sometimes turbulent social environments. Spilka explained that her volume was intended to take stock of accomplishments so far and to point the way for research in the coming years.

It is interesting to recall that just as academic researchers were becoming increasingly interested in bringing qualitative research methods to the puzzles and dilemmas of technical communication practice, so too were practitioners demanding new quantitative research efforts to demonstrate the value that technical communicators bring to their organizations. One result, of course, was the memorable February 1995 issue of *Technical Communication*, where two researchers, Dr Ginny Redish and Dr Judith Ramey, presented the results of their multifaceted investigation into the value of technical communication. Even here, however, our discipline could not function without the knowledge generated through qualitative methods. Redish wrote that "Numbers don't tell the whole story; process is also critical. They [technical communicators] may have to go beyond traditional corporate accounting systems to get credit for the value they add" (1995, 26). Ramey (1995) emphasized the importance of case studies and field studies (both of which are generally associated with qualitative methods) for gathering and analyzing data concerning the value of technical communication work.

Perhaps some technical communicators hoped that the study by Redish and Ramey would produce a definitive, quantitative answer to the persistent doubts and questions concerning the value of technical communication work. Judging by the extent to which questions of value continue to arise in our profession (see, for example: Carliner 1997; Hayhoe 2006; Henry 1998; Hughes 2002; Lanier 2009; Mead 1998), this hope has not been realized. However, recent research may be providing us with some clues about the sources of technical communication value. Hart and Conklin, republished in this collection as Chapter 5, found that the technical communicators who participated in their focus group study "showed a clear inclination to focus more on processes than products; we were told, for example, that people work mostly on teams, that they provide services as well as products, and that their value has more to do with the activity of their work than with the products they produce" (p. 130).

Thus, 25 years after Odell and Goswami brought us their ethnographies of workplace writing, 17 years after Spilka took stock of the ensuing research accomplishments, and 15 years after Redish and Ramey attempted to quantify the value of our manifold contributions to the quality and productivity of fast-changing workplaces, we offer the present volume—the first in over 15 years to bring together a representative sample of qualitative reports from a growing body of work, and to comment on the reasons for the extraordinary interest among technical communication practitioners and academics in qualitative research.

Why Qualitative Research?

Qualitative researchers see the research endeavor as a pluralist and exploratory enterprise that brings to light the results of a meaning-making process involving both researcher(s) and research participants. It does not generally seek unilateral answers to questions phrased as narrow hypotheses, but rather offers interpretations and perspectives about complex social phenomena that are grounded in rich narratives. Qualitative research confronts us with the messy and chaotic realities of specific social situations, and recognizes that different groups of people (writers, trainers, engineers, programmers, sales professionals, managers, customers, users, and so forth) experience the same social phenomena and technologies in different ways. Qualitative researchers find ways of depicting those realities that illustrate the underlying patterns and trends that provide us with a sense of coherence, identity, and purpose.

We believe that this collection is itself consistent with the qualitative enterprise. While five of the 15 articles contained in chapters 1 through 15 were written and published before 2008, seven articles originally appeared in the November 2008 special issue of *Technical Communication*, and the remaining three articles appear here for the first time. Authors of the more recent 10 articles were given broad guidelines, and were asked to bring forward reports of significant, structured qualitative research into various aspects of technical communication practice. We wanted to know what new insights researchers are generating about the working reality of today's technical communicators. We asked contributors to tell us how technical communicators are perceived and treated by managers and by colleagues from other disciplines. We sought examples of how technical communicators use qualitative methodologies—including ethnography, case study, focus groups, action research, grounded theory, and interview research—to strengthen their practice. The result is a rich harmony of perspectives, as diverse as the field of technical communication (or, indeed, qualitative research) itself.

Over the past 20 years, qualitative research has advanced from being seen as a way of revealing interesting "anecdotes" to being accepted as a significant method for generating new knowledge about human experience. Qualitative research designs are now found in all branches of social science inquiry, including health services research (once the exclusive territory of experimental designs using random samples). We include in our collection (in chapters 1 and 2) two early examples of qualitative research in technical communication—articles that helped to trigger the surge in the use of qualitative methods in our profession.

We assert that qualitative research is of fundamental importance to technical communication practitioners and academics, and its presuppositions and techniques have long been integrated into our practices. To gain a thorough understanding of an audience's needs and constraints, practitioners seek to understand the working realities of the users of technology. Thus, technical communicators have for years been absorbing qualitative perspectives and methods into their practice. Qualitative inquiry is derived from the idea that people construct their social reality through modes of talk and interaction (Lincoln and Guba 1985). By revealing how specific groups of people construct their social worlds, qualitative research can enrich our understanding of the user experience and of our own place in a communication dynamic encompassing scientists, engineers, business experts, and generalist audiences. Moreover, because technical communication is changing, exploration (a central purpose of qualitative inquiry) is needed to reveal patterns.

We hope that the essays in this collection will inform practitioners and academics alike about the breadth and strength of qualitative inquiry. We think that although qualitative techniques such as interviews and participant observation have become a natural part of the work of technical communicators, many practitioners may not have had the opportunity—or seen the need—to find ways of bringing more rigor to these techniques. In our call for contributions to this anthology, we challenged the authors to be clear about their data-gathering methods and analytic procedures, in the hope that their reports could enrich the work of practitioners. We believe that the resulting volume is responsive to that request.

The Audience for this Book

We designed this book for three audiences. First, we wanted to provide communication practitioners with useful information about how qualitative research can strengthen their practice. There are numerous excellent texts about how to conduct various types of qualitative inquiries (and most of the articles in the following chapters include lists of references that can be used to compile a bibliography on qualitative methods), so we elected to omit information on the procedures for designing and conducting a qualitative inquiry. Instead, we have focused on assembling a collection of articles that report the results of specific qualitative studies. The articles all include descriptions of the research methods that the authors used, so readers will be able to see for themselves how specific qualitative studies were able to provide compelling answers to complex research questions. This book addresses research methods that can be easily applied in the workplace to carry out essential practitioner tasks (such as audience analysis). Informed professionals will want to increase their effectiveness by learning about these methods.

We also believe that the book will be of interest to students of technical and professional communication. The book includes two articles (chapters 1 and 2) that raise important methodological issues, and 13 articles that report the results of specific studies. Students will be able to use the book to gain an appreciation for the range of qualitative research methods and the ways that they can be applied.

A third audience includes academics and practitioners who are interested in the way that technical communication is practiced today, and how that practice might change in the coming years. Technical communicators have shown a strong interest in taking stock of trends and changes in their own profession (see, for example: Conklin 2007; Giammona 2004; Hart and Conklin 2006; Lippincott 2003; Rainey, Turner, and Dayton 2005; Staples 1999). Qualitative methods are well-suited to explore social change, and many of the articles in this volume shed light on professional trends. In some ways, the findings reported in many of these articles might be viewed as a snapshot of the ways in which technical communicators are experiencing and making sense of their professional lives.

The Design of the Volume

The volume is divided into three parts. The first part consists of chapters 1 and 2, which are conceptual essays published in the 1990s that cover some of the more significant issues facing qualitative research. In Chapter 1, Sullivan and Spilka establish the importance of qualitative research for the profession of technical communication, and provide advice on how to design a rigorous qualitative inquiry. In Chapter 2, Blakeslee, Cole, and Conefrey focus on the issue of validity in qualitative research. They argue that the validity of research should be measured in terms of its practical value, and, like proponents of participative action research, suggest that research participants should be given a role in all phases of a research project. Each of these chapters is preceded by an introduction that presents the context and foregrounds the important themes of the article.

The second part of the volume, chapters 3 through 14, presents the results of specific qualitative research studies, encompassing a variety of qualitative methods. Again, each chapter is preceded by a brief introduction that explains the unique contribution that the article makes to the collection. These introductions also make special use of Giammona's article in Chapter 3, which deals with emerging trends and issues in technical communication. Giammona describes nine themes that emerged from her study:

- What is a technical communicator today?
- What forces are affecting the field?
- What is our future role in organizations?
- What should managers of technical communicators be concerned with?
- How do we contribute to innovation?
- How should we be educating future practitioners?
- What technologies are impacting us?
- Where do we go from here?
- Does technical communication matter?

Our introductions to each of these chapters indicate how each article sheds light on these themes. Table 0.1 provides a quick roadmap of how the articles address Giammona's themes.

Table 0.1 How the Articles Address the Themes

	What is a technical communicator today?	What forces are affecting the field?	What is our future role in organizations?	What should managers of technical communicators be concerned with?	How do we contribute to innovation?	How should we be educating future practitioners?	What technologies are impacting us?	Where do we go from here?	Does technical communication matter?
Chapter 4: Carliner	X		X		X		X	X	
Chapter 5: Hart and Conklin	X		X		X		X	X	
Chapter 6: Portewig	X				X				
Chapter 7: Breuch	X		X	X	X		X		
Chapter 8: Mogull			X		X	X		X	
Chapter 9: Willerton		X			X	X	X	X	X
Chapter 10: Thacker and Dayton					X		X		X
Chapter 11: Donker-Kuijer, de Jong, and Lentz						X			
Chapter 12: Vosecky, Seigel, and Wallace	X		X		X	X			
Chapter 13: Driskill and Watts									
Chapter 14: Hughes and Reeves	X		X		X				X

The third and final part of the volume consists of two concluding chapters that provide a synthesis and look ahead. Chapter 15 is a bibliographical essay by Davy and Valecillos that illustrates the importance of qualitative research for technical communication in recent years, and identifies the more common methods that researchers have employed and questions they have investigated. In the Conclusion, we provide our concluding thoughts on the subject and speculate on the future.

We conclude this introduction with a brief, final word on the importance of qualitative research. For the past five years, one of us has spent much of his time working as the evaluator of a health research transfer network. The purpose of the network is to connect health care practitioners, researchers, and policy makers in ways that ensure that scientists are creating relevant and useful knowledge that is needed by frontline caregivers, and that this useful knowledge is available and accessible to those who need it. Although the network's leaders have been concerned with demonstrating numerically the new value they bring to the health system, the network funders have made it clear that they have different expectations. "We want stories," they said emphatically at a 2008 meeting. "We want to be able to tell stories about the way our investments are improving the lives of citizens." In other words, they want qualitative narratives that allow people to see the value of the network in relation to human lives.

Qualitative research offers technical communicators vital and resilient tools with which to understand the needs of audiences and customers. Equally important, it allows technical communicators to explore and understand their own changing practice. We have learned that technical communication involves much more than the creation of well-written and carefully organized texts. Technical communication practice is a highly interactive process that requires the creation of relationships, conversations, and processes, as much as it requires the creation of texts and graphics. It is a social practice, and like other social phenomena it opens itself up to qualitative research.

In this book, we invite you to join us in taking stock of the current state of qualitative research in technical communication. We believe that you will be impressed—and perhaps surprised—by the rigor and diversity represented by the research presented here.

References

Brown, J. S., and P. Duguid. 2000. *The Social Life of Information.* Boston, MA: Harvard Business School Press.

Carliner, S. 1997. "Demonstrating Effectiveness and Value: A Process for Evaluating Technical Communication Products and Services." *Technical Communication* 44: 252–65.

Conklin, J. 2007. "From the Structure of Text to the Dynamic of Teams: The Changing Nature of Technical Communication Practice." *Technical Communication* 54: 210–31.

Doheny-Farina, S., and L. Odell. 1985. "Ethnographic Research on Writing: Assumptions and Methodology." In *Writing in Nonacademic Settings*, ed. L. Odell and D. Goswami. New York, NY: The Guilford Press, 503–35.

Faigley, L. 1985. "Nonacademic Writing: The Social Perspective." In *Writing in Nonacademic Settings*, ed. L. Odell and D. Goswami. New York, NY: The Guilford Press, 231–48.

Giammona, B. 2004. "The Future of Technical Communication: How Innovation, Technology, Information Management, and Other Forces are Shaping the Future of the Profession." *Technical Communication* 51: 349–66. (Reprinted here as Chapter 3.)

Hart, H., and J. Conklin. 2006. "Toward a Meaningful Model of Technical Communication." *Technical Communication* 53: 395–415. (Reprinted here as Chapter 5.)

Hayhoe, G. F. 2006. "Who We Are, Where We Are, What We Do: The Relevance of Research" (Editorial). *Technical Communication* 53: 393–94.

Henry, J. 1998. "Documenting Contributory Expertise: The Value Added by Technical Communicators in Collaborative Writing Situations." *Technical Communication* 45: 207–20.

Hughes, M. 2002. "Moving from Information Transfer to Knowledge Creation: A New Value Proposition for Technical Communicators." *Technical Communication* 49: 275–85.

Lanier, C. R. 2009. "Analysis of the Skills Called For by Technical Communication Employers in Recruitment Postings." *Technical Communication* 56: 51–61.

Lincoln, Y. S., and E. G. Guba. 1985. *Naturalistic Inquiry*. Beverley Hills, CA: Sage Publications.

Lippincott, G. 2003. "Moving Technical Communication into the Post-Industrial Age: Advice from 1910." *Technical Communication Quarterly* 12: 321–42.

Mead, J. 1998. "Measuring the Value Added by Technical Documentation: A Review of Research and Practice." *Technical Communication* 45: 353–79.

Odell, L., and D. Goswami, eds. 1985. *Writing in Nonacademic Settings*. New York, NY: The Guilford Press.

Rainey, K. T., R. K. Turner, and D. Dayton. 2005. "Do Curricula Correspond to Managerial Expectations? Core Competencies for Technical Communicators." *Technical Communication* 52: 323–52.

Ramey, J. 1995. "What Technical Communicators Think About Measuring Value Added: Report on a Questionnaire." *Technical Communication* 42: 40–51.

Redish, J. 1995. "Adding Value as a Professional Technical Communicator." *Technical Communication* 42: 26–39.

Spilka, R. 1993. Preface. In *Writing in the Workplace: New Research Perspectives*, ed. Rachel Spilka. Carbondale, IL: Southern Illinois University Press, vii–xl.

Staples, K. 1999. "Technical Communication from 1950–1998: Where Are We Now?" *Technical Communication Quarterly* 8: 153–65.

Taylor, J. R., and E. J. van Every. 2000. *The Emergent Organization: Communication as its Site and Surface*. Mahwah, NJ: Lawrence Erlbaum Associates.

Weick, K. 2009. *Making Sense of the Organization: The Impermanent Organization (Volume Two)*. Chichester, UK: John Wiley and Sons, Ltd.

Qualitative Research in Technical Communication
Issues of Value, Identity, and Use*

Patricia Sullivan and Rachel Spilka

Editors' Introduction

Although it was first published nearly 20 years ago, Patricia Sullivan and Rachel Spilka's "Qualitative Research in Technical Communication: Issues of Value, Identity, and Use" provides few clues of its age. Indeed, other than the opening anecdote about the NeXT Computer (a high-powered multimedia workstation developed by Steve Jobs during his exile from Apple from 1985 to 1997) and the publication years of its references, there is little here to betray its date of composition.

We have included this theoretical article in our collection for several reasons. It provides a solid introduction to the strengths and limitations of qualitative research, and it summarizes several frequently used qualitative methods. And in its exploration of qualitative research theory, it includes plenty of examples to help the unfamiliar reader understand the purposes and value of qualitative research. Finally, the authors believe that qualitative methods are highly appropriate for workplace research and accessible to practitioners. Moreover, the article's abundant examples will help technical communication practitioners see how a similar study might be applicable to their work environment. For all of these reasons, and particularly because our audience consists primarily of students, most of whom will soon become practitioners, as well as those already practicing in the workplace, we think that this article is particularly appropriate to this anthology.

Even today (although perhaps not as often as when Sullivan and Spilka wrote this article) the value and legitimacy of qualitative methods of research in technical communication are still questioned, especially by those whose backgrounds or experience involve training in engineering, natural science, or medicine, or who work with subject matter experts in those fields. Some are suspicious of the validity or reliability of qualitative studies, or of the fact that the results of qualitative research are not generalizable. Frequently, that skeptical attitude results from lack of understanding of qualitative methodology, so Part 1 of this article explores the usefulness of qualitative techniques by examining how they are used in educational and human–computer interaction research, and then turns to a consideration of goals for qualitative research in our field.

Spilka and Sullivan take most of their examples from ethnography or field studies. In technical, scientific, or professional communication, research involves extended observations of such phenomena as composing in the workplace, the interaction among technical communicators and subject matter specialists, or (from the perspective of the users of information products) the usefulness of documentation. Other approaches involve the analysis of communication artifacts.

Part 1 of the article concludes with a consideration of four typical goals of qualitative studies in our field: interpreting a phenomenon or event, exploring a situation to develop a broader understanding of it, developing a research perspective to explore innovative practices, and communicating with users about a phenomenon when they lack the vocabulary to respond to a survey or other quantitative instrument.

In Part 2, Sullivan and Spilka explore the usefulness of qualitative methods by analyzing their strengths and weaknesses. In discussing the strengths, the authors turn things around by raising five potential problems and considering possible solutions for each. So, for example, they suggest that research design and data collection techniques could be a cause for concern—collecting data for too short a time or from too limited a sample population, or excluding on-site observation when social interaction is a key part of the research question. They then suggest ways that the concerns can be transformed into strengths—through careful exploration of the research situation to identify all the key groups that need to be sampled, or through triangulation in data analysis. (By the way, we find the definition and examples of triangulation here to be among the best we've seen.)

As befits an article on qualitative research, Part 2 concludes with two extended examples of strong qualitative research. These studies, well known to technical communication researchers in the 1990s though perhaps lesser known today, are excellent models to illustrate the points made earlier in Part II.

In the third major section, Sullivan and Spilka discuss when and how to use qualitative research. Although they don't make overarching statements here, they provide two scenarios that help to demonstrate how the preceding parts of the article could be applied.

The article ends with half-page descriptions of the characteristics and strategies used in case studies, ethnographies, and field studies. Although these thumbnails are not sufficiently detailed to help a novice to construct a study, they suggest ways in which these techniques might be used in the workplace.

Although qualitative methods are more accepted today, at least within the community of researchers in our field (see Davy and Valecillos, "Qualitative Research in Technical Communication: A Review of Articles Published from 2003 to 2007," Chapter 15 in this volume), reservations about these techniques persist. Sullivan and Spilka do an excellent job of assuaging those doubts and suggesting ways of avoiding pitfalls that can trip up even experienced researchers.

A number of users who were learning to use a page layout program on the NeXT computer in an office experienced a "startle effect" when the tutorial began to talk to them. They jumped, fidgeted, exclaimed, swore, etc. Almost every one of them tried to find all the instances of voiced information in the tutorial. Most lost track of what they were trying to learn. Many brought their friends to see and hear it. Only the person responsible for report production learned the system and that person did not use the tutorial again.

(Sullivan 1991, 71)

Suppose you observed a situation such as this one, and it made you wonder whether some of the innovations in documentation that you and others were developing might be problematic for users. One solution to the product-specific question it raises would be to conduct a qualitative study of people trying to use the program you were documenting in their workplaces. If you were following the lead of Suchman's 1987 study of programmable copy machines, you might "hang out" by the copy machines and watch what happens. Such research could help you identify the problems for users, particularly the more serious problems.

But the observation could also raise far-reaching questions about how users interact with computers, and could then prompt formal qualitative research to study in depth the users of computers in an organization. Both responses would employ qualitative research methods.

Much has been said about the importance of, and need for, research in technical communication. For example, in 1983, Anderson, Brockmann, and Miller, analyzing what they then perceived as "the failure of research in technical and scientific communication" (1983, 8), speculated that improved research on workplace writing could have at least three positive effects for the field: building theory, supporting curriculum design, and informing the practices and problem-solving strategies of the professional (1983, 10).

In 1985, Frank Smith, in a much-quoted statement, agreed that "the more people we have trained to work in and do research in technical communication, the more likely we are to develop the recognition, appreciation, and standards [that] were lacking ten years ago" (1985, 7). Also, as recently as 1990, Mary Sue MacNealy, after surveying the types of research reported at the 1989 ITCC (International Technical Communication Conference) and IPCC (International Professional Communication Conference), recognized welcome improvements in both quantitative and qualitative research during the 1980s, but noted that "Given that empirical research is important to the growth of technical communication as a profession and discipline, the relatively small amount reported at the conferences is worrisome" (1990, 202).

There seems to be a consensus, therefore, that what is needed for further growth in technical communication is not only more research on workplace writing, but also more attempts to conduct scholarly, systematic studies of technical communication. Simply put, a body of research is needed for technical communication to be considered a profession.

In this article, we take up issues of the value, identity, and use of qualitative research methods from the perspective of incorporating qualitative research more fully into the growing profession of technical communication.

- First, we discuss the variety of meanings for qualitative research and its critical features in order to show that qualitative research has identity both as a formal and as a product-related activity.
- Then we discuss what makes a high-quality study—exposing the positives, the negatives, and the potential improvements—in order to equip readers of qualitative studies with a wider critical arsenal.
- Finally, we discuss how qualitative research might be applied to workplace situations in order to show its usefulness.

In doing so, we aim to argue that:

- Qualitative research findings contribute to the growing body of technical communication research by exploring situations, organizations, and cultures.
- Qualitative studies can be evaluated on the basis of their carefulness and rigor rather than routinely accepted.
- Formal research can be used to enrich thinking about analogous problems or to serve as a model for more informal work on a particular problem.

Part I: The Value and Identity of Qualitative Research in Technical Communication

The Value of Qualitative Research

In the context of technical communication the value of qualitative research, in particular, has received positive attention in the past decade. For example, Stephen Doheny-Farina and Lee Odell argued in 1985 that qualitative studies of workplace writing contribute substantially to the understanding of the composing process, of the characteristics of good writing, of how readers make meaning, and of the functions of writing. Some suggest that qualitative research, more than other types of research, has enormous potential to strengthen the field of technical communication.

In 1988, Jeanne Halpern argued that qualitative research enables us to "discover meaning" (1988, 30) and to

> [A]sk new kinds of questions about how people communicate at work and also about what we teach in our classes. . . . [These questions] offer new insights by suggesting incongruities or gaps between what we thought we knew and what is really the case. . . . [This type of research offers] the opportunity to explore new areas of interest . . . [to] get in deep . . . [to] allow us to construct the theoretical bases of business and technical communication.
>
> (1988, 38–39)

In 1990, MacNealy asserted that "the recent surge in interest in qualitative research in composition studies certainly suggests that technical communicators might also benefit from more work of this kind" (1990, 202). Yet despite increased interest in qualitative research among technical communicators, and the fact that some have recognized its potential to change our field in positive ways, there continue to be skeptics who question the rigor and validity of the enterprise. They often claim, for example, that qualitative researchers look at too few subjects, and too few settings, to be able to generalize.

We do not share this skepticism. In fact, we think qualitative research is the most accessible type of research to technical communicators. As the opening example illustrates, accounts of qualitative research usually focus on describing how and why people act as they do in the workplace, a quality that makes the accounts more compelling and arguably more useful to readers. We also believe this type of research, if done well, has great potential to lead to practical applications that can benefit the field and assist in its continued growth.

If such is the case, what will it take for more technical communicators to read, pay attention to, and benefit from the results of qualitative research of workplace writing? We are concerned by the fact that MacNealy found that just one-fourth of the research studies reported in the 1989 proceedings of the ITCC and IPCC used "ethnographic or case study methods involving a wide variety of procedures" (1989, 202), evidence that suggests that technical communicators might still be reluctant, or inadequately trained, to conduct systematic, high-calibre qualitative research studies.

Also, despite one study suggesting "that practitioners do value and use research" (Beard, Williams, and Doheny-Farina 1989, 193), we share Patricia Wright's more general concern "that the varied products of empirical research and the uses that can be made of these products are not always fully appreciated by those who are not themselves researchers" (1989, RT-3).

Many testimonials to the value of qualitative research can also be found in the social sciences. There, discussions of value often stress that qualitative research explores issues and works on real-life settings. For example, Lauer and Asher point out how a qualitative study can assist researchers in "closely studying individuals, small groups, or whole environments" (1988, 23) and can result in a "rich account" which shows interrelationships (1988, 55). They also note how this type of study can recognize "important variables" and "suggest new hypotheses" (1988, 45).

But this praise for the exploratory and explanatory nature of qualitative research is somewhat guarded, as social science methodologists are often as adamant in asserting that qualitative research cannot lead to generalizations about behavior— one of the central goals of the social sciences.

Discussions of qualitative research in the social sciences also stress its failure to study a problem or group in its natural setting. Such situated research can be claimed to "contribute uniquely to our knowledge of individual, organizational, social, and political phenomena" (Yin 1984). As Yin points out, the need for qualitative studies such as case studies "[arises] out of the desire to understand complex social phenomena; a case study allows an investigation to retain the holistic and meaningful characteristic of real-life events" (1984, 14).

This type of research is valuable, then, in helping organizations and their members come to understand how and why they act as they do.

What Makes Research Qualitative?

But what makes a research study qualitative? Although we find some disagreement about the answer to this question, we can identify a number of features that are often common to notions of qualitative research in various disciplines. This section reviews some of that literature, pointing to disagreements and gathering up agreements in order to assemble the reasonable goals and features for qualitative research in technical communication.

In the literatures of the social sciences, education, human factors, and technical communication, qualitative research does not refer to a single method, approach to analysis, or philosophy. A variety of approaches are labeled as "qualitative," and this diversity, in some cases, challenges the value of those methods—that is, judges them to be non-quantitative and hence non-scientific. In the social sciences, qualitative research has been used for decades but has been recognized just recently as an established field of inquiry, with courses and procedures to guide new researchers. Although some have seen it as a poor relation, with the ethnographic tradition in anthropology and the University of Chicago case studies in sociology as the most established models, its status is changing (Eisner and Peshkin 1990, 1).

The name of this research is not consistent, however. It is sometimes referred to as "qualitative," other times as "field," "naturalistic," "case study," "ethnographic," "focus group," or "descriptive" research. (See Appendix 1.1 for a thumbnail of key methods.) The differing terms sometimes refer to differing understandings of the goals as well as the conduct of this research. For example, "qualitative" could be said to focus on the type of data used in measurements and analysis (qualitative rather than quantitative); "field research" could be said to focus on where the data are gathered (the setting that is being studied); "descriptive" could be said to focus on the type of interpretation that is yielded (description rather than conclusions regarding causality).

There is some agreement that qualitative research involves contact with (some would say focuses exclusively on) the people being studied in their surroundings, and that it aims to explain what is happening to the community in focus. But methodologists disagree about the aims, the techniques, and the philosophical approaches of this research.

In *educational research*, for example, the common distinctions are between qualitative and quantitative research and between descriptive and experimental research. Disciplinary distinctions relate to whether the researchers get close to the situation when gathering their data; whether they stay closely involved with the situation while they are analyzing data; whether numerical analysis is used; and whether conscious manipulation of situations is important to the study.

The *Encyclopedia of Educational Research* (Alkin 1992), which is sponsored by the American Educational Research Association, has no entry on qualitative or quantitative research per se, relying on the entries of case study design, ethnography, experimental and quasi-experimental design, statistical methods, and survey design to cover the methodological territories.

The research taxonomy of Campbell and Stanley (1966) and their preference for experimental work are particularly telling in this discipline, as is the fact that the concept of the field study is missing because educational researchers do not view classrooms as the field.

That taxonomy does not take into account the naturalistic inquiry paradigm that Lincoln and Guba (1985) use to challenge the Campbell and Stanley taxonomy. Their work charges that researchers achieve their findings inside a specific worldview, and it creates a new taxonomy of qualitative research based on three paradigms: prepositivist; positivist; and postpositivist.

This debate about aims and philosophy demonstrates that meanings of "qualitative research" can differ substantially inside a discipline.

In a different arena, *human–computer interaction*, qualitative methods have been recently used in product-related research, particularly as an aid to interface design. This research serves as a contrast to the meanings for formal research recounted thus far, because it aims to use the formal methods for qualitative research to develop ways of improving a product. Building on the Whiteside and Wixon (1987) report about the necessity of testing products in the field (they reported on a computer that could not be assembled because all its boxes completely filled the room, leaving no space to move), human factors research involved with interface design has come to see value in qualitative research. Relevant methods include: participant observation; taped dialogs of users and evaluators reconstructing problems; thinking-aloud protocols during a work session; surveys followed up by interviews; and keyboard-activated tape machines allowing people to comment when they begin to use the program of interest.

Good (1989) names his method "contextual field research." It consists of interviewing a variety of customers, both while they are working (contextual interviews) and after they are finished working (summary interviews).

Campbell, Mack, and Roemer (1989) partition methods into quantitative (keystroke counts and performance times) and qualitative (interviews, thinking out loud, and video observation). They conclude that field studies can help to meet the measurable objectives of interface design, and that qualitative and quantitative methods can work together in answering questions necessary to the development of a good product. It is also clear from this work that product research can uncover questions that traditional, formal research might want to address as well.

In *technical communication*, we find considerable agreement that qualitative research is focused on studying issues of interest in detail, in their natural settings, with the aim of discovering the explanations for and the key patterns in the interaction of people with their environments. Although there have been no confrontations about the nature of qualitative research in technical communication, there are subtle—and not so subtle—differences in the perimeters of qualitative research in technical communication.

For example, Thompson and Cusella (1988) call it field research, Morgan (1988) calls it qualitative research strategies (including ethnography, case studies, and descriptive studies), Doheny-Farina and Odell (1985) call it ethnography, and Halpern (1988) calls it qualitative research.

In a comprehensive discussion of methods, Morgan (1988) asserts that ethnography focuses on the context; case studies focus on the inhabitant; and descriptive studies examine many instances of an event. *Ethnography* focuses on the context and on the relationships between people and their environment (for example, Mirel's 1991 study of why a manual was not getting used by the staff in a bursar's office). The *case study* focuses on identifying the features of a particular phenomenon through in-depth examination of one or more examples (for example, Lutz's 1987 study of students composing at the computer). The *descriptive study* (which others have named the *field study*) focuses on detailing a large-scale event to identify its features (for example, Ede and Lunsford's 1992 three-stage study of collaborative writing groups in various professions).

Other researchers demonstrate their awareness of how amorphous are the conceptions of qualitative research in technical communication by coining new terms for this type of research:

- Gould and Doheny-Farina, for example, coin "qualitative field research" to describe how to investigate the usefulness of documentation (1988). "Simply put," they say, "qualitative research involves investigating a few cases in great depth; field research involves investigations done in the natural environments of those under study" (329–30).
- Sullivan, in a similar move, names a type of usability study as a "longitudinal field study" in order to emphasize how it would differ from the dominant research approach in usability, the lab study (1989).

Reasonable Goals for Qualitative Research in Technical Communication

We agree with other researchers that qualitative research in technical communication should focus on discovery; on researching a problem, a product, or an issue in context; and on presenting the explanations developed in rich, descriptive detail. In technical communication, a qualitative study should relate important issues in the community studied to the field—for example, how the culture of a computer book company affects its editorial process (Simpson 1989), or how the relationship between written and oral discourse can influence the culture within a governmental department (Spilka 1993).

In general, the goals of qualitative research in technical communication are to interpret and understand situations in order to contribute to our understanding of how technical communicators function in various organizations. Its critical features are that it:

- Bases the choice of research design and methodology on research questions (that is, on theory and on the paradigms a researcher is operating in).
- Explains and defends the design of the research (and follows guidelines for the type of research as appropriate).
- Enters the culture (or its artifacts) and views the issue(s) in the context of that culture.
- Tries not to disturb the culture it studies.
- Gathers a variety of data (for example, oral, written, or recorded).
- Does not rely primarily on numerical analysis to produce its findings even though it looks for patterns in the data as a way to build an interpretation.
- Is careful to hold different viewpoints and perspectives toward the culture and the data (triangulation) in order to develop a reliable account of what it discovers (to achieve this goal, sometimes more than one researcher collects and analyzes data and often multiple research tools are used).
- Develops an interpretation that accommodates the data and shows evidence of considering other possible interpretations.

In technical communication, qualitative methods are particularly adept at addressing the following research needs:

1. *Interpreting a situation.* You encounter a phenomenon you don't understand or a problem you need to solve, so you gather information and build an explanation or description of the particular phenomenon or event.
2. *Exploring a situation.* You aren't sure what you are looking for, so you gather information to discover issues or problems. Or you gather information to develop a broader, more in-depth understanding of how a culture works or of a phenomenon's complexities.
3. *Developing a unique research perspective.* You are interested in learning more about an "uncommon story," about instances of innovation that would be silenced in studies focusing on what the majority of people think.
4. *Discovering a better way to communicate with users about research projects or products.* You want to know what others think about a particular problem, such as their opinions about graphic design in a magazine, but the language of graphic design has no meaning to the readers, so instead of conducting a survey with language foreign to the users, you decide to interview the users or run a focus group.

Clearly, we have tried to establish an identity for qualitative research that builds on the agreements among the disciplines about qualitative inquiry and the research needs and goals of technical communication as a field. This identity encompasses both formal research and the more problem-specific product investigations that technical communicators might also conduct. While we have been focusing on formal research, as we think we must for purposes of identity and evaluation, we also regard the more informal research as qualitative.

Part 2: The Usefulness of Qualitative Research

So far, this discussion has addressed what qualitative research attempts to do and why it is valuable to technical communication; but for qualitative research to be useful, it needs to present reliable and thoughtful findings about issues technical communicators are pondering. This section tackles two vital components of "usefulness": it discusses how to evaluate the quality of a particular study, and it suggests when and how qualitative findings can be applied to actual situations you might face as a technical communicator.

Evaluating Qualitative Research: What Makes Studies Strong?

We believe a strong qualitative study needs to account for the following types of research concerns.

Choice of this Type of Research

Potential problems:

- Choosing qualitative studies even when the situation calls for quantitative research.
- Deciding to conduct qualitative research without sufficient experience in technical communication (this refers to academics), or with a particular workplace culture, leading to problems in interpreting the findings.

Potential solutions:

- Basing the choice on the desire to answer *how* and *why* questions. For example, how do technical communicators address an audience they do not know (Huettman 1990)? While working on computer documentation, why do technical communicators need to consider how users learn (Mirel 1991)?
- Relying on one's own original research questions, and focusing on the phenomena one wants to observe, when choosing a research methodology.
- Conferring with technical communicators in several workplace cultures about the proposed research and spending time in a particular culture before studying it.

Choice of Research Design and Data Collection Method

Potential problems:

- Deciding to collect data for too short a time span, and then making conclusions on the basis of what could be an untypical situation.
- Deciding to limit observations just to the writers, even though original research questions pertain to all participants of the rhetorical situation, or to limit observations to a single organization, even though original research questions pertain to rhetoric across multiple organizations.
- Excluding on-site observations from research methodology, even when the original research question concerns social interactions or other on-site activities.
- Excluding some elements or characteristics of a particular phenomenon, such as observations of technology, even when technology is intricately involved with the process being observed.
- Deciding to obtain data from a single source, thereby risking the reliability and validity of the study. For example, if studying conflict in a group collaboration, consultation with only a single group member (or, for that matter, with only the group members and not with others concerned with the problem and the evolving document, such as document reviewers) will lead to a biased perspective.
- Deciding to adhere to the original research questions and relying on original hypotheses throughout the study; waiting until the end of the study to analyze and interpret data.

Potential solutions:

- Ensuring that choices of design and methodology are consistent with research questions. For example, if one is interested in how technical communicators analyze and address a multiple audience effectively, it would be necessary to plan to conduct a study long enough to enable observation of the reactions of a multiple audience to texts produced, and to observe this type of communication as it takes place across all organizational divisions (or across all organizations) that include members of the multiple audiences being addressed. If one is interested in studying social interactions, or the relationship between technology and social or rhetorical processes, it would be necessary to plan to observe these phenomena during the study.
- Exploring the research situation carefully in order to identify the proper informants for a particular study and the contexts in which these informants can aid the research.
- Planning for triangulation ("examining data from different perspectives to strengthen the validity of their conclusions" (Denzin 1970, 301–303)). Triangulation makes possible continuous comparison of data from different

sources, and of findings across sites, to allow for continuous reexamination of original research questions (Halpern 1988). Researchers can achieve triangulation in at least three ways (Denzin 1970):

1. Use a variety of theoretical perspectives when examining data (theory triangulation).
2. Involve multiple researchers instead of a single researcher (investigator triangulation).
3. Rely on an array of research tools from diverse sources when collecting data (data triangulation); for example, use printed records of meetings as well as informants' recollections of those meetings.

- Continually checking with on-site sources to validate inferences and conclusions made about data collected (this measure increases the reliability of a study's conclusions).
- Analyzing research data early enough in the data-collection process to open up opportunities for new hypotheses to emerge about the research problem and to make possible a reexamination and revision of original research questions.

Choice of Data Analysis Method

Potential problems:

- Deciding to analyze data without seeking patterns in the data. For example, you might have data about different strategies used by different organizational divisions for writing effective online documentation, but then analyze the strategies used by just one writing team in one organization.
- Deciding to analyze data without considering theoretical implications, or how findings challenge or build on previous hypotheses, or lead to new hypotheses. For example, you might have data concerning strategies for audience adaptation but then omit consideration of how those strategies challenge or build on existing audience theories or how those strategies might lead to a new model of how to adapt writing effectively to an organizational audience.

Potential solutions:

- Attempting to identify patterns in data collected from various research sources unless the primary research interest is in observing a unique, isolated type of behavior or activity; although providing thick description is important and commendable, it is important, also, to indicate which patterns in your data led to your overall conclusions (unless, of course, your key research interest is to examine an isolated phenomenon or "innovation" at the workplace).
- Taking the time to research, consider, and respond to previous theoretical approaches to your research issue—it is critically important to link data

analysis to theory, to provide a theoretical perspective to the research results; describing applications for practice without first analyzing theoretical implications of research results puts at risk the value and integrity of a qualitative study.

- Taking the time to understand fundamentals about technical communication, organizational behavior, or workplace cultures; or taking time (especially in early stages of a study) to conduct on-site observations of social and rhetorical behavior before attempting to collect and analyze data.
- Allowing sufficient time to collect and analyze data to ensure that one has detected workplace dimensions or phenomena of great importance to the research questions or conclusions.
- Continually checking your inferences and conclusions with study participants for their validation.

Choice of Data-Reporting Technique

Potential problem:

- Deciding not to describe and justify research designs and methodology while reporting a study at a conference, in a document internal to a company, or in external publications. For example, one might report on a study of flaming in e-mail by focusing on the tests and not giving a sense of when and how those texts were created; as a result, the report audience would wonder if it were really flaming or, if so, whether the flaming has social repercussions.

Potential solution:

- Providing a full description and rationalization of all major research decisions in a research report to convince the report audience of the validity of the findings; unless one can make a case for the type of research you are conducting, the integrity of the research will remain suspect.

Sources of Weakness

Critics of qualitative research have noted a number of potential problems with qualitative studies. They have cautioned researchers to guard against problems with generalizability, reliability, and possible misinterpretations of data, as well as to recognize the tendency of this type of research to be time-consuming, with an overwhelming amount of data to collect and analyze. However, as Chadwick, Bahr, and Albrecht (1984) point out, some of the potential weaknesses of qualitative research are also its strengths. For example, flexibility in data collecting can become both an advantage and a disadvantage, and although researchers might sacrifice some objectivity "in the process of obtaining a rich, intimate understanding of a people," it is also true that "too much attention to objectivity will rob the data of its imaginative, impressionistic feelings about how members of the group being studied define themselves and their world" (1984, 215).

Perhaps because of the diversity of research problems, some qualitative studies of workplace writing conducted in the past decade have suffered from a variety of weaknesses. We have noticed, for example, that some researchers choose to conduct qualitative studies even if their original research questions call for quantitative research. And unfortunately, some qualitative researchers of workplace writing decide not to describe (or justify) their research designs and methodology, so that it is impossible to tell how their research was conducted, let alone whether it was conducted well (MacNealy 1990).

We have also noticed problematic data-collection and data-analysis methods in recent studies. In some studies, researchers collect data in too short a time span, and then draw conclusions on the basis of what could be an untypical situation at a particular site; perhaps because of time, economic, or political constraints, a number of researchers do not observe phenomena over a sufficient time span to determine whether what is observed is the norm. Such practice is particularly troublesome when the target of observation is a complex situation requiring a longitudinal study.

In addition, some researchers limit their observations to the writers of documents, even though their original research questions pertain to all participants of a team (managers, engineers, marketers, etc.). Some of these researchers make no on-site observations but then draw conclusions about on-site social interaction. Other researchers limit observations to a single organization, even though their research questions pertain to rhetoric across multiple organizations, or to public issues of concern to multiple organizations. Some decide not to study technology, even if technology is intricately involved with the process being observed.

We have noticed, too, that qualitative researchers need to attend more to reporting observed patterns, rather than focusing on isolated instances. Sadly, some researchers even describe how their findings might change practice without first taking the time to consider the implications of applying their research results, a trend that puts at risk the value and integrity of their studies.

Finally, we feel a danger exists for academics who conduct this type of research but lack enough general experience with technical communication, or particular experience in a workplace culture, to understand what they see. There is a distinct possibility that these researchers will not detect certain workplace dimensions or phenomena of great importance to their research questions or conclusions.

Examples of Strong Studies

These concerns and cautions can be answered in a qualitative study. We now summarize the better qualities of two well-regarded qualitative studies of workplace writing as a way of demonstrating some strengths of recent qualitative research.

Example 1: Jennie Dautermann's "Negotiating Meaning in a Hospital Discourse Community" (1993)

In this qualitative study, conducted as her dissertation project at Purdue University, Dautermann's aim was to explore two questions about group collaboration within a hospital department of nursing:

1. Which strategies appear in collaborative writings on the job to address creative dissonances within the group and pressures from the community beyond the collaborators?
2. What relationship do the acts of composition have, in return, on the social contexts of writers? What (if anything) changes in the social context?

Part of what makes this study especially commendable is Dautermann's carefully planned and executed research methodology, as well as her consistent attempts to minimize bias in her data collection and analysis. She took care to defend all methodological choices and to link those choices to theory. During her almost two-year study, she functioned as a participant-observer of group collaboration at a hospital, thereby gaining an insider's perspective of the context, in which the nurses assembled to revise a nursing regulation system to make it more accessible to practicing nurses.

As Dautermann points out, this case demonstrates what can happen when writers (in this case, the nurses) attempt to effect significant changes in policy; as she puts it, "significant organizational change was attempted by writers with relatively little power within the hospital, but who were directly involved in the activities regulated by the documents at stake" (1993, 99).

During data collection, as she observed rhetorical and social patterns of behavior, Dautermann used caution to ensure that these observations would remain "tentative hypotheses which guided the collection of additional materials"; then, after further analysis, she made comparisons of data between research sources "to adjust the emerging picture" and to emerge with revised hypotheses (1993, 100).

To guard against researcher bias and to establish greater reliability in data analysis, Dautermann made special efforts throughout the study to collect observations from multiple community sources. To ensure accuracy and to verify inferences and detected patterns of decision making, she discussed her conclusions, inference, and detected patterns with study participants at different junctures of this two-year study. Also, Dautermann took an interdisciplinary, theoretical approach during data analysis to discover meaning from her data; in particular, she drew upon sociolinguistic and composition studies to help her understand the data she had collected.

Another commendable feature of Dautermann's approach was her open-mindedness to unexpected findings. As she entered the study, she was convinced that she would discover that writing was a way to make decisions, and that the activity of writing and rewriting would shape the community observed. However, while she found evidence of these phenomena, she was also able to detect interesting new patterns concerning oral planning.

Finally, in reporting the data, Dautermann allowed for "multiple voices"; that is, when making points about group collaboration and other phenomena observed, she included comments from various people to corroborate those points.

We think Dautermann's continuous cautions in data collection, analysis, and reporting add to the validity and worth of her study, making her findings particularly credible and her study as a whole particularly valuable.

Example 2: Susan Kleimann's "The Reciprocal Relationship of Workplace Culture and Review" (1993)

Kleimann studied the process of document review at the Government Accounting Organization (GAO; renamed the Government Accountability Office in 2004) as part of a dissertation project at the University of Maryland. In this 18-month study, Kleimann was interested in exploring the reciprocal relationship between the review process and organizational culture, especially in how one or more cultures can shape composing processes and behavior. Her main research questions were:

1. How do organizational and divisional cultures affect the nature of review comments?
2. In what ways do reviewers reflect the organization's culture in the style and content of their comments?

To explore these broad questions adequately and to determine whether different cultures within the GAO influenced the review process in different ways, Kleimann decided that it was important to spend more than a year at the GAO and to conduct multiple case studies in different GAO divisions. Initially, her plan was to follow the evolution of seven reports in three divisions of the organization; however, because each team observed produced between 13 and 24 drafts of each report, Kleimann decided later to limit her data analysis to three assignments in two GAO divisions.

This decision concerning research design resulted in an excellent study and led to her detection of some important research patterns linked directly to her original research questions. By limiting her reporting to what happened in two GAO divisions (instead of three), Kleimann was able to detect significant differences between the two division cultures that influenced the review process in different ways, while also providing a more in-depth description of what happened in the three assignments observed.

Also, by ensuring that her study lasted 18 months, she was able to draw conclusions about the long-term nature of the divisional management perspective in the review process.

Kleimann was careful to use methodological triangulation in data collection, a decision that enabled her to build reliability by accretion, and to detect significant patterns concerning the roles of oral exchanges. In particular, by using a variety of research methods, Kleimann was able to discover that

Because review involves negotiating a balance of values and constraints, oral exchanges are central to the success of a review process. Frequent exchanges help staff members navigate the political waters . . . help members learn the organization and secure agreement on an approach to a report long before they have committed a great deal of time and effort to writing . . . [and] help staff members achieve buy-in to their institutional product; overall, they increase the chance that people communicate with, rather than past, each other.

(Kleimann 1993, 68)

Kleimann's methods included the following:

- attending formal and informal meetings;
- conducting open-ended interviews after meetings to confirm observations or elicit responses to the meetings;
- conducting structured interviews with staff to record differences in perceptions about the same events;
- collecting communication logs (sheets that require participants to record discussions, meetings, and phone calls pertaining to the assignment) when she was not present;
- conducting discourse-based interviews, informal and often unplanned interviews, throughout the process;
- conducting open-ended (unstructured) interviews as each report was completed.

This study is also commendable in its use of both qualitative and quantitative research methodology. Besides relying upon the previously listed qualitative measures during data collection, Kleimann organized the written review comments with a modified GAO computer program to help in the analysis of those comments. In addition, she relied on four analytic techniques:

1. She used SPSS (Statistical Package for the Social Sciences) to create cross-indexing of various measures, for analysis of details about the comments.
2. She created a "complexity chart" that illustrated a report's review process and served as a useful visual analytic tool that provided a picture of the entire review process for each report.
3. She conducted an analysis process enabling detection of patterns both within and across the three cases observed.
4. She asked participants frequently, at the end of meetings or interviews, to verify observations, and adjusted the data when necessary for accuracy.

As in Dautermann's case, Kleimann's decisions regarding research design, data collection, and data analysis led to a well-regarded, credible study. Studies such as these demonstrate that writing in organizations is being studied qualitatively, carefully, and well.

The Brass Tacks: When and How Qualitative Research is Helpful to Technical Communicators

Let's return to the example we posed at the start in order to frame a discussion of how to apply qualitative research. A writer's chance observation of people having trouble with sound in a tutorial eventually led the observer to consider the problem of how sound in a computer tutorial can make that tutorial uncomfortable for people in a crowded office. It is unlikely that the discomfort would have been noticed in a laboratory test of the tutorial's usefulness; the on-site observation was necessary to uncover a problem that could inhibit the learning of the software.

If the tutorial were the only one of the educational materials that delivered information critical to learning the software, then users who were self-conscious about running that tutorial in "public" would miss important information. Obviously, the qualitative research that yielded this finding is valuable to the technical communicators working on the tutorial in question.

But that research is also suggestive to other technical communicators who are designing manuals for use. It invites technical communicators to imagine how the users will employ a manual as they work and then make decisions that will aid them in those uses.

Clearly, we can find people who speak of the value of research in general and qualitative research in particular: earlier in the discussion we quoted some such testimonials. But what is more important is when and how qualitative research can help one to do a job more effectively or to learn more about technical communication. We discuss this question in light of two hypothetical scenarios. Our purpose is to identify some ways that qualitative inquiry might enrich the analysis of various specific problems that technical communicators may face—to demonstrate that qualitative research is not just something that people do in colleges and research institutes.

As you read through these scenarios, consider what questions we might frame that would lead you to do qualitative research or to look for relevant research to read and apply. Remember that qualitative research is not the only response we can and should have to these situations. If we view the scenarios as requiring statistically defensible answers, for example, we will not use qualitative research in any way. But when we need to explore situations, build interpretations, develop a unique story, or discover answers to "how" and "why" questions, then we need to consider qualitative methods.

Scenario 1: Establishing a Documentation Department's Value

Suppose the computer company you work for is planning a reorganization and your documentation department of 16 people (managers, designers, editors, usability testers, and writers) has been asked to present its case for the value of having a documentation department as a separate unit, as opposed to assigning the various writers to projects, with final editorial and production being housed in some service function. The request is a serious one, not a way to cover a prior managerial decision.

Suppose, too, that the idea has come in part from your department's success with getting writers involved with the design teams: Some of the design teams want the writers assigned to them. This is problematic because earlier in your company's history, writers used to be housed with design teams, but they were not valued for their contribution because it was thought that anyone could write documentation.

Your department has been successful in establishing the professional status of writers, and you see the proposed reorganization as a primary danger for it. How can you show the work your department does and demonstrate its merit? You have resources: quarterly productivity reports; customer feedback on the documentation; data about product support; marketing surveys that include information about documentation; and usability reports about the products and their documentation.

How might qualitative research contribute to the case you must build to establish the value of your department? Certainly you should look for qualitative studies (or any other research) of other documentation departments inside your industry that are similar in size and function. If you find qualitative studies, you need to examine their design and implementation carefully enough to be confident that they are reasonable research. You may well find case studies of automation, publications management, or productivity of workers, and such reports may provide evidence, arguments, or models for your thinking.

Of course, the problem poses many types of questions: For example, how does the department operate in its current set-up? Why does it make sense to house those involved with the publications process in one unit? What kinds of useful autonomy does a writer gain from working with a team but being supervised in another unit? Some of the questions can be answered using qualitative research, but the problem cannot be comprehensively addressed without the use of quantitative analysis as well. You are also working under such time pressure that no new research can be initiated; you need to rely on your own resources and what models you can locate in previous research.

Scenario 2: Investigating the Merits of Recorded Versus Paper Documentation

Your company is planning to market a programmable phone-answering machine to elderly customers and can either revise the existing paper documentation or develop a new recorded tutorial that talks customers through the programming tasks. Both the recording option and the audience are new to you. While the idea is intriguing, and could potentially be done in a way that keeps all or part of the paper documentation as well, you think you need to know more about elderly customers in order to make an informed decision.

You have some data from focus groups (which means you already have some results of qualitative research) on which functions are more important to elderly consumers, and it is clear from those data that their ideas change after functions have been explained; the names of the functions are not clear to them.

Some questions arising from this scenario might be:

- Will they be reluctant to play the recording?
- What language and style should be maintained for elderly customers? How do you explain concepts succinctly without seeming pejorative or too informal for their tastes?
- In a natural setting, how difficult will it be for consumers to actually follow the recorded directions?

Clearly your decision making can profit from your own qualitative research or from bringing in someone else to do that research. The results of the focus group point to the need for such research; a survey, for example, probably would not yield a reliable solution to your problem, since the consumers in the focus group did not understand the meanings of their options. Observation and interviewing are needed to identify the feasibility of using a new approach to documentation for this group. As always, the key will be time. You will need to focus your efforts to complete research that answers your product-specific question and abides by the principles of good design.

Conclusion

We consider qualitative research a rich means for exploring important issues in technical communication in depth and breadth, including such critical global concerns as usability, total quality management, technology transfer, and office automation. Regular reading of this type of research can yield strategies for improved communication in the workplace; for example, you might discover new strategies for communicating effectively to a multiple audience situated in multiple organizations, or for dealing with conflict during collaboration in the preparation of in-house documents.

The findings of qualitative studies can also reinforce your arguments; for example, by persuading supervisors to prolong planning time in order to allow for some opportunity to resolve conflicts before writing begins, or by encouraging fuller participation in document development by concerned people who would otherwise find themselves excluded from key decision-making processes.

By consistently reading reports of this type of research, you can gradually develop increased sensitivity to what makes a study commendable or weak. With this increased sensitivity, it is easier to evaluate whether a particular study has produced relevant findings that could be cited in debates about important company issues.

With increased knowledge of this type of research, you or some of your colleagues might decide to conduct qualitative studies at your own worksite, or to seek ways of convincing management to bring in outsiders to conduct this type of research with the goal of identifying ways to improve communication practices locally.

You might also find it more tempting, as well as easier, to participate actively in discussions with other technical communicators about how recent qualitative research might apply to issues of mutual concern—for example, at the annual meetings of the Society for Technical Communication.

Tuning into qualitative research will also help you to be a more sensitive observer, often on the prowl for a story that captures "how things work around here." We think that you will soon see opportunities for research wherever you look; at least, that is our hope.

Appendix 1.1: Thumbnails of Frequently Used Qualitative Methods

Case Studies

A "case study" can be simple or elaborate. Often data will be gathered without much fanfare and called a "case." Case studies gather descriptive information about a phenomenon of interest. But case studies are often quite elaborate as well, as Yin asserts when he claims that case studies are particularly good strategies for situations where "a 'how' or 'what' question is being asked about a contemporary set of events, over which the investigator has little or no control" (1984, 23). Yin goes on to define a case study as "an empirical inquiry that:

- Investigates a contemporary phenomenon within its real-life context; when
- The boundaries between phenomenon and context are not clearly evident; and in which
- Multiple sources of evidence are used." (Yin 1984, 21)

Yin distinguishes the case from ethnography on the basis of time. A case study starts with pointed questions, so the case study is much more directional in its inquiry than the ethnographic study. A case study is easier to design than an ethnographic study for several reasons:

- The case study does not necessarily take a long time.
- The case study design can handle, and indeed it is dependent on, a variety of data. Where much of the ethnographic study depends on the researcher's participant observation, the case study needs to have a range of materials that serve as corroboration of the interpretation proposed. Yin points to six sources for case study data:
  1. documents;
  2. archival records;
  3. interviews;
  4. direct observation;
  5. participant observation;
  6. physical artifacts.
- The elaborate case study can also serve as a good problem-solving tool, because it can work from very pointed starting questions.

Ethnographies

An "ethnography" is the study of culture from the stance of a researcher who enters and participates in that culture. Because it comes from anthropology, it is more focused on the researcher joining in with—even merging with—the context. Hammersley and Atkinson give ethnography a loose definition:

> Ethnography (or participant observation, a cognate term) is simply one special research method. . . . The ethnographer participates, overtly or covertly, in people's daily lives for an extended period of time, watching what happens, listening to what is said, asking questions; in fact collecting whatever data are available to throw light on issues with which he or she is concerned.
>
> (Hammersley and Atkinson 1983, 2)

They acknowledge that ethnography is practiced in somewhat different ways in differing fields:

> There is disagreement as to whether ethnography's distinctive feature is the elicitation of cultural knowledge, the detailed investigation of social inter-action, or holistic analysis of societies. Sometimes ethnography is portrayed as essentially descriptive, or perhaps as a form of storytelling; occasionally, by contrast, great emphasis is laid on the development and testing of theory.
>
> (Hammersley and Atkinson 1983, 1)

It is true that some ethnographies describe while others analyze and develop theories, but all ethnographies include culture as a focal element. An ethnographic study is quite difficult for product studies because:

- The primary researcher must become a participant and observer at the study site for a substantial period of time; preferably one to two years. This process calls for considerable job flexibility and for a relationship of trust with another company.
- The primary researcher must also have special training and experience in ethnographic research.
- The ethnographic researcher comes to a culture with some ideas in mind, but expects the culture to provide the true focus and thus concentrates on being open to the true nature of things. Usability research is normally more problem centered and directed than is a typical ethnographic study.

Field Studies

"Field study" is a general classification for research that collects its data in the subjects' environment. It can be applied to case studies and to ethnographies, indicating that the focus of the data collection was in the field, and blurring the philosophical goals behind a particular study. A variety of measures can be used: interviews, surveys, observation (both participant observation and detached observation), and collected documents. In general, the data produced are qualitative data used to describe the "field" and to interpret the attitudes and actions taking place in the field, though quantitative data can be used as well.

Erickson (1986) articulates some of the questions that fieldwork answers:

1. What is happening, specifically, in social action that takes place in this particular setting?
2. What do these actions mean to the actors involved in them?
3. How are the happenings organized in patterns of social organization?
4. How is what is happening related to happenings at other system levels?
5. How do the ways everyday life in this setting is organized compare with other ways of organizing social life?

(Erickson 1986, 121)

Product studies are often called field studies.

Acknowledgment

*This article was originally published in *Technical Communication* (1992), 39: 592–606.

References

Alkin, M. C., ed. 1992. *Encyclopedia of Educational Research*. New York, NY: Macmillan.

Anderson, P. V., R. J. Brockmann, and C. R. Miller, eds. 1983. Introduction, *New Essays in Technical and Scientific Communication: Research, Theory, Practice*. Farmingdale, NY: Baywood.

Beard, J. D., D. L. Williams, and S. Doheny-Farina. 1989. "Can Research Assist Technical Communication?" *Technical Communication* 36: 188–94.

Campbell, D. T., and J. C. Stanley. 1966. *Experimental and Quasi-Experimental Designs for Research*. Boston, MA: Houghton.

Campbell, R. L., R. L. Mack, and J. M. Roemer. 1989. "Extending the Scope of Field Research in HCI." *SIGCHI Bulletin* 20 (4): 30–32.

Chadwick, B. A., H. M. Bahr, and S. L. Albrecht. 1984. *Social Science Research Methods*. Englewood Cliffs, NJ: Prentice-Hall.

Dautermann, J. 1993. "Negotiating Meaning in a Hospital Discourse Community." In *Writing in the Workplace: New Research Perspectives*, ed. R. Spilka. Carbondale, IL: Southern Illinois University Press.

Denzin, N. 1970. *The Research Act in Sociology*. London: Butterworth.

Doheny-Farina, S., and L. Odell. 1985. "Ethnographic Research on Writing: Assumptions and Methodology." In *Writing in Nonacadmic Settings*, ed. L. Odell and D. Goswami. New York, NY: The Guilford Press.

Ede, L., and A. Lunsford. 1992. *Singular Texts/Plural Authors: Perspectives on Collaborative Writing*. Carbondale, IL: Southern Illinois University Press.

Eisner, E. W., and A. Peshkin, eds. 1990. *Qualitative Inquiry in Education: The Continuing Debate*. New York, NY: Teachers College Press.

Erickson, F. 1986. "Qualitative Methods in Research on Teaching." In *Handbook on Research on Teaching*, 3rd edn, ed. M. C. Wittrock. New York, NY: Macmillan.

Good, M., ed. 1989. "Seven Experiences with Contextual Field Research." *SIGCHI Bulletin* 20 (4): 25–32.

Gould, E., and S. Doheny-Farina. 1988. "Studying Usability in the Field: Qualitative Research Techniques for Technical Communicators." In *Effective Documentation: What We Have Learned from Research*, ed. S. Doheny-Farina. Cambridge, MA: MIT Press.

Halpern, J. 1988. "Getting In Deep: Using Qualitative Research in Business and Technical Communication." *Journal of Business and Technical Communication* 2 (2): 22–43.

Hammersley, M., and P. Atkinson. 1983. *Ethnography: Principles and Practice*. London: Tavistock Publications.

Huettman, E. 1990. "Writing for the Unknown Reader: An Ethnographic Case Study in a Business Setting." PhD dissertation, Purdue University. AAT 9116406.

Kleimann, S. 1993. "The Reciprocal Relationship of Workplace Culture and Review." In *Writing in the Workplace: New Research Perspectives*, ed. R. Spilka. Carbondale, IL: Southern Illinois University Press.

Lauer, J. M., and J. W. Asher. 1988. *Composition Research: Empirical Designs*. New York, NY: Oxford University Press.

Lincoln, Y. S., and E. G. Guba. 1985. *Naturalistic Inquiry*. Beverly Hills, CA: Sage.

Lutz, J. A. 1987. "A Study of Professional and Experienced Writers Revising and Editing at the Computer and with Pen and Paper." *Research in the Teaching of English* 21: 398–412.

MacNealy, M. S. 1990. "Moving Toward Maturity: Research in Technical Communication." *IEEE Transactions on Professional Communication* 33: 197–204.

Mirel, B. 1991. "Comparing MIS and User Views about Task Needs." SIGDOC '91 *Proceedings*. New York, NY: Association for Computing Machinery.

Morgan, M. 1988. "Empirical Research Designs: Choices for Technical Communicators." In *Effective Documentation: What We Have Learned From Research*, ed. S. Doheny-Farina. Cambridge, MA: MIT Press.

Simpson, M. 1989. "Shaping Computer Documentation for Multiple Audiences: An Ethnographic Study." PhD dissertation, Purdue University. AAT 9018906.

Smith, F. R. 1985. "Twenty-Five Years Later." *Technical Communication* 36: 5–7.

Spilka, R. 1993. "Moving Between Oral and Written Discourse to Fulfill Rhetorical and Social Goals." In *Writing in the Workplace: New Research Perspectives*, ed. R. Spilka. Carbondale, IL: Southern Illinois University Press.

Suchman, L. A. 1987. *Plans and Situated Actions*. London: Oxford University Press.

Sullivan, P. 1989. "Usability in the Computer Industry: What Contribution Can Longitudinal Field Studies Make?" International Professional Communication Conference *Record*. Piscataway, NJ: IEEE.

——. 1991. "Multimedia Computer Products and Usability: Needed Research." In International Professional Communication Conference *Proceedings*. Piscataway, NJ: IEEE.

Thompson, T. L., and L. P. Cusella. 1988. "Field Research in the Organization." In *Conducting Research in Business Communication*, ed. G. Campbell Patty, T. Housel, and K. O. Locker. Urbana, IL: Association for Business Communication.

Whiteside, J., and D. Wixon. 1987. "Discussion: Improving Human–Computer Interaction—A Quest for a Cognitive Science." In *Interfacing Thought*, ed. J. M. Carroll. Cambridge, MA: MIT Press.

Wright, P. 1989. "Can Research Assist Technical Communication?" In *Proceedings* of the 36th International Technical Communication Conference. Washington DC: Society for Technical Communication.

Yin, R. K. 1984. *Case Study Research: Design and Methods*. Beverly Hills, CA: Sage.

Chapter 2

Evaluating Qualitative Inquiry in Technical and Scientific Communication

Toward a Practical and Dialogic Validity*

*Ann M. Blakeslee, Caroline M. Cole,
and Theresa Conefrey*

Editors' Introduction

In this conceptual article, first published in 1996, Blakeslee, Cole, and Conefrey propose a reframing of the problem of validity, and discuss the challenges and benefits of sharing authority among researchers and participants.

Blakeslee and colleagues are reacting to the concepts of internal and external validity as held by most quantitative social science researchers. Internal validity in quantitative research is seen as the extent to which a research study brings to light the intended social phenomenon. Suppose a research team is trying to determine whether a set of installation documents promotes the *efficiency* of teams of installers who are configuring and commissioning a new information system. A research study that asks installers whether the documents helped them to become more efficient may have somewhat limited validity because we may doubt the ability of the installers to arrive at an accurate conclusion about their own efficiency. However, a research study that asks for the installers' opinion, that asks for the opinions of others who observe the installers' work, and that also compares the work of the installers using the documentation with the work of installers who are not using the documentation may be seen to have more validity. In this latter case, the research results are considered more likely to shed light on the efficiency of the installation process and thus have a higher level of internal validity.

Quantitative researchers see external validity as the extent to which the results produced by a research study can be generalized and applied to other situations. Many quantitative researchers are interested in revealing the general laws that govern social action. In the above example, quantitative researchers may wish to formulate general laws or rules governing the efficient implementation of the new information system in all possible customer locations. They may believe that if they can arrive at those laws, they will be able to create a "recipe" for installing information systems that will allow most installation teams to succeed at their task in an efficient manner.

Many believe that these conceptions of validity cannot be applied to qualitative research. In their watershed publication, *Naturalistic Inquiry* (1985), Lincoln and Guba pointed out that the concept of internal validity assumes a correspondence theory of truth: for every social phenomenon, there is a corresponding single best description and conceptualization to apply. Qualitative research, however, derives from a constructionist view of social reality, which suggests that there can be multiple, relevant experiences of the same social phenomenon, and thus many relevant and justifiable interpretations of the same phenomenon. Returning to our example, a qualitative researcher might say that the efficient installation of the new information system may be more problematic than is evident at first. The installers may view efficiency in relation to the speed of the installation; however, the speed of the installation may produce serious work interruptions for the users of the new technology. Thus, an efficient installation from the installers' perspective may be a highly inefficient installation from the users' perspective.

Moreover, qualitative researchers tend to argue that social phenomena are fundamentally situated in a local social context, with specific characteristics and agents. Instead of trying to generalize specific research findings into universal social laws, Lincoln and Guba (1985) argue that qualitative researchers describe their research results through a "thick" (or detailed) description of their study, so that readers can judge for themselves whether the results are transferable to other social contexts.

Blakeslee and colleagues offer an elegant and compelling solution to this polarizing debate between quantitative and qualitative researchers. They suggest that qualitative inquiry in technical and scientific communication should be evaluated in terms of the research's *practical validity*. They argue that qualitative research is valid to the extent that it is meaningful and useful for research participants and researchers. A valid qualitative study is one that brings improvements to frontline practice, while also meeting the needs of academic researchers to produce new knowledge and publish their results. Consequently, they advocate in the strongest terms for the involvement of research participants in all phases of the research process—including the interpretation of findings.

Importantly, Blakeslee and colleagues clearly state that they are not advocating for a bottomless relativism. They want the views and interpretations of research participants to be prominent in our research reports, and they are comfortable with the idea that research conclusions may be ambiguous, with disagreements between participants and researchers. At the same time, though, they insist that while research findings can and should be shared with and commented on by research participants, findings must always be based on evidence, and inter- pretations must be justified by logical arguments. They also believe that a qualitative report must explicitly acknowledge the biases and viewpoints of both researchers and participants, so readers can judge for themselves the congruence between methods, findings, and interpretations.

Happily, their article includes practical advice on and examples of how we might involve research participants in the interpretation of qualitative data. They advise us to be open to disagreement, to accept the contingent nature of research knowledge, to view research as an emergent process of learning, and to recognize that a single research endeavor may give rise to a variety of interpretations and narratives. They recommend that we give participants a chance to review our interpretations and that we highlight (rather than conceal) differences in interpretation.

In our view, this article is an important and ground-breaking conceptualization of qualitative research, and it bears noteworthy similarity to the more recent work of social psychologist Chris Argyris (2004), who argues in favor of an "implementable validity." Argyris states that social scientists ought to pursue research that ultimately promotes improved performance. To accomplish this goal, he says that it is often appropriate to gather data and interpretations from research participants, but then those meanings must be examined to see whether they are based on verifiable data or reflect the biases, assumptions, and values of research participants. Like Blakeslee and colleagues, Argyris conceives of research as promoting actionable knowledge. The ultimate purpose of social science research is to enhance the value of the human experience, and hence a valid research study is one that results in improvements to our social and task systems.

We notice one change between the time this essay was published and today. The influence of the "dominant paradigm of the academy" (by which they mean a positivist and quantitative worldview), which the authors refer to toward the end of their article, has relented somewhat today. The paradigm wars are over, having fallen prey to the indifference of new generations of researchers. Qualitative methods have now entered the mainstream.

References

Argyris, C. 2004. *Reasons and Rationalizations: The Limits to Organizational Knowledge.* Oxford, UK: Oxford University Press.

Lincoln, Y. S., and E. G. Guba. 1985. *Naturalistic Inquiry.* Beverley Hills, CA: Sage Publications.

Recent developments in qualitative research—in line with postmodern, poststructuralist, and feminist thinking—raise questions about how to evaluate the credibility of our work. How do we answer these concerns and still produce accounts that are adequate and credible? Or, as Steven Athanases and Shirley Brice Heath ask, "Is anything we do either fully adequate or ideologically safe?" (1995, 278). Students enrolled in a graduate seminar on qualitative methodology led by one of us (Blakeslee) have wondered how they can make their accounts multi-vocal, self-reflexive, representative of and accountable to their participants, and credible at the same time. In other words, the practices that postmodern, poststructuralist, or feminist stances on qualitative inquiry suggest can make such inquiry seem impossible. However, accountability, multi-vocality, and self-reflexivity on the one hand, and credibility on the other are not necessarily mutually exclusive and, in fact, can be very compatible. The challenge becomes one of sorting out and understanding these practices and avoiding the feelings of paralysis they can evoke. In this article, we address this challenge by suggesting a way to approach and evaluate our qualitative inquiry that is sensitive to these recent developments and that is also reasonable. In presenting our ideas, we foreground concerns that we believe are particularly important to research in technical and scientific communication. Like David Altheide and John Johnson, we believe that we must set forth the goals of our inquiry before we can satisfactorily judge it (1994, 489). Thus, we stress that one important goal of such research is to improve our understanding of the settings and individuals we study through accounts that describe the rhetorical practices of our participants in ways that are meaningful and useful to them and to ourselves.

In this article, we are concerned, most generally, with the question of what makes qualitative research in technical and scientific communication good, valuable, or useful. We argue that an important criterion for determining this value and usefulness is whether researchers and participants believe in the findings of the research and are willing to act on them. In other words, we argue for judging how meaningful and worthwhile our accounts are, based on how well they inform practice and on what they teach us. Using such a criterion will require certain actions on the part of the researchers, which we elaborate in our discussion below. The actions we focus on include:

- Foregrounding and problematizing interpretive stances in our work—our own as well as those of our participants—instead of disguising and shrouding those stances. Also, acknowledging and embracing the discrepancies that may exist in these stances and engaging participants in dialogue as a way of identifying and understanding these discrepancies.
- Continuing to move our inquiry toward greater reflexivity and self-awareness, both for ourselves and for our participants.

These actions, and the evaluative criteria we propose, are consistent with suggestions some scholars have made in other recent discussions of how to evaluate qualitative work. Elliot Mishler, for example, advocates that we abandon the formal procedures and abstract rules traditionally used for evaluating such inquiry and that we look instead to criteria for evaluation that are relevant, meaningful, and within our grasp (1990, 418). Along similar lines, Martin Packer and Richard Addison (1989) argue that a valid or good account is that which is coherent, fits with external evidence, is convincing to others, and has the power to change practice. Patti Lather calls for a "reconceptualized validity that is grounded in theorizing our practice" (1991, 3), and Stephen Doheny-Farina argues for a more practical validity that is determined through the responses of our audiences: "validity is determined through a range of readings by audiences located within the researcher's discipline, as well as those located within the research sites" (1993, 261). Our belief that qualitative research in scientific and technical domains should be judged according to its usefulness and its ability to redirect practice is consonant with these scholars' views. In addition, by being responsive to and by implicating informants in these judgments of quality, we make our research more participatory and egalitarian, another concern many scholars now share.

Before considering the two actions articulated above, we discuss ways to reconceptualize validity that are consistent with our views of such research and with our ideas for evaluating it. We then explore how engaging the interpretive stances of our participants—particularly ones that may be contrary to our own— can enhance our understanding of the practices we study and make our work more meaningful. We also examine how greater reflexivity can make our work more accountable and useful. Underlying our perspectives on these issues is the belief that, as Renato Rosaldo has said, researchers must "grapple with the realization" that our objects of analysis are also "analyzing subjects who critically interrogate ethnographers—their writings, their ethics and their politics" (1989, 21).

Reconceptualizing Validity in Qualitative Research

Recent shifts in attitudes toward qualitative inquiry in general, and the criteria we suggest here for judging such inquiry in particular, suggest a need for reconsidering the notion of validity. By focusing on just one criterion for evaluating our qualitative studies of technical communication—that is, whether we find our accounts meaningful and capable of redirecting our scholarly and professional practices—and by examining ways to satisfy this criterion, we seek in this article to make the problem of evaluation in qualitative inquiry more manageable. We are not seeking to eliminate the notion of validity altogether, nor are we claiming to offer a comprehensive solution to the problems it poses. Instead we propose reconceptualizing validity so as to broaden it and make it more flexible. We argue, in particular, for making validity a more fluid notion that disperses authority, instead of situating it solely with the researcher.

Making Validity More Fluid

Because of the fluid and dynamic nature of most qualitative studies, we believe that we need to view validity as being less static and more generative than traditional usage of this notion suggests. Validity should be conceptualized in a way that accounts for and values the shifting and variable character of our research. In other words, we need to view validity as being more than a matter of determining whether, in fact, we are measuring what we think or say we are measuring, which is how many scholars continue to define validity (Goubil-Gambrell 1992, 587; Grant-Davie 1992, 280; Lauer and Asher 1988, 140).

If the phenomena we observe are, in fact, unstable, we can never ascertain the validity of our observations with any degree of certainty. Therefore, we should view validity, as Lather suggests, as "multiple, partial, [and] endlessly deferred" (1991, 5). Lather posits that validity is not about epistemological guarantees, because in the absence of value-free interpretations, or "truth," there are no such guarantees (1991, 5). Other scholars make similar arguments, which are frequently rooted in reactions to positivist epistemologies and to conceptions of validity implicated by them—that is, objectivist stances toward reality suggesting that observations can be somehow codified and/or measured, and that certainty and truth are attainable. In response to such stances, Sandra Harding points out that in qualitative inquiry there is no truth to be assessed, no interpretation-free facts, no ultimate external reality, and no Archimedean point (1991, 59). Similarly, Packer and Addison argue, "Those who seek fixed validity criteria are requiring something of interpretive inquiry that, in actuality, not even natural science can provide" (1989, 288). If we take these scholars' concerns seriously, we will make validity in qualitative inquiry a more fluid and generative notion that better accounts for the shifting and contingent nature of the phenomena we study. We will also embrace the tentativeness of our work rather than expect exactitude. We will make our accounts more honest, thus opening them to broader and more creative interpretations. In short, reconceptualizing validity as fluid and generative will help us to better account for the contingency and variability of our research sites, of our informants, our data, our findings, and our interpretations.

Dispersing Authority

We also believe that validity should be reconstructed to disperse or redistribute authority between researchers and the participants in their studies. Usually validity is thought of in terms that are fairly abstract, removed, and impersonal, and that tend to reify traditional subject/object distinctions. By engaging participants in judgments about the quality and usefulness of our work, we can minimize these distinctions and make validity a shared and more egalitarian notion; a viewpoint that other scholars seem to share. For example, Anne Herrington defines validity as an accountability to the situation being studied that can be judged by readers of the research and/or by the people who are actually in the situation (1993, 66). Altheide and Johnson likewise encourage researchers to obtain participants' perspectives on the research, and to show in their accounts where their own voices are located in relation to those of their participants (1994, 490). Finally, Lather suggests that we think of validity as the generation of counterpractices to authority that are grounded in the crisis of representation (1993).

In attempting to make validity a more fluid and egalitarian notion, we are not suggesting that researchers relinquish their authority entirely or that they bend completely to the perspectives of their participants. Instead, like Geoffrey Cross, we believe that a valid account is one that seeks to balance, rather than to privilege input from participants, the research community, the data, or the researcher (Cross 1994, 121). In other words, although we advocate taking our accounts back to our participants and valuing their responses to them, we do not suggest that we replace our perspectives with theirs. Neither are we advocating a phenomenological approach to validity—as suggested by Stephen North (1987)—in which our participants possess the authority to approve or disprove our analyses and interpretations. Instead, we agree with Doheny-Farina that it is a myth of internal validity to assume that our findings are somehow more valid if they match the perceptions of our participants (1993, 260). We also agree with Cross, who says that an important constraint with such approaches to validity is that researchers may be prevented from presenting those interpretations that, for some reason, are not agreeable to their participants, even when those interpretations are reasonable and respectful (1994, 120). As an example of this constraint, one of this paper's authors (Blakeslee) could have been prevented from publishing her research findings on the role of social interaction in scientific writing had her participants been given final say and authority in approving her interpretations (see "Continuing the Dialogue: One Researcher's Experiences," below). Thus, we believe that we should give equal consideration to our own and to our participants' perspectives, and that we should engage our participants in a dialogue in order to understand their concerns better and to improve the chances that our accounts will inform their practices, as well as our own. In the next section we discuss ways to solicit our participants' perspectives, and we address how we can use these perspectives to make our research more useful and meaningful.

Foregrounding and Problematizing the Interpretive Stances of Researchers and Participants: Learning from Contradiction and Dialogue

If we want to reevaluate validity and base our judgments about the quality of our inquiry on the believability of the findings we present, as well as on our own and our participants' willingness to act on them, we need to demonstrate in our inquiry greater responsibility to and concern for our research participants. Specifically, we need to acknowledge and account even more for their perspectives on our findings, and we need to be sensitive and responsive to these perspectives, as well as to any discrepancies that may occur in them. In this section we suggest ways to achieve these aims.

Failing to Acknowledge Participants

Although researchers have increasingly expressed concerns about their responsibilities to their participants and to the settings they study, they still rarely make explicit in their accounts the roles their participants played in their work, or the influences they may have had on it—even though such acknowledgments may be crucial for evaluating the research. Herrington, for example, addresses how ethnographic research has attempted to involve participants more, but that

> Still, the participants' role remains relatively passive, limited to offering information—being the "informants"—and serving in some way to validate emerging interpretations. The assumption is that the researcher will, through

analysis, interpretation, and writing of the account, take care to represent the multiple realities of the participants or informants.

(1993, 45)

Herrington goes on to say that too often we opt for studies in which our participants remain passive while we, as researchers, speak about and for them. However, "if our aim is to understand and represent the experiences of others, we need to involve them more centrally in shaping that knowledge" (1993, 51). Gesa Kirsch similarly addresses how we have failed to address dimensions of the researcher–subject relationship sufficiently in our field:

> The researcher–subject relationship is an issue that cuts across all research methodologies but has received little attention in composition studies. In fact, most discussions that address the researcher–participant relationship are designed to document the researcher's lack of interest in and interaction with research participants.
>
> (1992, 261–62)

In essence, although we say we are concerned with making our participants and their voices more pronounced in our work, we continue to adopt privileged and authoritative rhetorical stances—we foreground one story or version of our work despite the many other stories we could tell. We especially do this when we interpret and present our findings. Seldom do we acknowledge or interpret and present our findings and interpretations at various stages of our work. We also seldom indicate whether and over what issues our participants have disagreed with us, or the outcomes of those disagreements (whether the researcher reconceptualized or reconsidered interpretations because of them, or even acknowledged and addressed them). In fact, sometimes we avoid the practice of sharing our work with participants altogether, perhaps because we make ourselves vulnerable when we expose our work and our writing to the "Other(s)" whom we study. Virginia Olesen has noted how even in feminist research, where concerns with participants are usually central, the practice of taking accounts back to participants is widely discussed but seldom followed (1994, 166).

Alternately, those researchers who do follow such practices, and who acknowledge in their accounts that they shared their findings with their participants, seldom elaborate any outcomes from these actions, perhaps because they fear that contradiction will undermine the credibility of their accounts. They present their work, instead, in a manner that makes their findings appear seamless and unified, implying that the research participants agree with the findings and that the interactions between authors and participants have proceeded smoothly and agreeably. Such failure to seek or to acknowledge input from our participants is unfortunate, especially given the valuable insights we might gain from such practices. Kirsch, for example, notes how feminist scholars argue that the researcher–subject relationship can provide researchers with additional insights

(1992, 262). John Thompson likewise reminds us that findings of qualitative studies "stand in a relation of potential feedback with the very subject-object domain about which the results are formulated, in a way that has no direct parallel in the natural sciences" (1990, 376). Therefore, our failure to actively seek and acknowledge this feedback, we believe, constrains the usefulness of our accounts and of the research process more generally.

Encouraging and Valuing Contradiction

In regard to studying human behavior, Margery Wolf says: "When human behavior is the data, a tolerance for ambiguity, multiplicity, contradiction, and instability is essential" (1992, 129). She also maintains, "As ethnographers, our job is not simply to pass on the disorderly complexity of culture, but also to try to hypothesize about apparent consistencies, to lay out our best guesses, without hiding the contradictions and the instability" (Wolf 1992, 129). In the remainder of this section we concern ourselves with what such contradictions may reveal and what we can learn from them. We also argue for and highlight ways in which researchers can seek and learn from the interpretive viewpoints expressed by their participants, especially when they differ from the researchers' viewpoints.

We believe that researchers should seek and acknowledge the interpretive stances of their participants, even those that are discrepant and contradictory. In fact, such discrepancies should be viewed as productive occasions for learning. As Michael Jackson argues, interpretation becomes the practice of "actively debating and exchanging points of view with our informants. It means placing our ideas on a par with theirs, testing them not against predetermined standards in rationality but against the immediate exigencies of life" (quoted in Schwandt 1994, 132). Thus, rather than viewing any differences in perspective as an impasse, researchers should approach these situations with openness and interest and view them as productive occasions for furthering understanding. Such exchanges also present opportunities for sharing authority with the participants in our research, allowing us to recognize and acknowledge, as Paul Rabinow suggests, that there is no privileged position and no absolute perspective in qualitative inquiry (1977, 151). As Kirsch suggests, then, we should open up our research agendas to our participants, listen to their stories, and allow them to participate actively, as much as possible, in the design, development, and reporting of our research (1992, 257). We should view the researcher–participant relationship as flexible, cooperative, and mutually beneficial. We should also know that our research agendas can be grounded in participants' experiences and, most importantly in light of the concerns we present in this paper, that our findings can be relevant to both participants and researchers (Kirsch 1992, 263–64).

Studies of professional and scientific discourse, we believe, are particularly conducive to the dialogic processes we are advocating, partly because of the characteristics of the researcher–participant relationship in such studies. After all, our participants in these studies are usually professionals and practitioners in specialized fields or in academic disciplines. As such, they offer us informed insights into their professional activities. Sometimes we even share an understanding of these activities because of our own experiences and practice in settings similar to those we are studying—for example, the professional writer turned graduate student who studies communication in a corporate setting for thesis work, or the academic who studies other academics, though perhaps in different institutional or disciplinary locations. Failure to solicit, listen to, and seriously consider the perspectives of our participants may limit our studies and cause us to overlook important data and to miss important understandings. Thus, we believe that we should listen to and value the perspectives of our participants and give them opportunities, not unlike the opportunities we repeatedly give ourselves, to reflect on and to offer insights and commentary on their professional practices. We should also strive in our inquiry to arrive at an understanding not only *of* our participants and their actions but also *with* them. By so valuing the voices and authority of our participants, we believe that we can make our inquiry more accountable to them and more responsive to their needs, practices, and concerns.

Valuing the Interpretations of Participants: Realistic Approaches

Engaging in exchanges in which we encourage our participants to share their perspectives on our findings and on our interpretations requires that we embrace the interpretive and contingent nature of our work and view our inquiry as an open system characterized by dialogicality and recursiveness, rather than as a closed system that depends on a linear enactment of research, data analysis, and writing. Such exchanges also necessitate that we value the voices and perspectives of our participants, particularly in regard to their judgment of our findings and of our interpretations. Along these lines, Herrington suggests that researchers exhibit a commitment to listening to other perspectives and to trying to understand the assumptions underlying them (1993, 65). Thus, fostering an ongoing exchange with our research participants requires that we adopt certain dispositions toward them that support such interactions. In particular, researchers:

- Need to admit and avoid a tendency to seek agreement only. It is too easy and tempting to ask participants for their perspectives on our work and then to ignore those perspectives if they do not cohere with our own thinking and interpretations. Such actions constrain and limit our studies.
- Need to admit and work with the knowledge that there is no single, correct interpretation of any event, nor is there ever an absolute or pre-defined standard against which to measure one's interpretations. Consonant with such perspectives, researchers must also recognize and accept that disagreements or discrepancies in perspective need not undermine their authority, expertise, and credibility. Disagreement need not result in an impasse.

- Should view occasions for continuing their exchanges with their participants as opportunities for learning more about their participants and their practices, and as occasions for productively reconsidering and possibly even rethinking interpretations of those practices.
- Must accept that there are different ways to view and make meaning from situations and must view interpretations as multiple and fluid. Researchers must also remember and acknowledge that the theoretical and conceptual frameworks that they use influence the choices they make and the stories they tell in their accounts. Likewise, the theoretical and conceptual frameworks that participants possess influence how they tell their stories.

Given the dispositions addressed above, some of the strategies researchers can use to foster exchange and dialogue with their participants include:

- Asking participants to review interpretations and/or written accounts portraying their practices and experiences. We should invite and acknowledge their participation in and contributions to our work, and we should foster a mutual exchange of information and knowledge with them throughout or at particular stages in our inquiry. We should also avoid assuming, at any point in our research, that we know how our participants feel about or view our work.
- Highlighting and differentiating in our accounts, where appropriate, our own and our participants' viewpoints, perspectives, voices, and presence. We should strive to highlight in our texts our own, as well as our participants', experiences, presence, and contributions to the inquiry—writing both ourselves and our participants into our accounts.
- Describing how we arrived at and formed our interpretations and the roles that our participants played in these activities. We should also indicate, when appropriate, how we negotiated or renegotiated our interpretations with our participants.
- Being more reflexive in our accounts about our approaches to research, our interpretations, our relationships with participants, and our positioning and authority in relation to them. We should also reflect on our participants' interpretations, viewpoints, positions, and authority (see "Approaching Our Inquiry and Writing Our Accounts Reflexively," below, for further elaboration on this final strategy).

We are not arguing here for adopting a relativistic stance through such activities, where one reading of our work is as good as any other. Instead, we agree with Kirsch, who points out that in dialogic, reflexive, and participatory research we still need to provide evidence for our findings, to develop arguments for our interpretations, and to convince readers that our interpretations are feasible (1992, 259).

Another important factor we need to consider when we seek the input and perspectives of the participants in our inquiry is that our relationships with and our roles in relation to those participants may shift throughout our studies. These shifts may lead us, along with our participants, to view and interpret our findings differently at different points during our inquiry. For example, a novice researcher using the tools and strategies of qualitative research for the first time may view a set of findings one way at the beginning of the study and another way further along, with the benefit of greater familiarity and facility with using such approaches to inquiry. The perspectives of more experienced researchers may change too, as they become more familiar with and accustomed to the settings they are studying. Likewise, participants' perspectives may change as they learn more about the work and become more experienced at and more accustomed to being participants. Rabinow addresses this latter issue when he discusses how informants learn to explicate their cultures and practices, to become self-conscious about them, and to objectify them (1977, 152). It would be naïve for us to assume, then, that our participants never reflect on or consider their roles, behavior, and comfort as participants; that they do not somehow change during our inquiry or are not somehow altered by it.

Shifts in our professional roles and status (for example, when a graduate student becomes an assistant professor) can also influence researchers' and participants' views toward and responses to the inquiry. For example, when exposure to the work increases through publication, and more is at stake with the work, participants may become more self-conscious and critical of how researchers represent and portray them (see "Continuing the Dialogue: One Researcher's Experiences," below). Thus, researchers need to acknowledge their evolving relationships and potentially shifting roles in relation to participants as factors that may influence how they and their participants view and interpret their findings. Sharing interpretations and findings with participants, and listening to their concerns throughout the inquiry, can mitigate the potential problems and misunderstandings that may occur as a result of these shifts and changes.

Benefiting from Considering Alternative Perspectives

All of the strategies addressed in this section in some way situate our participants in our inquiry and make such inquiry more responsive and accountable to participants, who still remain conspicuously absent from many qualitative accounts. Using these strategies can help us to eliminate this absence, as well as accomplish several other aims and objectives in keeping with more participatory conceptions of qualitative inquiry. In particular, researchers:

- May gain new insights in their work by checking findings and interpretations with participants. Such exchanges may lead to a dialogue that extends the work and contributes new understandings and meanings.
- Can increase the credibility of their studies and foster greater trust and respect in their relationships with participants by not casting the practices of participants solely in abstract, theoretical terms.

- Can give participants a greater stake in the research and in its outcomes by inviting participants to contribute more actively to the work, which, in turn, may increase participants' engagement in and their commitment to the research.
- Can begin to close the gap between our scholarly research in technical and scientific communication and the day-to-day work and practices of those we study by engaging participants in a dialogue about the findings of the inquiry. In essence, researchers and participants can learn with and from each other by sharing their interests and concerns with understanding and improving practice.

In regard to this final benefit, Thomas Pinelli and Rebecca Barclay have argued similarly that we need mechanisms to link researchers and theoreticians with technical communication practitioners and to translate the results of research into practice (1982, 531). They propose that technical communicators in universities should collaborate more closely and frequently with the directors and managers of technical communication programs in government and industry. Such cooperation, they argue, could provide opportunities for field research that would produce findings vital for solving professional problems (Pinelli and Barclay 1982, 531). Their recommendations are closely connected to and in accord with our concerns with making our inquiry more accountable to practice.

Seeking the outcomes described above can make our work more conscientious, responsible, and ethical, as well as more credible—in ways similar to those described by Doheny-Farina (1993, 261) and Altheide and Johnson (1994)—whereas failure to seek them may result in our reinforcing and valuing our authority as researchers over that of our practitioner participants. This failure may in turn lead us to valorize our own understandings and interpretations of our participants' experiences at the expense of their understandings and interpretations, which otherwise could inform and enhance our understandings.

Below we offer an example of how one of the authors of this paper (Blakeslee) used some of these strategies in her own work, but only after she learned that one of her participants disagreed with her interpretation of one of his actions. (Because this example was written as a personal reflection by Blakeslee, it is presented here in the first person.) Through the discussion that resulted from this disagreement, the researcher and her participant arrived at a broadened conception of collaboration. What we hope to demonstrate through this example is how discussions with our participants can generate additional understanding and knowledge for both researchers and participants. Such exchanges, we contend, can help us to enhance our accounts of those situations and settings.

Continuing the Dialogue: One Researcher's Experiences

In 1991, as part of my dissertation work, I conducted a nine-month qualitative study of three theoretical physicists as they conducted research and wrote two papers on the subject of molecular simulation (Blakeslee 1992). My primary sources of data for this study included observations of meetings at which the physicists reviewed drafts of their papers, interviews in which I followed up on the discussions in the meetings, and analyses of all of the drafts they produced. I also conducted periodic interviews with two specialist informants who were not members of the research group. One was an experimental physicist who worked in the same field as the primary participants (condensed matter physics), and who was familiar with their work. The other was an experimentalist from a different field (high energy particle physics), who was unfamiliar with the physicists' work, but who served as an expert informant on rhetorical practice in his area of physics. During the study I periodically checked my research findings with my primary participants, and I also discussed my observations with the two specialist informants. After completing my study I continued to engage in these discussions in order to verify my observations and to improve my understanding of the technical subject matter with which I was dealing. At this stage of my work I also gave my participants chapters of my dissertation to read.

As I reflect on my experience of sharing my initial and then some later accounts of my work with my participants, I realize that my own and my participants' shifting roles at various stages of the study influenced their responses to my work. For example, when I was still a graduate student my participants viewed me, at least partly, as someone they were helping. They often said that they liked that they were helping me obtain my degree. The postdoctoral fellow and the graduate student, in particular, read my work in a manner not unlike that in which I read theirs; with interest and curiosity and with the intention of being helpful. These participants also seemed intrigued with being involved in my study and being written about. I have often suspected, however, that this intrigue may have supplanted their critical stance toward my work. They seldom, if ever, questioned or criticized my interpretations of their activities in any of my accounts.

The professor in the group, Robert Swendsen, was also interested in and intrigued by my work and by being a participant in it. However, Swendsen's status as a professor and as a member of my thesis committee seemed to make him more willing to question my interpretations and to ask for elaboration of various issues. Still, he never questioned the legitimacy of anything I said about the physicists. I believe that our respective roles, along with the document I was writing, my thesis, had something to do with this. First, as a graduate student I was in a subordinate role in relation to my professor participant. Second, because I was only writing my thesis, my work had a limited audience—my committee, other faculty members in my department, and perhaps other graduate students. However, after I left graduate school and started to publish my work, my audience expanded, and it was at this point that Swendesen's views on and responses to my work seemed to change. In particular, he became more self-conscious about and more critical of my interpretations of his actions.

The first publication that I shared with Swendsen was an article on collaboration that appeared in the Winter 1993 issue of *Technical Communication*

Quarterly (Blakeslee 1993). Because I extracted the material for this article from my thesis, which Swendsen had approved without censor, I saw little reason to ask him to review it a second time. Therefore, I waited until after it was published to send him a copy. However, in a subsequent e-mail message Swendsen conveyed his dissatisfaction with the article. He said that the argument I had made about the collaborative nature of his interactions with three scientists who read and responded to one of the physicists' drafts misrepresented what had occurred in this situation. He also said that my use of the term "collaboration" was too inclusive and, therefore, not useful. (I had called the group's interactions with these readers a collaboration because of the extent to which the feedback the readers gave the physicists influenced and ultimately altered their text.)

In response to Swendsen's concerns, I reexamined the article, interviewed Swendsen, and constructed a response. My interviews with Swendsen revealed something I had overlooked and failed to consider in my earlier interpretations; namely, Swendsen had a fixed conception of collaboration that was tied to his disciplinary experiences and perspectives. I, too, had a conception of collaboration that stemmed from my own theoretical and disciplinary perspectives. These conceptions differed. While I took as collaborative any direct or indirect influence of another on the form and/or long-term meaning and consequences of a text, Swendsen defined collaboration in terms of a direct commitment to solving a scientific problem. Further, for Swendsen, the notion of collaboration appears to hold an important place in the reward and responsibility system in science. A collaborative commitment for Swendsen entails particular kinds of stances, activities, and responsibilities in regard to a text. In other words, whereas I took account in my conception of the complex social processes through which textual artifacts and meanings emerge, considering the overall dialogicality of text, Swendsen had in mind specific roles, along with specific kinds of rewards, blame, and personal responsibility in regard to collaborations.

As a result of continuing my discussions with Swendsen, I was able to improve my understanding of his activities and of his perspectives on them. In elaborating why he views collaboration the way that he does, Swendsen helped me to better understand the roles and functions of various collaborative-like relationships in scientific activity, especially as they relate to attributions of reward and responsibility in science. I, in turn, helped him to understand better my conceptions of and perspectives on collaboration from a rhetorical perspective. Both of us, I believe, arrived at an expanded understanding of the notion of collaboration, or at least a broadened sense of our varied perspectives on it. We also discovered how our respective disciplinary locations and concerns influenced how we defined and viewed collaboration. We learned that we had different, but still legitimate reasons and motivations for holding our respective viewpoints. Had I ignored Swendsen's objections to my arguments, or failed to articulate them in one of my subsequent accounts (Blakeslee 2001), which admittedly was tempting, I would have missed an opportunity to understand his viewpoints better, and he mine. I would also have prevented others from hearing his viewpoints. Instead, Swendsen's objections caused both of us to reflect further on our understandings and interpretations. Addressing the tensions that arose in our relationship allowed us to clarify and differentiate our positions and perspectives and to learn from them. It also prompted us to adopt a productive critical stance toward our interpretations, ultimately making those interpretations more meaningful to ourselves and, hopefully, to others.

Continuing the Dialogue: Other Examples

In addition to the example of these practices offered in Blakeslee's work, other scholars who study technical and scientific communication have also made efforts to acknowledge and address the perspectives of their participants in their accounts. One notable example is provided by Dorothy Winsor's (1996) book, which describes the experiences of four engineering co-op students as they wrote at work over a five-year period. In her book Winsor includes sections titled "Backtalk" after each of the chapters in which she discusses her participants. In these sections, Winsor presents letters written to her by the students that recount their experiences as participants in the study and their reactions to those experiences and to reading about themselves. Winsor's concern in including these sections is wedded to a concern with her inability as a researcher to convey the true complexity of her subjects in her account. In discussing this constraint she says, "The people they [our research accounts] are about are flattened out to become research 'subjects,' a term that is curious because it really means that they have become objects" (1996, ix). Winsor refrains from offering commentary on these letters because she wants them to convey the unmediated voices of the students: "I wanted them to have the last word" (1996, ix–x). However, she also notes that, "despite my best intentions, my words surround theirs and affect how they are read" (1996, ix–x).

In addition to being surrounded by Winsor's text, another factor that may mediate the students' letters, though indirectly and perhaps unintentionally, is the fact that they were requested by and addressed to Winsor. Each of the writers makes frequent reference to Winsor and to the interactions in which they engaged with her. Each also is very complimentary of the book and of Winsor's roles as researcher and author. Yet despite these potential imitations, the letters do in fact serve as a forum for reflection for Winsor's participants. Two of the four students, in particular, used the letters to reflect further on their experiences during the inquiry and to elaborate their perspectives on those experiences and on the research more generally. The insights the students articulate in these letters offer an additional source of information and understanding to Winsor's account. One student even concluded his letter by saying, "I thought that I would give you a little more information on how I see my writing in the workplace without being asked the questions" (1996, 103).

One other example we would like to offer here is provided by Gregory Clark and Stephen Doheny-Farina's 1990 account of a case study about a student writing in a school and in an organizational setting. In this account the authors describe how their analysis of their subject's experiences was problematized when they considered alternative interpretations of the case (1990, 456). The responses of the reviewer, editors, and others who read their account, they said, caused them to realize that their reading and account of the situation ended up concealing as many interpretations and insights as it revealed (1990, 479). They concluded that the best thing to do in this situation was to further the discussion of the case; therefore, they shifted the focus of their work from the experience of their subject to their own experience in making sense of it (1990, 479–80).

Clark and Doheny-Farina argue in this article that the best way to build theory in our field is by exchanging and examining diverse interpretations and arguments (1990, 480). They argue that theory building in composition needs to be viewed as a collaborative process that entails ongoing critical exchange (1990, 457). In other words, they suggest that we must situate theory building in an ongoing exchange of analyses and, in so doing, confront conflicting interpretations of how writing is done and how it should be taught (1990, 460). They also address in their work how the ways in which we choose constructs and interpretive frameworks in our qualitative inquiry are limited and unstable, and they note how in this situation the alternative interpretations caused them to acknowledge and examine this instability (1990, 457, 478).

We offer Clark and Doheny-Farina's account as an example here because it argues for practices not unlike those we are suggesting in this paper, particularly in regard to entertaining alternative perspectives on our work. The difference between their argument and ours concerns whose alternative perspectives we entertain. Clark and Doheny-Farina address in their work the contradictory perspectives offered by scholars in their own disciplinary field—editors, reviewers, and colleagues. We argue for also considering contradictory perspectives of participants in research. We believe that we have as much, if not more responsibility to the participants in our research as we do to our colleagues. We also believe that we stand to gain a great deal by involving and engaging our participants even more in our inquiry, particularly in regard to our understandings of their activities and our credibility in portraying those activities and making them meaningful.

Understanding What Underlies Our Own and Our Participants' Interpretations: A Caution

Before concluding this section, we want to offer a caution to scholars in regard to using these practices. Namely, we must remember that, as Rabinow suggests, our participants' interpretations are as mediated by history and culture—and by other life factors and circumstances—as our own (1977, 119). In some cases, our research participants may hold perspectives and viewpoints that seem irreconcilably counter to or at odds with our own. Such discrepancies may be attributable to differing ideologies or other deeply held beliefs or viewpoints. In some, though not all cases, articulating these contrary perspectives may impede our understanding or prove counterproductive to our work. Also, our participants at times may have difficulties grasping our conceptual or theoretical frameworks. Therefore, as researchers we must consider and explore all of these possibilities and seek to understand what underlies them. In other words, we should not accept either our own or our participants' perspectives at face value. Rather, if our aim is to improve our own and our participants' practices, we should consider the various perspectives that may exist on such practices thoughtfully and reflexively. Doing so may guard against potential errors in interpretation and against misunderstandings that may prevent us from achieving our larger aims. In the next section we address the importance of engaging in such reflection as well as approaches for and considerations in doing so.

Approaching our Inquiry and Writing our Accounts Reflexively

Many recent critiques of qualitative inquiry have encouraged greater reflexivity in regard to the processes and products of such inquiry, greater self-consciousness on the parts of researchers, and the recognition that all experience is mediated. Kathleen Jones defines reflexivity as a concern with scrutinizing the beliefs and practices of the researcher insofar as they contribute to her findings (1993, 200). Other scholars address how reflexivity is implicated in a reconceptualized validity. Harding, for example, writes, "Currently paradigmatic uncertainty in the human sciences is leading to the re-conceptualizing of validity. . . . Our best tactic at present is to construct research designs that demand a rigorous self-reflexivity" (1991, 66). Similarly, Donna Haraway argues for "politics and epistemologies of location, positioning, and situation, where partiality and not universality is the condition of being heard to make rational knowledge claims" (1988, 589). According to these scholars, reflexivity should lead researchers to adopt a politics of position and to infuse their accounts with a sense of situatedness.

Using Reflexivity in a Manner Consistent with our Evaluative Criteria

We, too, are concerned with what reflexivity can contribute to accounts of our qualitative inquiry. The question with which we are concerned most is: How can we use reflexivity in a manner that is consistent with our concerns for evaluating our inquiry on the basis of how well it informs and redirects our practices? Doheny-Farina argues that research on writing in nonacademic settings is shaped significantly by researchers' own rhetorical agendas (1993, 266). Similarly, Tyler Bouldin and Lee Odell address how our images of the world and our assumptions play a large role in how we view and interpret our research findings (1993, 277). Such images and assumptions, we would argue, need to be acknowledged and articulated in our work, as do the assumptions and values of our participants, whose images of the world and knowledge play a role in how they view us, our research, and their own practices. In other words, we need to remember that the participants in our inquiry are no more disinterested and impartial than we are. Therefore, in addition to our own, we must bring our participants' biases, positions, and perspectives to light in our accounts. Addressing these perspectives and biases will allow readers to see and to understand better what underlies the various stances researchers and participants take toward the inquiry.

Being Reflexive: Suggestions and an Example

Anne Herrington (1993) provides suggestions that may help in the endeavor to achieve these reflexive ends. Herrington argues for the need for researchers to unmask ourselves in our accounts and to acknowledge our situatedness. She suggests that such reflection sheds light on the assumptions that underlie what we do, on the factors and circumstances that influence our problems and questions, on the views and voices represented in our work, and on our audiences and purposes (1993, 41, 57). Reflection, she says, can also lead to a heightened awareness of how oneself and one's research decisions shape what is seen (1993, 52). She argues, in particular, for the need to acknowledge that ideology and values are implicated in all research (1993, 58–59), and that we should examine and articulate our values, background, and involvements, bringing to our own and our readers' awareness how these shape the meanings we make: "Because the 'data' we select and make some sort of meaning of are shaped by those personal factors, being 'accountable to the data' requires such self-examinations. Being accountable to others requires examining our social conscience" (1993, 64). Herrington also reflects, retrospectively, on her study, conducted in the 1980s, of the writings that occurred in two engineering classes. She admits that while she was engaged in this research she was not very aware of the assumptions implicit in the practices she followed. She also considers how this work might have been different had she approached it more reflexively (1993, 42, 50).

Herrington articulates several questions that she believes we should pose if we are to conduct qualitative inquiry reflexively (1993, 42). These include:

- Whose views are included in the research and who represents them?
- What roles do participants play throughout our studies?
- How do researchers recognize and/or address their own roles and the influence of those roles in shaping their findings?
- How do researchers recognize and/or acknowledge the functions of their ideologies and values in their studies?

In addition to Herrington's questions, we believe we also need to ask:

- How do researchers recognize and/or address their participants' roles and the influence of those roles in shaping their findings?
- How do researchers recognize and/or acknowledge the functions of their participants' ideologies and values in their studies?

Addressing these questions can help us reveal and reflect on our own, as well as our participants' positions in and perspectives on our research. Such reflection, we believe, can also enhance the value and usefulness of our work, for our participants and for ourselves.

Acknowledging the Risks of Reflexivity

Although the call for greater reflexivity in our inquiry is an important and much needed one, it also carries some risks. In particular, researchers must be aware that the self-reflexivity characteristic of some recent ethnographic forms can sometimes become problematic. For example, if carried too far, writing in a more reflexive style, or "I-Witnessing" as Clifford Geertz calls it (1998, 78), can crowd out our participants and lead to what Geertz calls "the diary disease" (1998, 90). Doheny-Farina similarly cautions that reflexivity can move us too far inward, with one consequence being that it may lead us away from the issues we need to study (1993, 267). Because too strong an imposition of ourselves and of our disciplines may eclipse the sociohistorical contexts of the settings we study and the experiences and practices of our participants, we must, as Doheny-Farina suggests, balance such self-consciousness with attempts to engage in our inquiry systematically and perceptively (1993, 267).

Sharon Traweek's (1988) account of high energy particle physics, we think, offers a good example of reflexivity used reasonably. Traweek moves easily between her rhetorical positioning in the text as narrator and her positioning as a researcher in the field. Traweek was concerned in her study with the interactional dynamics among large groups of collaborating physicists that occurred at two particle accelerator sites—one in the US and the other in Japan. In her account she acknowledges her status as subordinate to her participants. She highlights the importance of the relationship between a researcher and her participants, and she conveys in her account a sense of proximity to her participants. There are also acknowledgments throughout the account of reciprocity—a mutual exchange of information and knowledge between herself and the physicists.

Throughout her book Traweek uses strategies consistent with our goals for more reflexive, as well as more participatory approaches to situated inquiry. Traweek is reflexive about her positioning and authority in relation to that of her participants, and about her gender and nationality—and in connection with this, about the role of gender and race in her research, particularly in relation to her mostly male participants and to her Japanese participants. Traweek also highlights her own presence throughout her work, frequently writing in the first person. She asserts that it is essential for researchers to "write themselves into their accounts and to describe the conditions under which they formed their interpretations" (1988, ix). She describes how she accomplished this goal in her work by "[choosing] a form of this book that underscores the interaction between me, the physicists I study, and the readers of the book" (1988, x). She claims that she discusses "what counts to the physicists as interesting" (1988, x), and that her work "is an account of how high energy physicists see their own world" (1988, 1). She assures us that "the anthropologist no longer has the last word in the dialogue of fieldwork," saying, "when I submit a manuscript for publication, those who are asked to review its merits always include physicists" (1988, 6). Traweek's use of reflexivity is in accord with our beliefs that accounts of how we arrive at our interpretations and perhaps even renegotiate them with our participants can be interesting and informative to our readers, and that such accounts can make our work more meaningful and authentic.

Addressing Factors that may Constrain our Practices

As much as any of us may agree with and wish to adopt the practices we have addressed in this article, we still face many methodological, institutional, and pragmatic constraints in doing so. For example, because of our disciplines' expectations for accountable, credible, and supposedly objective research, there may be little room or opportunity for doing what is needed to make our accounts more meaningful and useful to our participants and to ourselves, such as acknowledging discrepant perspectives and being more reflexive in regard to our findings. In other words, despite our increasing acknowledgment of the contingent, partial, and subjective nature of all research, we still face institutional pressures to exclude elements from our accounts that suggest their partiality and to exclude, especially, any discussion of how our research and/or our findings may have been influenced by our personal agendas and biases. Patricia Sullivan addresses how the dominant paradigm in the academy suggests that such agendas and biases are not to influence observations and analysis if the researcher is to arrive at an objective and undistorted picture of reality (1992, 55). Doheny-Farina suggests that because of this paradigm researchers still aspire to a certain scientific pretense in their work (1993, 257). Researchers and journal editors alike, he says, are apt to reject self-reflexive practices since they admit subjectivity (1993, 262). These ossified academic structures, as Olesen calls them (1994, 167), deter researchers from acknowledging discrepant perspectives and from reflecting on the problems and limitations of their work, even though such practices can contribute something to the work.

Another constraint we may face in trying to carry out the practices we suggest in this paper is posed by the authority we possess as researchers. Judith Stacey, for example, notes how

> ethnographic method appears to (and often does) place the researcher and her informants in a collaborative, reciprocal quest for understanding; but the research product is ultimately that of the researcher, however modified or influenced by informants. With very rare exceptions it is the researcher who narrates, who 'authors' the ethnography. . . . [A]n ethnography is a written document structured primarily by a researcher's purposes, offering a researcher's interpretations, registered in a researcher's voice.
>
> (Stacey 1991, 114)

Joan Acker, Kate Barry, and Johanna Esseveld also conclude that "it is impossible to create a research process that erases the contradictions (in power and consciousness) between researcher and researched" (1991, 150). In other words, we can never totally relinquish or escape our authority as researchers. However, such authority need not be unproductive or negative. Both researchers and participants, we believe, bring perspectives and viewpoints that can inform the inquiry, and we should acknowledge and value the authority of each in our work. In our view authority becomes a problem only if and when we use it to overshadow, appropriate, or exploit our participants. We must engage, instead, in research that is self-reflexive and dialogic, as we have just discussed.

Conclusion

In this article, we have explored a way to assess qualitative inquiry in technical and scientific communication in a manner consistent with the aims of such inquiry to improve and redirect practice. We have argued for the need to consider and embrace the interpretive perspectives and stances of our informants, even when those stances differ from our own. Such discrepancies, we believe, can be a source of additional understanding and insights, for us and for our participants. In other words, we believe that by acknowledging and embracing disagreement, and by engaging in a dialogue with our participants, we can enlarge the scope of our inquiry and of our understanding of it. As professionals interested in understanding discursive activity in technical and scientific contexts, we should value and acknowledge the various viewpoints our inquiry evokes and seek ways to examine and better understand those viewpoints. Engaging in such activities can improve our understanding of the situations and practices we study.

Acknowledgment

*This article was originally published in *Technical Communication Quarterly* (1996), 5: 125–49.

References

Acker, J., K. Berry, and J. Esseveld. 1991. "Objectivity and Truth: Problems in Doing Feminist Research." In *Beyond Methodology: Feminist Scholarship as Lived Researched*, ed. M. M. Fonnow and J. A. Cook. Bloomington, IN: Indiana University Press.

Altheide, D. L., and J. M. Johnson. 1994. "Criteria for Assessing Interpretive Validity in Qualitative Research." In *Handbook of Qualitative Research*, ed. N. Denzin and Y. Lincoln. Thousand Oaks, CA: Sage.

Athanases, S., and S. B. Heath. 1995. "Ethnography in the Study of the Teaching and Learning of English." *Research in the Teaching of English* 29: 263–87.

Blakeslee, A. M. 1992. "Inventing Scientific Discourse: Dimensions of Rhetorical Knowledge in Physics." PhD dissertation, Carnegie Mellon University.

——. 1993. "Readers and Authors: Fictionalized Constructs or Dynamic Collaborations?" *Technical Communication Quarterly* 2: 23–35.

———. 2001. "Sorting Out Social Influences: Distinguishing Authorial, Audience, and Other Roles in Scientific Work." In *Interacting with Audiences: Social Influences on the Production of Scientific Writing*. Mahwah, NJ: Erlbaum.

Bouldin, T., and L. Odell. 1993. "Surveying the Field and Looking Ahead: A Systems Theory Perspective on Research on Writing in the Workplace." In *Writing in the Workplace: New Research Perspectives*, ed. R. Spilka. Carbondale, IL: Southern Illinois University Press.

Clark, G., and S. Doheny-Farina. 1990. "Public Discourse and Personal Expression: A Case-Study in Theory-Building." *Written Communication* 7: 456–81.

Cross, G. A. 1994. "Ethnographic Research in Business and Technical Writing: Between Extremes and Margins." *Journal of Business and Technical Communication* 8: 118–34.

Doheny-Farina, S. 1993. "Research as Rhetoric: Confronting the Methodological and Ethical Problems of Research on Writing in Nonacademic Settings." In *Writing in the Workplace: New Research Perspectives*, ed. R. Spilka. Carbondale, IL: Southern Illinois University Press.

Geertz, C. 1988. *Works and Lives: The Anthropologist as Author*. Stanford, CA: Stanford University Press.

Goubil-Gambrell, P. 1992. "A Practitioner's Guide to Research Methods." *Technical Communication* 39: 582–91.

Grant-Davie, K. 1992. "Coding Data: Issues of Validity, Reliability, and Interpretation." In *Methods and Methodology in Composition Research*, ed. G. Kirsch and P. Sullivan. Carbondale, IL: Southern Illinois University Press.

Haraway, D. 1988. "Situated Knowledges: The Science Question in Feminism and the Privilege of Partial Perspective." *Feminist Studies* 14: 575–99.

Harding, S. 1991. *Whose Knowledge? Thinking from Women's Lives*. New York, NY: Columbia University Press.

Herrington, A. J. 1993. "Reflections on Empirical Research: Examining Some Ties Between Theory and Action." In *Theory and Practice in the Teaching of Writing: Rethinking the Discipline*, ed. L. Odell. Carbondale, IL: Southern Illinois University Press.

Jones, K. B. 1993. *Compassionate Authority: Democracy and the Representation of Women*. New York, NY: Routledge.

Kirsch, G. 1992. "Methodological Pluralism: Epistemological Issues." In *Method and Methodology in Composition Research*, ed. G. Kirsch and P. Sullivan. Carbondale, IL: Southern Illinois University Press.

Lather, Patricia. 1991. *Getting Smart: Feminist Research and Pedagogy with/in the Postmodern*. New York, NY: Routledge.

———. 1993. "Fertile Obsession: Validity after Poststructuralism." *The Sociological Quarterly* 34: 673–93.

Lauer, J. M., and W. J. Asher. 1988. *Composition Research/Empirical Designs*. New York, NY: Oxford University Press.

Mishler, E. 1990. "Validation in Inquiry-Guided Research: The Role of Exemplars in Narrative Studies." *Harvard Education Review* 4: 415–42.

North, S. M. 1987. *The Making of Knowledge in Composition: Portrait of an Emerging Field*. Upper Montclair, NJ: Boynton/Cook.

Olesen, V. 1994. "Feminisms and Models of Qualitative Research." In *Handbook of Qualitative Research*, ed. N. Denzin and Y. Lincoln. Thousand Oaks, CA: Sage.

Packer, M. J., and R. B. Addison, eds. 1989. *Entering the Circle: Hermeneutic Investigation in Psychology*. Albany, NY: State University of New York Press.

Pinelli, T. E., and R. O. Barclay. 1982. "Research in Technical Communication: Perspectives and Thoughts on the Process." *Technical Communication* 39: 526–32.

Rabinow, P. 1977. *Reflections on Fieldwork in Morocco.* Berkeley, CA: University of California Press.

Rosaldo, R. 1989. *Culture and Truth: The Remaking of Social Analysis.* Boston, MA: Beacon Press.

Schwandt, T. A. 1994. "Constructivist, Interpretivist Approaches to Human Inquiry." In *Handbook of Qualitative Research,* ed. N. Denzin and Y. Lincoln. Thousand Oaks, CA: Sage.

Stacey, J. 1991. "Can There Be a Feminist Ethnography?" In *Women's Words: The Feminist Practice of Oral History,* ed. S. B. Gluck and D. Patai. New York, NY: Routledge.

Sullivan, P. A. 1992. "Feminism and Methodology in Composition Studies." In *Methods and Methodology in Composition Research,* ed. Gesa Kirsch and Patricia Sullivan. Carbondale, IL: Southern Illinois Univerrsity Press.

Thompson, J. B. 1990. *Ideology and Modern Culture: Critical Social Theory in the Era of Mass Communication.* Cambridge, MA: Polity.

Traweek, S. 1988. *Beamtimes and Lifetimes: The World of High Energy Physics.* Cambridge, MA: Harvard University Press.

Winsor, D. 1996. *Writing Like an Engineer: A Rhetorical Education.* Mahwah, NJ: Erlbaum.

Wolf, M. 1992. *A Thrice Told Tale: Feminism, Postmodernism and Ethnographic Responsibility.* Stanford, CA: Stanford University Press.

Chapter 3

The Future of Technical Communication

How Innovation, Technology, Information Management, and Other Forces are Shaping the Future of the Profession*

Barbara A. Giammona

Editors' Introduction

Barbara Giammona's "The Future of Technical Communication: How Innovation, Technology, Information Management, and Other Forces are Shaping the Future of the Profession" is part mirror, part time machine. The article reflects the current state of the profession (not so different today than when she wrote the piece in 2004) and also suggests what technical communication might become in the not too distant future—both for good and for ill.

We included this article in our anthology for several reasons. First, it is a superb example of qualitative research in its design, execution, and reporting. Using the data gathered from questionnaires and interviews, it creates a "rich" or "thick" description of the current and future states of our profession. Giammona combines frequent quotations and paraphrases from the open-ended responses to her questionnaire as well as the comments of interviewees in a masterful narrative that helps the reader understand where we are and where we are headed, as seen through the eyes of the 28 technical communicators who contributed to her study. And these participants are representative of the wider profession in terms of the countries where they live and the kinds of work that they do.

Another reason for including this essay in the collection is the skill that the author uses in briefly detailing her research methodology (supplemented by the complete text of her extensive questionnaire in Appendix 3.1) and in working a thorough review of the literature into the text of the article.

Finally, we added this article to the collection because we will use it as a touchstone to discuss and analyze the articles in chapters 4 through 14 in the collection that follows it. We will examine these other articles in terms of their exploration of the same themes that Giammona raises here:

- What is a technical communicator today?
- What forces are affecting the field?

- What is our future role in organizations?
- What should managers of technical communicators be concerned with?
- How do we contribute to innovation?
- How should we be educating future practitioners?
- What technologies are impacting on us?
- Where do we go from here?
- Does technical communication matter?

Although other articles in the book do not take the same comprehensive approach to answering these questions that we see in Giammona's essay, most of them do consider at least one of them, either explicitly or implicitly.

This article is also notable in another way. We remember that Sullivan and Spilka commented in Chapter 1 that qualitative research is important in technical communication because it is more accessible to practitioners than quantitative studies—not only in terms of comprehending it but also in terms of performing it themselves. Practitioners will have no problem understanding Giammona's article, from the simple elegance of its research design to the way that she has woven together the responses of her participants to frame answers to each of the questions she poses. But many practitioners will respond, having read this article, "I can do that." Although the essay is derived from a paper the author wrote to fulfill one of the degree requirements in a master's program, it doesn't feel "academic" in its tone or in its exploration of real-world problems. Instead, it makes the practitioner reader comfortable with its subject as well as its approach.

The basic message here? Good research doesn't have to be intimidating. On the contrary, as Blakeslee and colleagues insist in Chapter 2, good research is useful.

In December 2003, I completed a master's degree at Polytechnic University in New York City in the management of technology (MOT). The following research was conducted as one of the requirements for that degree. This article looks at the future of technical communication from the point of view of many of its most seasoned and influential practitioners. And it wraps that point of view around the themes of the MOT program— innovation, global concerns, managing technical leaders and practitioners, the impact of new technologies, and the future role of technologists in organizations. It concludes by providing a series of recommendations for the future direction of the profession.

Introduction

Technical communication is the translation of the language of the expert into the language of the novice.

(Kistler 2003)

In May 2003, at the Society for Technical Communication's 50th Annual Conference in Dallas, Texas, closing session presenters made some startling and eye-opening statements. According to De Murr, then the manager of the technical publications department for Walt Disney Imagineering and STC Fellow, who attended the session,

> the speakers were presenting a doom and gloom view. I spoke to people afterwards who heard the same message. There was a negativity that was saying that if you didn't get an MBA, you might not be employable with your current skills. They were implying that the technical communicator of five years ago would not survive five years from now—that the whole profession would go away.
>
> (Murr 2003)

The response to the presentation was lively. Many walked away wondering if there was a future for our profession at all.

The STC itself has seen the need to revitalize and "also recognizes that the organization must evolve to be more competitive moving forward." To achieve this evolution, the STC formed a Transformation Team charged with "developing and refining the future state vision and charting the operational course to transforming STC from its current state to that future vision" (Society for Technical Communication 2004).

So it seems that we are indeed at some sort of crossroads. What does that mean for the profession? Where are we going? Will we disappear? Or will we, as often in the past, reinvent ourselves to face new requirements? What do we need to do differently or need to be thinking about to survive?

Methodology

To discover what is the tide of thinking in the field today, I conducted a series of interviews in October and November 2003 with a group of experienced practitioners. I also sent a questionnaire covering similar topics to a separate group of participants worldwide. The 28 participants included senior-level professionals, in large corporations and small, located across the United States and in international locations including Canada, the UK, India, Austria, Japan, and the Republic of China. The group included recruiters, managers, consultants, hands-on practitioners, academics, and individuals who, from their participation in professional society activities and/or prominence in the field, are recognized leaders in our profession.

I formulated the interview questions around what I knew to be key themes in the field today and also to satisfy the requirements for my degree program. (See Appendix 3.1 for the questionnaire used for all participants.) Thus, I will be discussing the responses in light of themes that are impacting the management of technical communicators as a component of the larger notion of the management of technology. These themes include such ideas as skills and the future role of the technical communicator, challenges facing the managers of technical communicators, the importance of our function and our role in innovation, our concerns with global business and technology issues, the education of our future practitioners, and some of the technologies that affect us today and will affect us going forward.

The responses from the questionnaires and interviews were analyzed and distilled into the set of action points and recommendations presented at the end of this paper.

What is a Technical Communicator Today?

Technical communicators are the anthropologists of the technical world for their ability to be participant-observers in the efforts in which they are involved.

(Decatrel 2003)

A technical communicator, as the STC defines the role, is anyone whose "job involves communicating technical information" (Society for Technical Communication 2004). This may mean anyone whose professional job title or daily job role includes technical writing, editing, illustration, Web design, or any number of related functions. In more sophisticated environments, he or she might be testing usability of products and contributing to the design of such products, especially those elements that are used to convey information.

I presented my participants with a list of skills that are common to practitioners in our field and asked them to identify the five they thought were most important (see Appendix 3.1, Question 7). The temptation of most participants was to want to check them all off—implying that all the skills shown are important. Most struggled to select only five.

But the one common denominator was writing—everyone agreed that a technical communicator must, at the core, be able to write. "I wish I would not have to check off writing—that I could take that for granted," says George Hayhoe, professor and director of the master of science program in technical communication management in the School of Engineering at Mercer University. "Today's students are really into online communication and multimedia communication. They find writing and editing to be less interesting. They think they won't need these skills" (Hayhoe 2003a). But the lack of interest in it does not eliminate its importance.

Cheryl Lockett-Zubak, president of Work Write, Inc., author and user assistance specialist, notes that "technical skills may initially get you the job. But technical skills are not what will make you excel at the job. Writing is. You have to be able to do both" (2003).

A significant skill that I did not list is one adeptly identified by Jack Molisani, a former STC chapter president, whose business includes outsourcing and placement of technical communicators. He labeled this skill "the evaluation of importance," the ability to recognize what is important in a situation or in a set of information. I have been searching for a name for this trait for years, for it is a core skill that I have seen lacking in many senior-level professionals. It's the proverbial ability to "see the forest for the trees" and to still know which individual trees matter. It's the ability to ask the right questions of a subject matter expert so as to "cut to the chase" and not waste a busy person's time. It's the ability to know what to include in and exclude from an explanation to make it accurate and complete, without being overly detailed. And it is absolutely crucial to success in our field.

Some additional skills raised by the participants included:

- being proficient at instructional design;
- being big-picture oriented and project oriented;
- having a user orientation;
- understanding readers and reading, and understanding how people read, use, and absorb documents;
- being a quick study;
- having strong interpersonal skills;
- being good at team building.

The role and importance of the technical communicator has been changing over the last few years in most organizations. We are becoming more than writers. We are also becoming product usability experts. Whitney Quesenbery, user interface designer and consultant, says, "What I have seen is that documentation departments are beginning to see that they cannot operate in a 'ghetto' just focusing on manuals. Communication with users takes place in the [user interface], in documentation, but also on extranets, Web sites, and a broad spectrum of different channels" (2003). And the technical communicator has been moving into all of these channels in recent years.

But no definition of the technical communicator today would be complete without a mention of the persona of the technical communicator. I have been hiring and managing technical communicators for 15 of my more than 20 years in the field. We are, as a breed, an eclectic and unusual group of people. Participants in this study agreed that people drawn to this field are often introverted, smart, artistic, creative, perfectionist, rigid, and fascinated with details of writing and technology. Saul Carliner, assistant professor of educational technology at Concordia University, Montreal, Canada, summed it up well: "We need to learn to be flexible and get along. Although no formal study has been done, there is a preponderance of evidence on our traits. We are clerk-oriented and rule-bound. While periods matter in our jobs, they don't matter in people's lives" (2003).

This, coupled with our complex skillset, makes for a complex personality that puts us in our unique place in the professional world.

What Forces are Affecting the Field?

> I've gone W-2 for the first time in 20 years. I couldn't pay myself out of my own business for a year. At the point of survival, I'm taking a job.
>
> (Glick-Smith 2003)

Judy Glick-Smith, owner of her own technical communication business, took her first full-time job in over 20 years in October 2003. Her former business, which specialized in technical communication placement, projects, and out-sourcing, closed its doors on September 30, 2003. She has now started a new business, which she will get off the ground while supporting herself with her new full-time job. Her new business will have an entirely new focus.

"For technical writers, writing is only 15–20 percent of the job," she explains. "In my new business, I'm now selling to the 80 percent of the job, which is the analysis piece." In her new company, she will focus on selling services to an entirely different audience—not documentation or development managers, but presidents of companies who need the kind of skills and intelligence she can offer. She says, "If we don't recognize quickly what the impact of the current environment is going to be, we could die out, and the business analysts and product designers will take over. All of today's technical communicators will be working at Wal-Mart and driving buses" (Glick-Smith 2003).

Reduced IT Spending and the Economy

Glick-Smith's story is just one of many I heard in the course of my research about how hard times have been in our profession for the last three years. It's thought that as many as one-third of us have lost our jobs; these mostly senior-level people (Lockett-Zubak 2003). Membership in the STC is down by as much as 20–25 percent (Hayhoe 2003a; See 2004).

Peaking at a high of over 25,000 members in 2001, the STC saw a 9 percent drop in membership in fiscal year 2002, the largest single-year drop in 30 years (See 2004). "In the past," says George Hayhoe, "membership usually increased during a downturn as people invested in it as a means of forming contacts and assisting in their professional development. Some people I spoke to at the STC conference are seeing their third recession since the early 1980s. But this time, business is not coming back" (Hayhoe 2003a).

"The economic slowdown has hit our community particularly hard. Budgets have been slashed. The technical communicator is not seen as essential. Technical communication departments are being shut down or cut down to the bare bones. So the work that is being done is also bare bones," writes Donna Timpone, STC chapter president and owner of her own technical communication business, UserEdge (2003).

Even at large companies that are heavily dependent on documentation, the effects are being felt. Amy Logsdon Taricco, user assistance manager for the Tablet PC project at Microsoft, has felt the impact.

Reduced IT spending has a huge impact. The budgets are severely tight. We have to justify our resources like crazy and do more with fewer people. We have to continually justify the value added in design, user resources, and user experience. I need to justify that this position for a technical writer is more important than another position they might want to add for a researcher. It's very difficult, especially compared with years ago.

(Logsdon Taricco 2003)

The Importance and Role of the Documentation Function

All too often, the technical communication function is not seen as vital to an organization. George Hayhoe notes, "The role is generally less significant than it was five years ago. Companies are offshoring it and downsizing it. That indicates that technical communication is not seen as strategic—not a core business function" (Hayhoe 2003a).

This lack of a strategic orientation is rooted in our often being seen as a service function and not as a bottom-line contributor in most firms, leaving the manager of technical communication with little or no political clout in the organization.

JoAnn Hackos, director of the Center for Information-Development Management and well-known author in the field of documentation management, says she believes that managers are challenged to take information development organizations into the mainstream of their companies' work rather than being seen as a "service" function:

There is a constant battle to show value, which means that managers have to develop a clear notion of what value means in a larger corporate sense. I see too many cases where managers tell me that they aren't permitted to interact with customers. I don't see how we can possibly advance our standing without customer contact. Otherwise we are working in a complete vacuum. Work done without customer understanding is easily outsourced off shore.

(Hackos 2003).

But strides have been made to make that shift to the mainstream. The model for information development at IBM is an example. At IBM, technical writers are equivalent to other technologists in title, and their influence is growing.

IBM's Andrea Ames calls herself an information strategist—"a captive information architect." She was one of the presenters at that controversial closing session of the STC Annual Conference in 2003. She is a cross-divisional internal consultant at the corporate level. "I tell people what I do all day is attend meetings and influence people." She says that the role of the technical communicator at IBM is a fairly good one. "It's significant. And is getting more significant. A lot of people have been there for 10, 15, 20 years and have been writing manuals. They are about the 'Commodity' role that I mention[ed] in my presentation. We have work groups and councils that are working to create consistency across the organizations, including looking at reuse and tools." Ames is a highranking professional at IBM, and thus has tremendous strategic influence. "So when I push the strategic piece, they don't push back!" (Ames 2003a).

Our Persona, Once Again

And once again, we cannot escape the force of our overarching personalities. Vici Koster-Lenhardt, who manages technical publications for Coca-Cola in Vienna, Austria, knows us well.

> We are in an industry of whiners. . . . I don't include myself in that group! But we are seen as very sensitive, not very business focused. It's not just political savvy I am talking about. It's earning the right to be able to say to others on our projects that the documentation was not correctly included in the project plan—without whining. Then we will be heard and will make a difference.
>
> (Koster-Lenhardt 2003)

The tough economic environment coupled with our personalities adds to our struggle. Jack Molisani concurs, saying, "Technical writers tend to focus on perfection of information instead of on good enough information. You need to be delivering what the customer wants—whether the customer is inside or outside the firm—that's the most important thing" (Molisani 2003).

What is our Future Role in Organizations?

> The next big thing is figuring out who we are. Teamwork is central to the health of our profession. We can't just sit in our cubes anymore. We have to be seen as critical to the development of a product.
>
> (Lockett-Zubak 2003)

As I noted in the introduction, our profession today seems to be at a crossroads. Interestingly, I believe that this crossroads was created in part by the intersection of economic factors and the personality of our profession. Thus, our future role needs to be one that represents new economic value to those who pay us and one that represents new technical, service, and social value to those who work with us. Neil Perlin, consultant and STC Associate Fellow, sees it like this:

> I think that our role has to change as the environment changes. To do so, we need to get out of the cubicle to shake the artsy introvert image, become more technical, become more involved in company operations and even strategic direction setting, and start thinking "content" rather than "technical writing." I think this will be one of the greatest challenges for those who entered the field before the mid-1980s and thus before the explosion of technology that we see today.
>
> (Perlin 2003)

Many of the participants in this study expressed the sentiment that the new role of the technical communicator is going to be much more closely linked with that of the product developer—and that if we do not seek out this relationship, our function may become extinct. "We need to be proactive and take part in the product development process. We need to be perceived as an important part of the technology," says Cheryl Lockett-Zubak. "We shouldn't just expect this as some sort of gift—it should be true because we earned the position" (Lockett-Zubak 2003).

In my interviews and in the questionnaire, I asked each participant to tell me what they thought the "next big thing" was going to be for our profession. George Hayhoe agreed with Cheryl Lockett-Zubak that the next big thing is where we go as a profession. Hayhoe reflected back on Andrea Ames's presentation in Dallas, as well. He said, "Andrea's continuum is [on target]. If we don't break out of the conviction that to be a technical communicator is about knowing all the tools, if we don't demonstrate how we add value, that we are a strategic part of the business, we are doomed" (Hayhoe 2003a).

The "continuum" to which Hayhoe refers (see Figure 3.1) is Ames's depiction of the direction in which our profession and its practitioners need to move in order to ensure our future role in organizations. It is an image that shows us moving from a role of low value and low leadership—the "commodity role" that many in our profession are in today—to a role of high value and high leadership; a "strategic contributor role" that we need to seek out (Ames 2003b).

Says Ames, "We need the ability to drive profitability and strategy and product usability and adoption though information soft skills. The writing needs to be a

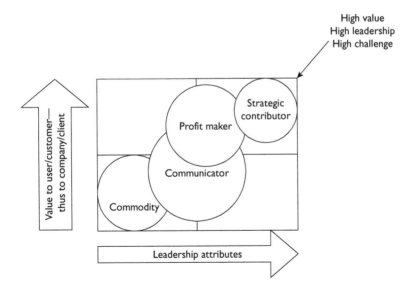

Figure 3.1 Andrea Ames's Value Continuum for Technical Communicators

no-brainer—a baseline skill. We now need expanded skills: Negotiation, driving strategy. We need to be moving technical communication into the realm of user experience design. We need to look at our jobs more expansively, taking a broader perspective" (Ames 2003a).

Donna Timpone summed the same idea up nicely: "To ensure our future role, we need to be flexible and not resist change. We need to think outside the box. We need to be sources of new ideas. We need to present a different approach and keep up on new tools and capabilities. We need to keep ourselves and our skills [constantly in front of] management" (Timpone 2003).

In other words, we need to reinvent ourselves, and that responsibility begins with our managers.

What Should Managers of Technical Communicators be Concerned With?

> In the workplace, we don't demand that much from our technical com-municators. Employees who get too wigged out about italics and bolding should be fired after a warning! We, as a profession, need to learn how a business is run, so that management won't say we are comma chasers.
>
> (Carliner 2003)

There was much talk throughout the interviews and questionnaires about the uncertainty of our future as a profession, but little doubt about how to make a difference in that future: The change has to begin with the leadership in our profession. JoAnn Hackos expressed her concerns that only organizations with strong and inspired leadership will survive. "That leadership is politically savvy, aware of arguing for the right values, focused on usefulness for the customers, willing to reduce costs and increase efficiencies, and able to stand with other middle managers," she says. Without such leadership, she feels that jobs will continue to be reduced and outsourced. "In a way, we priced ourselves out of the market in the 90s, earning high salaries for work that provided dubious corporate value" (Hackos 2003). In her view, we must change the fundamental ways we frame our business to survive.

How do we do that? We need to look for opportunities to use our unique skills to show value. But the seeking of opportunity and capitalizing on it is one of those skills that seems to be outside our common skillset. "It's the looking for the opportunities that the typical introverted technical writer doesn't do!" says Vici Koster-Lenhardt. "The managers or the extraverts are going to have to be out there creating these opportunities for us" (Koster-Lenhardt 2003).

One of the ways we can do that is for our managers to focus on acquiring more business acumen. This may mean both standing up for ourselves in ways we never have, and educating ourselves on aspects of business that we have never before concerned ourselves with. Jack Molisani says, "We are ignorant of how business works. We need to be able to justify our costs against the bottom line. We need to learn to create business cases. We need to speak 'CEO'" (Molisani 2003).

Andrea Ames's path to the role of strategic contributor requires a rise on the leadership axis of her chart:

> It's not about a specific technology. It's a way of thinking about our jobs. It means gaining business skills so that we can communicate with the bean counters who evaluate projects from a managerial perspective. Having business skills, demonstrating value, pushing strategy, showing ROI, showing how we contribute to profitability, all based on principles of usability, being a customer and user advocate. This needs to be true for our industry, or we may not have an industry anymore. Being the commoditized technical writer just isn't going to work anymore.
>
> (Ames 2003a)

Hiring

Not only do our leaders need to develop themselves, they also need to take a fresh look at the people they are selecting who will become our future leaders in the profession. This means taking a fresh look at the hiring process.

Jack Molisani is outspoken on the subject of hiring and has published articles on the subject in the STC's magazine *Intercom* and elsewhere (Molisani 1999). When I interviewed him, he harkened back to those same basic required skills: communicating using the written word, being a quick study, being a good investigator, and having the ability to learn tools quickly. The key point is that tools are the last skill on the list. "Today, I would say the ability to learn quickly and adapt, a tolerance for change, hands-on technical skills appropriate to what you are documenting, experience in the industry in which you are writing, and communication skills are key" (Molisani 2003).

But tools have often been our main criterion when looking at candidates to hire. Saul Carliner says, "If we hire based solely on tool skills, we are setting the agenda for our entire field. Writing is more important than technical [tool] skills. Technical [tool] skills are perishable. We've screwed up our profession by focusing on perishable skills and taking the focus away from the strategic, intellectual skills" (Carliner 2003).

This idea of raising the bar in hiring for our profession has been the theme of my own speaking and writing for some time. When hiring, I suggest that a manager look for a solid overall professional package, and not just a set of specific tools and experience. In evaluating that overall package, the hiring manager should be asking the following questions (Giammona 2001):

- Does the candidate have a professional demeanor?
- Does the candidate write well?
- Does the candidate have good project management skills?
- Does the candidate have good self-management skills (time management, multi-tasking, appropriate separation of work and personal issues)?

- Is the candidate a good verbal communicator?
- Is the candidate a team player?
- Is the candidate technically qualified for the opening (or teachable if specific skills are not present)?
- Would the candidate contribute to or enhance the positive image of your department in the firm?
- Is the candidate a person you and your team would want to spend every day with for the next two to five years?

If you cannot answer yes to all the above questions, the person you are considering is probably not the kind of person who can handle the challenges our profession is facing today and can grow to become a future leader.

Outsourcing

One issue managers are facing today is outsourcing. It is an issue that we need to be prepared to discuss intelligently with our management as we look for ways to save money and show our value. Most of us in management will be asked to consider the option or will face the reality of it in the near future—whether it means our entire function is sent offshore or whether certain work is sent outside our firm. When approaching outsourcing, the study participants raised several useful points for consideration.

First, we need to be able to articulate for management when outsourcing makes sense. Jack Molisani observes that outsourcing can be successful for discretely defined projects. "It's hard to lend a hand as an outsourcer. If you just want someone to come in and help out with a heavy workload, you are better off hiring a temporary onsite person than working in a relationship with an outsourcer" (Molisani 2003).

Outsourcing is also helpful if it is viewed as a means of boosting the capabilities of the current staff to equip them to go forward, preparing them to add value in the future without the ongoing support of the outsourcer. Cheryl Lockett-Zubak feels that it is important that outsourcing not remove work from the current staff: "We should develop partnerships that support the existing staff, not remove someone's job" (2003).

Finally, use of unqualified resources can endanger the integrity of the work done by our function. This can especially be an issue overseas (an issue I will address later in this article). JoAnn Hackos voiced concerns that some overseas firms "are run by shrewd business people who don't know much, if anything, about our field. They do not contribute in any way to the advancement of the field. They simply do what they're hired to do, often by people who don't understand information development and delivery" (2003).

How Do We Contribute to Innovation?

> Innovation occurs when people from different backgrounds come together to try to make something new. Can technical communicators contribute to innovation? Absolutely. Do we? Not often enough.
>
> (Lockett-Zubak 2003)

To a large extent, managing technology is all about managing innovation. So what is the technical communicator's role in this process?

The fact is, we often don't have one. Suzanne Sowinska, user assistance training manager at Microsoft, feels that our job requirements often exclude us from the innovative process. "Often, the skill that makes one a good technical writer does not qualify someone to be a true innovator. Most technical writers are acting as service or production professionals, strapped to meet deadlines, and are not asked to innovate" (Sowinska 2003).

But the fact that we are not asked does not necessarily mean that we are not qualified. We bring some unique gifts to the table. For example, "A writer can slice right through a discussion to reach a new understanding of things. That is one way we can contribute," says Cheryl Lockett-Zubak (2003).

De Murr sees it this way:

> One of the many gifts we bring to the table is the unique ability to see all the trees and see the forest. We ask questions when we join a project team. The people on the project may only be seeing the tree. You ask how the tree fits into the forest. Sometimes these are uncomfortable questions. We can actually save a lot of money with this kind of questioning. This gift of understanding how pieces fit into the whole is how we add value to the innovative process.
>
> (Murr 2003)

Today's environment offers opportunities that exploit these unique gifts. "In the past, technical communicators have only been asked to think about small to midsize problems. Now there's no place for small thinking," says Andrea Ames. "When I look at the work I was doing 10 years ago versus the influence I am having today, it's amazing. It's not easy, but the environment today is primed for this! And the value I am adding is being recognized and rewarded" (Ames 2003a).

But we cannot make this contribution by sitting alone in our cubicles. We need to get involved in a project early on, says Amy Logsdon Taricco. "You need to develop skills in your current people to be involved in design or hire new people who can go in early to a project," she says. This may mean hiring people with different skills going forward. "You need an instructional design background, plus courage and interpersonal skills to go in there and face these opinionated design people" (Logsdon Taricco 2003).

What about innovation that advances our own profession? Cheryl Jenkins, a former Hewlett Packard employee who now manages projects for Microsoft, cited work at Microsoft that involves categorizing information with content types. She says that customer-driven content is a next wave in our field, engaging the customer through chat rooms and community groups so that the customers write, or help to write, the content themselves, as is done in open source settings today (Jenkins 2003).

In the end, our contribution to innovation consists of the information products we produce. Says Vici Koster-Lenhardt, "The products [that] we deliver and that . . . contribute to the firm's success [are] our contribution to innovation" (2003).

What are the Global Concerns in Our Field?

Along with innovation, the global nature of business is another important theme facing technology managers today. The technical communicator will need to be flexible to move with that strategy, so as to be considered vital to operations.

JoAnn Hackos feels that a global focus is essential and will mean working closely with information developers in many languages and from different cultural perspectives. "The offshoring [of technical writing] is most likely to continue through globalization if not only through cost savings," says Hackos. This will mean taking a global perspective on information development. "In particular, it means extending our customer analysis reach outside the US, something we have rarely seen in the past," she says. "We have to be very careful about expecting everyone to act and write like Americans, or being convinced that we always know best. We need to promote professionalism worldwide and work for higher wages in the field everywhere" (Hackos 2003).

Part of the development of professionalism will be the increasing need for the technical communicator to become project manager. Marian Newell, a consultant based in the UK, explains.

> I suspect that the [technical communication] field will shrink within the developed countries, as manufacturing and large sections of service industries continue to migrate to cheaper locations. I have prioritized project management in the skills list because I know several people managing teams of technical staff (often through outsourcing arrangements) in less developed countries and I expect the same trend to affect technical communication.
>
> (Newell 2003)

De Murr agrees with the changing role. "More and more of us here in the US may become project managers as . . . [work] moves abroad. We will need to be editors, to bring the work back . . . to the US" (Murr 2003).

Offshoring of Technical Communication Work

Certainly the top global concern for technical communicators in the United States is the notion of sending technical communication work offshore. Most participants in this study were concerned about the production of English-language materials outside of English-speaking countries. Cheryl Lockett-Zubak's comments were representative of many others.

> I hear grumblings about offshore work. It's a concern. There is this idea that by using resources from other countries you are going to make your product more feasible—not necessarily better, just more feasible. And of course, we just lose jobs. It could become a problem because the main audience of most documents is US-based.
>
> (Lockett-Zubak 2003)

While there were many critics of the potential downsides of offshore arrangements, some practical upsides were noted as well. "A vendor model is in place for basic writing and editing," says Suzanne Sowinska. "The benefits are low cost and 24-hour work cycles. Content created overnight abroad can be edited during the day. I've heard that companies have started to do this" (Sowinska 2003).

Regardless of the upsides or downsides, it is clear that movement of technical writing from United States-based companies to offshore locations has begun. Specific locations mentioned included Ireland, Chile, and India.

Technical Communication in India

As with every other technical profession, the move of technical communication from the US to India has begun. Technical communication is a small but growing profession in India. Layla Matthew, senior technical editor with Cisco Systems in Bangalore, reports on the changing nature of the profession there:

> Initially in India, technical communicators were viewed as some type of secretaries! This was because technical communicators had access to word processors and were seen as people who "prettied up" documents for the engineers. Fortunately, that has changed. Now, to be an effective technical communicator, the person needs a degree/diploma from a reputable computer institution. Ideally, the technical communicator should have an engineering degree, but not many engineers are willing to be technical communicators.
>
> (Matthew 2003)

What is the quality of the work that comes out of India? Matthew says, "So far, since all the writers at my company and I are fluent in English and have worked in the US for some time, we do not have any translation/localization problems. Sometimes, however, some 'Indianisms' tend to creep into our writing. As an editor, I try to spot these and correct them" (2003).

Judy Glick-Smith has spoken of work being done for Alcatel in India:

> The work that comes back from there is pretty good. As a result, I don't think that this kind of job [will be] coming back to the US. In India, [as in] the US

in the 1980s, the technical writers and programmers are job hopping for more money. Billing rates will go up. When rates get high enough there, the US will start sending work to China.

(Glick-Smith 2003)

But today, they are working in India, and for wages that so far undercut US wages that they cannot help but be attractive options to United States companies. Donna Timpone also reported that she recently heard of technical writers in India being paid as little as US$6 to US$8 per hour (Timpone 2003). As a result, outsourcing work to India is likely to increase for United States companies in the coming years.

That being the case, we as a profession need to be prepared for the project management and editorial challenges this kind of work will demand, and build the relationships needed to succeed. Saul Carliner explains:

I visited India and that sparked an incredible relationship with the India STC chapter. They do write well. Yes, there are some problems. But their work is more than adequate and they want to do a good job. By all means, send the "coding" to India—the production of bulk manuals. And over here in the US, we need to learn to be more productive. If you are providing communication services in the US, show how you are adding value.

(Carliner 2003)

How Should we be Educating Future Practitioners?

Technical communication has emerged as a distinct academic major at many universities. One can pursue a degree at the bachelor's, master's, or PhD level, or obtain a certificate through an abbreviated course of study. But views on the value, content, and structure of degree programs varied greatly between study participants.

Jack Molisani takes an extreme position:

You don't need a degree in technical writing. There is not that much to learn. Get a degree in computer science or engineering and take a class in technical writing. If you understand a concept—like circuit design—you can understand all kinds of circuits. I don't think academics are preparing people to be technical writers.

(Molisani 2003)

Some argue that there is a middle ground, somewhere between a dedicated degree and having no academic preparation at all. Vancouver-based consultant Darren Barefoot described his approach:

If I were creating an academic program, it would be half in the engineering or computer science department and half in the writing department. [Then] technical writers would have as much technical writing as possible. To be viable and above average, you need to have business savvy, or get some industry experience, or have a technical background. The days of the arts major going into technical writing are numbered.

(Barefoot 2003)

The majority of those whom I interviewed or who filled out questionnaires felt that a technical writing degree was a valid pursuit—and a large number of the participants were themselves involved on either a full or part-time basis in the education of technical communicators. "The program that I teach in," says Andrea Ames, "helps with the visibility of our professionals and provides people with the skills they need to be strategic contributors" (Ames 2003a). Technical communication programs are being taken more seriously by many universities. According to Stan Dicks, a professor at North Carolina State University, "Technical communication is becoming more important at my school. We currently offer an MS degree in technical communication. In the last five years we have hired two new faculty in the field, and we currently have a proposal on the table for a new PhD program" (Dicks 2003).

What shape should these programs take? There was a general agreement that more emphasis on business skills is important—that the programs should better prepare students for work in the "real world." Barry Batorsky of DeVry Institute says that in recent years his technical writing course has been absorbed into a more general business writing course, but that it has, in the last year, been resurrected as technical writing, with a student population from both technical and management programs (Batorsky 2003).

JoAnn Hackos agrees with this trend:

> I'd like to see the training programs be much more business oriented than they are today. I don't believe that graduates are well prepared for the demands of the job. I would also like to see a focus on more extroverts than introverts. We should not be attracting people to the programs who are so introverted that they have difficulty speaking up for themselves. Students need to be prepared to make a business case for their work, to hold their own with senior managers, not be afraid of their shadows.
>
> (Hackos 2003)

But while the matter of business skills, especially project management, was a popular theme, there was one point of general agreement: programs still need to emphasize writing. Cheryl Lockett-Zubak sums it up: "At first technical writing programs were all writing. Then they focused on electronic publishing and writing. Now they look at HTML, information design, publishing, and writing; but they should always be something plus writing" (2003).

Saul Carliner cited several common problems in technical communication programs that need be addressed (2003). These include:

- PhD programs that do not prepare the participants to be technical communication leaders in the academy and industry.
- Lack of distinction between bachelor's and master's programs.
- Lack of professional experience and understanding by professors, and university hiring practices that do not offer tenure-track positions to people with nonacademic backgrounds to encourage movement of seasoned professionals into teaching.

Some schools are actively seeking new alternatives to make their programs more relevant to today's marketplace. Karen Schnakenberg, director of professional and technical writing at Carnegie Mellon University, indicated that her program is now interacting with the School of Design to create a joint master's in communication planning and information design. "Half of the students come from a writing background and half from a design background," she explains. "Writing is taught from a design perspective—writing as a design process" (Schnakenberg 2003).

In the UK, the flavor of education for technical communicators is also changing. Marian Newell explains.

> In the UK, we are seeing a steady shift from City & Guilds (trade) qualifications to more academic alternatives, specifically through Sheffield Hallam University's distance learning MA course. An attempt to set up national vocational qualifications stalled, apparently due to issues with funding and industry interest. I expect more technical communicators to pursue an academic qualification to increase their professional standing and distinguish themselves from their peers. I expect the City & Guilds modules to be withdrawn at some point.
>
> (Newell 2003)

In India, the education of technical communicators is also evolving. Layla Matthew notes, "Earlier, we had people with journalism degrees and degrees in English applying for technical communication jobs. Now we realize that a technical communicator needs strong technical skills, Web design skills, and interviewing skills. Future training for technical communicators will be based on sharpening these skills" (2003).

The curriculum for today's programs also needs to be forward-looking in terms of the technologies and changing business models affecting our field. Stan Dicks indicated that his program is increasing the focus on translation and localization (Dicks 2003). Other study participants indicated the importance of including subjects such as content management, single sourcing, information management and architecture, user awareness, and design of information for mobile and voice devices—all areas that are commonly encountered in technical communication literature, conferences, and practice in firms today.

What Technologies are Impacting on Us?

> Single sourcing should have impacted us but it has not. . . . There's a lot of talk and there are a lot of conference sessions on single sourcing, but not a lot of people are doing it.
>
> (Molisani 2003)

The list of technologies that impact on the technical communication profession could be endless—especially if you consider that every technology that can be documented is a technology that potentially impacts on us. But in the course of this study, there were definitely several technology themes that consistently surfaced.

Information Management—Content Management and Single Sourcing

Single sourcing is "using a single document source to generate multiple types of document outputs" (Williams 2003). And a content management system might be used to store content in a database separately from the templates that control its appearance, so content can be updated or reused in various contexts without anyone changing the content (Darwin Executive Guides 2003). The two together are meant to empower publishers of information, saving time and redundancy, and making information easy to access and publish. And while both of these issues are widely discussed, their impact on the industry seems to be slow in coming.

Gavin Ireland, former president of the Institute for Scientific and Technical Communicators (ISTC) in the UK, says, "Single sourcing is a concept that I'm still trying to introduce and so far is being treated with disbelief. Content management has just been introduced and is having a great effect already. The initial setup was very expensive in terms of time, but we are already seeing benefits in time, quality, and management" (Ireland 2003).

Cheryl Lockett-Zubak indicates that it's difficult for these concepts to break into companies from the technical communication function:

> it is a hard process for a company to take on. To make it successful, you can't start on a large scale. Pick the places that make the most sense to start and add in the rest later. In our profession, unless you have the whole company on board, you can't afford to do it. It's too expensive. It's mostly large companies that are doing it.
>
> (Lockett-Zubak 2003)

In fact, Vici Koster-Lenhardt believes that momentum for these systems and processes will end up coming from outside our area rather than from within: "The next big thing is likely to be a tool that so simplifies content management that it becomes mainstream and that people in the business outside the documentation area will be asking why we aren't using it" (2003).

As a result, a new role will become increasingly important—the information architect who designs the information that goes into the content management system. Suzanne Sowinska elaborates:

> The ability to become a content administrator is going to become more important. We are going to be needing a person to run the content server who also has the Web skills to make the content talk to the server. This is a specialized role requiring an analyst's mindset and skills in editing, Web development, server administration, indexing—the document storage business. Soon we are all going to be moving content from paper file cabinets to storage servers for long-term access.
>
> (Sowinska 2003)

The Internet

Clearly, the impact of the Internet has only just begun to be felt. But it has profoundly changed our profession already. The following observations were characteristic of the many comments about the importance and impact of the Internet:

> The Internet has absolutely changed the way we work. You used to have a writer, editor, illustrator—no more. Now you have to do it all. The technology itself has changed. Now we have to use more online publishing than before. The media we publish to has changed. And the rate of change has dramatically changed. We moved from the arrival of the Internet to HTML to XML in two years.
>
> (Molisani 2003)

> One appealing aspect of the Internet is that there are no printing vendors involved. You can make changes on the fly. The collaborative aspects are even more important. The relationship with the user can be more of a two-way street as a technical writer and as a company. You can get immediate, constant user feedback on products and documents. It's a fluid environment. It's OK if there's a typo and that the document has not been reviewed three times. It's likely to change anyway. We become more a compiler and distributor of information.
>
> (Barefoot 2003)

> Since my company is a networking company, all our documentation is on the Web. All our manuals are on our Web site, and the users do not need to get printed books. Since the manuals are on the Web, making corrections/updates is quicker than reprinting the manual.
>
> (Matthew 2003)

Wireless

There is no denying that wireless technology, like the Internet, is having a huge impact on society. However, the technical communicators that I surveyed vary greatly in their viewpoint on the importance of this technology. Some see wireless as nothing more than another platform on which to apply our skills—in fact, many see it that way. Others see it is a place where we can have a special impact.

> There is huge potential in wireless devices. There is an economy of information with these products—either auditory or visual. Knowing how people use information will be increasingly important. . . . Technical communicators will be a great ally to the developer in these kinds of products.
>
> (Murr 2003)

Amy Logsdon Taricco, whose team worked on the Tablet PC project for Microsoft, has a large stake in the future of these devices. She believes that we will have to be smarter about how to package the information for these small products, determining the minimum information that will be needed and creating simplistic information designs. "We'll need to make it friendly and get rid of the obtuse language we use. Computers are way too hard to use today. Wireless will impact information architecture and raise the bar for computer ease of use," she says (Logsdon Taricco 2003).

What Else is New?

In response to interview requests to identify "the next big thing," many respondents offered comments that centered on technology. Table 3.1 shows a list of the technologies that were thought to be important trends for the technical communicator and a list of related skills that will be needed to support those technologies.

Table 3.1 Important Technologies and the Skills Needed to Support Them

Technology trends	Skills needed to support these trends
Content management systems	Content administration
Single sourcing	Minimalistic writing
Embedded help systems	Interface design
E-learning	XML/database technologies
Speech recognition	Instructional design
Voice-to-print technology	Usability
Large-format touch screens	Information architecture
Wireless technologies	Design of visual information
Gaming technologies	Standard meta-language
Standards	Shepherding community writing efforts
	Business analysis
	Human factors

Not all of these technologies and skills are new to technical communication, but when they are fully in place, they will represent a significant shift in how we do our jobs.

The gaming industry is an interesting example. It is significant not only for the wide audience of young people who are invested in it today, but for the large audience of future adults whose style of interaction and expectations about information delivery are going to be shaped by their gaming experience. Suzanne Sowinska observes that "Today's 13-year-olds are going to expect the kinds of information exchange that they see in gaming to be everywhere by the time they grow up. This visual orientation will get translated into technical content through things like 3D modeling" (2003). Also from the gaming industry and the open source world will come the concept of user community-created documentation.

As many industries are standardizing to readily share information, technical communicators will need to be looking into standards of their own to handle the sophisticated information exchanges in our future. "I believe the next big thing will be the definition of internal standards and the adoption of both internal and external standards," says Neil Perlin. "This will be a difficult task for many writers, who approach technical communication as an art rather than a science, but it's the only way that our content can be made flexible enough to move in whatever direction the technology ultimately goes" (Perlin 2003).

In the end, there is no doubt that the technical communicator will continue to need to be technical.

Where do We Go from Here?

> It's time to stop whining about the past and look at the future. . . . Stop thinking about manuals and start thinking about communication.
>
> (Quesenbery 2003)

As part of the interviews, respondents were asked, "What can we do to ensure our unique role going forward?" But as Andrea Ames responded, "Today our challenge is going to be to ensure any role, not just a unique role!" (Ames 2003a). So where do we go from here? There appear to be some obvious first steps. Here are the key steps our profession should be taking.

Become Part of the Development and Innovation Processes

The first step is to make a strong move toward inserting ourselves in the development and innovation processes for the products we support. There needs to be a clear connection in the minds of our employers between the sources of a firm's revenue and our contribution. That may mean acquiring some new skills. Cheryl Lockett-Zubak explains.

If we aren't designing products, I don't know how many of us will be here [in the future]. We need to pick up additional skills. It will be [assumed] that we know how to take part in teams, that we have adaptable communication skills. We have to be technologists.

(Lockett-Zubak 2003)

Jack Molisani agrees:

In 5 to 10 years, we will be more tightly integrated with development teams. We will have to fight for that arrangement. The benefits will be that we will get in on projects earlier and have more say in the development of products. We would become [user interface] specialists as we [become] more involved in the design phase.

(Molisani 2003)

This change means reexamining our core competencies and carefully defining them for those who pay for our services. According to Saul Carliner, "Everyone offers an idea of their core skills. We are not, as [a former] STC president was fond of saying, 'an umbrella profession.' We need to differentiate ourselves" (Carliner 2003).

By being more involved in product development, we will become more easily identifiable as contributors with a unique set of skills to offer. We will also be involved in innovation again, contributing actively the specialized knowledge that our skills make us well suited for. As Donna Timpone observes, "In 5 or 10 years, I hope that we will find our focus to be back on usability issues, providing more innovation in the areas of improving performance" (2003).

Launch a Public Relations Campaign for Our Profession

The time has come for our profession to make itself better known and understood in the wider world of technology and business. George Hayhoe underscored this point in his editorial in the August 2003 issue of *Technical Communication*. He said, "The fact that our profession is not widely known among management and the general public is no one's fault but our own. If we don't speak for ourselves, can we expect anyone else to do so?" He suggests that it is time to launch an active PR campaign in schools, with executive management, and in popular periodicals to make ourselves better understood and more widely known (Hayhoe 2003b).

Judy Glick-Smith agrees, saying, "Technical communicators can play a huge role in the future if they get out there and let people know what they can do. We need to let people know we want to be involved" (Glick-Smith 2003).

Participants outside the United States also felt the same way. "First and foremost, we need to raise awareness of the profession by highlighting the benefits of using technical communicators, and the risks and costs of not using them," says

Iain Wright, an editor for British Telecom (Wright 2003). Gavin Ireland of the ISTC agrees. "We need to keep going forward with technology and make sure we can sing the praises of technical communicators everywhere. Our biggest hurdle is that people aren't aware of us and the value that we can add to their products and services" (Ireland 2003).

How do we do this? In part, by taking better advantage of our native communication skills and by reaching out. That approach requires some of the additional steps described below.

Improve Our Professional Societies

As mentioned earlier, STC has examined how to transform itself to better meet the needs of our profession. Some specific suggestions have been offered. In particular, we need to be sure that those leading us in the society are exercising the kinds of sound business sense that we are going to be expecting from our managers in the future. According to Saul Carliner, "We have zero business acumen. And we choose leaders in our professional societies with no business acumen" (2003). This is often difficult when dealing with a volunteer organization. But it is up to the members themselves to ensure that they elect to office those among them who can be the best ambassadors for our profession.

The global growth of our profession is also going to be key as more work moves overseas. We need to do all we can to encourage international membership in the STC. As Saul Carliner observes, "Canadians earn three-quarters of what we earn in the States. In India, they earn one half. And India is our biggest growing market in . . . STC" (2003).

Where should the strategic value of the society come from? From focusing on the kinds of forward-looking concerns for our profession that I am citing here. Others agree. Ginny Redish, in her comments in the 50th anniversary issue of *Technical Communication*, described the forward-looking perspective that STC needs to take: "Document design, information design, user experience design, usability, information architecture with content management, special needs, international focus, teaching, and many other aspects of communication all fit within STC" (Redish 2003).

Become Better Business People and Managers

For the future, technical communication requires managers who are more professional in their management roles. This change is absolutely vital to our survival. We need to be able to sit at the table with the heads of technology functions in our organizations, as well as with those on the business side—with those in manufacturing, marketing, sales, customer service, human resources, and with senior management—to pitch our services, make a business case for our functions and deliverables, and delineate eloquently the value we provide. The fact is that many of us—because we rise through the ranks from our traditional contributor's role and persona—have trouble making this transition. But, ironically, because of our unique role, we are often well prepared to make a significant management contribution. As George Hayhoe sees it,

> Some of us . . . have a better understanding of the businesses we work in than virtually anyone else in our organizations because we are savvy and we have access to information that most people don't have access to in the normal course of our jobs. If we can see connections, we can know more than anyone. We can use that knowledge for the benefit of our businesses because we have other skills that allow us to make sense of things. Few others can do this. Our potential is huge.
>
> (Hayhoe 2003a)

With that potential, we can see the rise of the technical communicator through the ranks, not only of the wider technical organization, but of the corporation itself. "It will be a great day, when we see a technical communicator on the board of a company. We need to get the technical communicators out there on the golf course—seriously! We need to be interacting with the upper levels of the firm," says Vici Koster-Lenhardt (2003). It's not impossible. My manager in my first documentation job 20 years ago is today the executive vice-president of a software firm. But to rise to that role, not only did he exploit his communication skills, he also had to be willing to learn everything about the business he was in.

We can play this role of contributor to the business of the firm not just from the management level, but from an individual level as well. Darren Barefoot says that he tells young people entering the profession to "think of yourself not as an employee of your company but as a consultant for all the areas of the company. Make yourself more valuable to the company by learning what the company does" (Barefoot 2003).

The days are gone when we could sit in our corner cubicles and churn out documents. We now need to prepare ourselves to compete in the world of business as well.

Repackage Ourselves for the Future

Throughout this article, we have been looking at the future role of the technical communicator. But our profession is not the only one standing at a crossroads today. As we look ahead, a whole new generation of future professionals is rising to adulthood. To prepare to be in the workplace with this next generation, and to prepare to produce information products to meet their needs, we need to be looking at the complete skillset that all professionals will need for the future, not just the skills that are unique to our own line of work.

The Partnership for 21st Century Skills has outlined what it believes will be the skillset needed for the workplace of the future (see Figure 3.2). Interestingly,

Skills required for success in the 21st century		
In the digital economy, one U.S. technology company expects current and prospective employees to bring this set of skills to the workplace:		
Set business direction	**Align and motivate others**	**Deliver results**
• Business acumen • Customer focus • Financial acumen • Strategic agility	• Build effective teams • Develop direct reports • Hire and staff • Motivate others	• Command skills • Deal with ambiguity • Drive for results • Intellectual horsepower • Integrigy and trust

Figure 3.2 Skills Required for the 21st Century

this skillset is not so very dissimilar to the one identified in this paper, including the need for business acumen, working better with others, and delivering more significant and complex contributions (Partnership for 21st Century Skills 2003).

What can prevent us from developing and exercising these 21st-century skills? Nothing perhaps, except, again, our core personas. "Technical writers are a lot of MyersBriggs INFJ types, who lack soft skills and like to sit in their offices. The good ones, [the ones] who keep up with change, have strong interpersonal skills. It's the quiet ones who stay off in the corners who don't get ahead," says Vici Koster-Lenhardt. Will there be room for the quiet ones going forward? Perhaps not in our profession or in any profession. Koster-Lenhardt continues, "Those interpersonal skills need to be a core skill of our profession— someone who grows, changes, goes out and tries new things. Introverts can learn to be extroverts when they need to!" (Koster-Lenhardt 2003).

This is a call to change for our practitioners. "My message to people," says Judy Glick-Smith, "is that you've got to keep learning and figure out how you add value. That is an ongoing process. It's how you keep being able to do what you love. You constantly need to repackage yourself—that's the trick!" (Glick-Smith 2003).

Going forward, technical communication will have a broader definition, which means our core skill, writing, will take on more flavors than it has today. Staying employed may depend on this flexibility at our core.

Darren Barefoot says, "For me, increasingly, I am more Web-oriented. I learned the administration of a Web site. I gained applied industry knowledge, and I added some marketing skills. I'm not doing much user documentation anymore. I'm doing online demos, white papers, collateral materials. That's how I've reinvented myself" (Barefoot 2003).

Vici Koster-Lenhardt continues the theme, underscoring our core skill:

> But it still matters that you can write. That won't change. The writing
> deliverables might become broader. We evolved from user documents,
> installation guides, API documents, online help, information for online use,
> information mapping, and now perhaps [marketing communication]. Cuts in
> other writing positions will end up sending all kinds of writing in the company
> our way. So the ability to write different styles across the writing realm will
> be important. Not everyone has all of these kinds of writing in their bag of
> tricks. It will become a necessity for employment.
>
> (Koster-Lenhardt 2003)

Does Technical Communication Matter?

There was a lot of debate in 2003 about the importance of information technology
in general. Nicholas Carr, in his controversial article, "IT Doesn't Matter" in
Harvard Business Review, argued that perhaps IT has become such a commodity
that it is no longer a source of strategic value (Carr 2003). Is that true for technical
communication? Are so many of us in the bottom left corner of Andrea Ames's
value continuum for technical communicators as to be adding no value to the
organizations we serve, and therefore making ourselves prime candidates for
outsourcing or offshoring?

We need to be concerned that if we do not act quickly to dispute the growing
perception by our management that we are not strategic, our profession as we know
it will be in jeopardy. In some respects we already have begun the change, as the
name "technical writer" drifts further from describing the core functions we
provide. Whatever we are called, we will matter, and we will be making significant
contributions in a new marketplace—we just may look a little different going
forward. Judy Glick-Smith offers some options:

> The technical writer name is dead. It's not called technical writing or
> technical communication anymore. We have to call it something else.
> Requirements analysis or business analysis, or enterprise architecture or
> information architecture, or product design. Usability is hot today, but I
> believe it will become a no-brainer as things are more and more usable.
> Quality assurance is an option for us. Training and instructional design is
> another—becoming e-learning developers. All of these are options for our
> future.
>
> (Glick-Smith 2003)

Is the technical writer dead? In name, perhaps, but it appears that the role that
was once known as "technical writer" may have even broader future possibilities
than ever previously imagined. It will be up to the leaders and practitioners in
the profession to exploit those possibilities by taking the action steps outlined

here—repackaging ourselves for a new marketplace, becoming more influential and savvy to the businesses we serve, supporting ourselves from within through vital professional societies, and making the world in general more aware of who we are and what we can do, starting with technical teams in our own firms who sit down the hall from us. With these actions to guide us, technical communication will continue to matter—more than ever—in the increasingly complex age of information.

Appendix 3.1: Interview Questionnaire

The Future of Technical Communication: How Innovation, Technology, and Information Management are Shaping the Discipline

Please answer all the questions in Section 1. Then, based on your role and your experiences, please answer as many of the remaining questions as seem relevant to you or on which you would like to express an opinion.

SECTION 1: BACKGROUND

1. Your comments may be quoted in the context of my paper. Your name will be attributed to your comments, unless you indicate that you wish to remain anonymous.
   - n You have my permission to use my name
   - n I wish to remain anonymous
2. What is your current primary role in the Technical Communication field?
   - n Writer
   - n Editor
   - n Manager
   - n Consultant
   - n Educator
   - n Other: Enter your response here
3. How long have you been in the field?
   - n 1–5 Years
   - n 6–10 Years
   - n 11–20 Years
   - n More than 20 Years
4. Are you a member of a professional organization? (Check all that apply.)
   - n STC (Society for Technical Communication)
   - n ISTC (Institute for Scientific and Technical Communicators)
   - n IABC (International Association of Business Communicators)
   - n IEEE Professional Communication Society
   - n ACM SIGDOC
   - n UPA (Usability Professionals Association)
   - n ISPI (International Society for Performance and Instruction)

- n   Other: Enter your response here
- n   I do not belong to any professional societies

5. What is the primary way you stay current with changes in the Technical Communication industry? (Check only one.)
   - n   Professional Society Meetings/Conferences
   - n   Professional Society Publications
   - n   Networking with Peers
   - n   Technical Books and Periodicals
   - n   Online Publications/Internet Sites
   - n   Mail/User Groups
   - n   Other: Enter your response here

6. Please provide in the box below or attach in a separate file a brief one- to two-paragraph career summary. I may be placing these in an appendix to my document to identify my contacts and references.

Enter your response here.

SECTION 2: TODAY'S LANDSCAPE

7. Which of the following skills would you say are the most important to the success of the technical communicator? Select five from the list below.
   - n   Writing
   - n   Editing
   - n   Organizing Information
   - n   Project Management
   - n   Authoring/Publishing Tools
   - n   Web Design
   - n   Print Document Design
   - n   Usability/GUI Design
   - n   Information Architecture
   - n   Research
   - n   Interviewing/Listening
   - n   Political Savvy
   - n   Programming or Hands-On Technical Skills
   - n   Business/Industry-Specific Experience
   - n   Other: Enter your response here

8. How significant a role does Technical Communication play in your company/industry? Has that changed in the last five years?

Enter your response here.

9. How has the Internet changed the way you perform your work?

Enter your response here.

10. What impact are single-sourcing and content management systems having on your organization? Is the technical communicator playing an active role in information management and knowledge management issues beyond your department and in your firm at large?

Enter your response here.

11. How has reduced IT spending impacted on your technical writing organization?

Enter your response here.

12. Are you using outsourcing for the technical writing function? Why or why not? Is it successful? What are some of the greatest issues you face with outsourcing?

Enter your response here.

SECTION 3: THE FUTURE OUTLOOK

13. What do you see as the "Next Big Thing" for technical communicators?

Enter your response here.

14. What is the technical communicator's role in the realm of wireless devices (that is, providing information related to small, mobile, or ubiquitous devices)?

Enter your response here.

15. What will technical communicators be doing five years from now? Ten?

Enter your response here.

16. Looking at the same list of skills from question 7 above, which five skills would you select as being the most important to the success of the technical communicator of the future?
    - n Writing
    - n Editing
    - n Organizing Information
    - n Project Management
    - n Authoring/Publishing Tools
    - n Web Design
    - n Print Document Design
    - n Usability/GUI Design
    - n Information Architecture

- n Research
- n Interviewing/Listening
- n Political Savvy
- n Programming or Hands-On Technical Skills
- n Business/Industry-Specific Experience
- n Other: Enter your response here

17. What can technical communicators do to ensure that our unique role in organizations continues to exist going forward?

Enter your response here.

SECTION 4: THE COMMUNICATOR AND THE INNOVATIVE PROCESS

18. What is the technical communicator contributing to your firm's innovative processes?

Enter your response here.

19. What is the most innovative thing you and your team are currently doing?

Enter your response here.

SECTION 5: GLOBAL PERSPECTIVES

20. If your organization is global, how centralized or decentralized is your documentation operation?

Enter your response here.

21. How important are translation and localization to the work you are doing?

Enter your response here.

22. What collaborative tools and methods are you using to connect multi-location teams?

Enter your response here.

23. Do you work in a global or multinational organization? If yes, in how many are technical communicators on staff in-country?

Enter your response here.

SECTION 6: MANAGEMENT ISSUES

24. As a manager of technical communicators, what is one of your key current challenges?

Enter your response here.

25. Is your team connected to strategic initiatives (for example, customer products or documents for technical staff) or in a supporting role (for example, policies and procedures or regulatory documents)?

Enter your response here.

26. How much political clout do you feel you have in your organization?

Enter your response here.

27. How have alternative work arrangements affected your role as technical communicator/manager?

Enter your response here.

28. How do you develop your staff? With limited resources, what would you most prefer to spend money on toward the development of your staff:
    - n   Conferences
    - n   Tools Training
    - n   Professional Skills Training
    - n   Project Management Training
    - n   Technology Training
    - n   Industry Training
    - n   Design Skills Training
    - n   Other: Enter your response here

SECTION 7: ACADEMIC ISSUES

29. How will the training of technical communicators change in the future, based on your understanding of how the profession may change?

Enter your response here.

30. The primary focus of my teaching today is on (rank from 1 to 7, with 1 being the most important focus):
    - n   Authoring/Publishing Tools
    - n   Programming/Technical Tools
    - n   Information Architecture

- n GUI Design
- n Graphic Design
- n Writing/Editing
- n Professional Skills
- n General Business Skills
- n Other: Enter your response here

31. What is the hardest skill to teach a student in Technical Communication?

Enter your response here.

Acknowledgment

*This article was originally published in *Technical Communication* (2004), 51: 349–66.

References

Ames, A. 2003a. Personal interview, October 22.

———. 2003b. "Transforming Your Career . . . With the Economy and the Industry: Moving from Commodity to Strategic Contributor." STC 50th Annual Conference presentation.

Barefoot, D. 2003. Personal interview, October 27.

Batorsky, B. 2003. Questionnaire completed October 28.

Carliner, S. 2003. Personal interview, October 27.

Carr, N. G. 2003. "IT Doesn't Matter." *Harvard Business Review* (May): 3–10.

Darwin Executive Guides. 2003. *Content Management*. Boston, MA: IDG.

Decatrel, M. 2003. Personal interview, October 22.

Dicks, S. 2003. Questionnaire completed November 10.

Giammona, B. 2001. "Successful Hiring: Tips for Finding the Best and the Brightest." *Best Practices. A Publication of the Center for Information Development Management* 3: 78.

Glick-Smith, J. 2003. Personal interview, October 22.

Hackos, J. 2003. Questionnaire completed November 17.

Hayhoe, G. 2003a. Personal interview, October 14.

———. 2003b. "Who Speaks for our Profession?" *Technical Communication* 50: 313–14.

Ireland, G. 2003. Questionnaire completed October 28.

Jenkins, C. 2003. Questionnaire completed November 3.

Kistler, K. 2003. "What is Technical Communication? Definitions by Members." *MetroVoice* (November/December). http://stcnymetro.org/metro_voice/mv_122003/mv_122003.htm

Koster-Lenhardt, V. 2003. Personal interview, November 4.

Lockett-Zubak, C. 2003. Personal interview, October 31.

Logsdon Taricco, A. 2003. Personal interview, October 28.

Matthew, L. 2003. Questionnaire completed October 31.

Molisani, J. 1999. "Tools or Talent? What it Takes to be a Good Technical Writer." http://www.writerssociety.com/ writers/molisani2.html

——. 2003. Personal interview, October 28.

Murr, D. 2003. Personal interview, October 9.

Newell, M. 2003. Questionnaire completed October 29.

Partnership for 21st Century Skills. 2003. *Learning for the 21st Century: A Report and Mile Guide for 21st Century Skills.* http://www.21stcenturyskills.org/downloads/P21_Report.pdf, p. 7.

Perlin, N. E. 2003. Questionnaire completed October 28.

Quesenbery, W. 2003. Questionnaire completed October 28.

Redish, J. 2003. "Reflections @ 50." *Technical Communication* 50: 42.

Schnakenberg, K. 2003. Personal interview, October 30.

See, E. 2004. Exchange of e-mails, March.

Society for Technical Communication. 2004. STC Web site. http://www.stc.org

Sowinska, S. 2003. Personal interview, November 18.

Timpone, D. 2003. Personal interview, October 9.

Williams, J. D. 2003. "The Implications of Single Sourcing for Technical Communicators." *Technical Communication* 50: 321.

Wright, I. 2003. Questionnaire completed November 10.

Chapter 4

Modeling Information for Three-Dimensional Space

Lessons Learned from Museum Exhibit Design*

Saul Carliner

Editors' Introduction

In "Modeling Information for Three-Dimensional Space: Lessons Learned from Museum Exhibit Design," Saul Carliner explores the relationship between Web design and the design of museum exhibits. The article reports some of the results from a study using a grounded theory design, and his data-collection methods included interviews, observations, and document analysis. These methods were supplemented by a literature review, observations of visitor behavior, visits to over 200 museums around the world, and participation in conferences and events for museum exhibit designers.

A qualitative method is clearly needed for this study. Carliner is not interested in exploring correlations or causal relationships. He wishes to explore the phenomena of design in museums, and to consider whether his findings might be used to strengthen certain aspects of technical communication practice.

We include this article in part because it illustrates one of the more popular and rigorous qualitative research designs, grounded theory, including the technique of constant comparative analysis as practiced by Strauss and Corbin (1990, 1994). Readers may wish to note that since Carliner's work was published, grounded theory has been revisited and, to some extent, reconceived by Charmaz (2006). Many qualitative approaches have been influenced by the rigor of grounded theory, and have adopted the technique of using a structured coding and theming approach, and of moving iteratively between gathering data and analyzing data as a way of determining when data saturation has been achieved.

We also like this article for the exemplary manner in which it reports the study findings. Carliner's presentation of themes is elegant and useful, with each theme discussed in terms of what the author observed and heard, and the lessons that he gleans for Web designers. He allows the voices of his participants to be heard through numerous quotations, and he is careful to separate his descriptions from his interpretations. Carliner's article might be seen as an example of the sort of qualitative study advocated by Blakeslee and colleagues earlier in this volume, a study that seeks to justify itself in terms of its usefulness for practitioners—though here, Carliner is not focused so much on benefiting the study participants (designers of museum exhibits) as on helping technical communicators see how they can apply best practices of museum design to Web design.

Today, despite the dizzying changes that have occurred in online technologies, the article continues to read like a primer on fundamental principles for Web design. The linkage between the design of Web sites and museum exhibits is startling, and should remind us that communication and learning are fundamental human traits that can be encountered and studied in many different social situations.

One wonders whether Carliner still believes, as he says in his closing thoughts, that "online worlds are ultimately artificial ones"—or whether the growing pervasiveness of online technology has blurred the boundaries between the virtual and the real. His enduring lesson, that Web designers "must take responsibility for their roles" and must bring leadership to their design teams, is both provocative and intriguing, and perhaps points toward a new research project to be designed and implemented in the near future.

References

Charmaz, K. 2006. *Constructing Grounded Theory: A Practical Guide through Qualitative Analysis.* Thousand Oaks, CA: Sage Publications.

Strauss, A. and J. Corbin. 1990. *Basics of Qualitative Research: Grounded Theory Procedures and Techniques.* Newbury Park, CA: Sage Publications.

—— and ——. 1994. "Grounded Theory Methodology: An Overview." In *Handbook of Qualitative Research,* ed. N. K. Denzin and Y. S. Lincoln. Newbury Park, CA: Sage Publications.

Perhaps these concerns sound familiar:

- Visitors complain that they cannot find information of interest. One observes, "I know there's information about that type of robotics here, but darned if I can find it."
- Visitors enter the site but don't stay particularly long. Some might even express an interest in the subject; let's say it's modern art. But they leave almost as quickly as they enter without paying much attention to the artwork that the designers painstakingly displayed.
- Other visitors spend hours at the site but never seem to notice particular sections. For example, a visitor might be thoroughly familiar with the content on radios but oblivious to the section on industrial hardware.

These observations could describe visitors to Web sites, none of which are more than 10 years old. Actually, these observations describe museum visitors. As a type of institution, the museum has existed for nearly three centuries, and these concerns are nothing new to museum exhibit designers. Since the first research in the late 1920s and early 1930s, museum professionals have observed visitor behavior and, in response, transformed exhibit design practices (Chambers 1999). These practices were further refined between the 1960s and 1980s as museums redefined their mission, from warehouses of artifacts to institutions of informal learning (that is, learning without a predetermined outcome) (Bloom and Powell 1984).

I systematically observed current exhibit design practices as part of an extended study. The primary purpose of that study was to see how practices from my primary field of study, instructional design (whose primary focus is on formal learning with predetermined outcomes in the classroom, through workbooks, and online) transferred to the design of informal learning in museums.

An interpretation of these observations yielded a more flexible perspective on instructional design (Carliner 1998). It also yielded a number of communication practices that could be transferred from the community of museum exhibit designers to the community of information designers. Sharing that second set of interpretations is my purpose here.

Following a brief description of the research project, I share eight lessons, or categories of practices, that I observed. For each lesson, I first describe in detail what I observed in museums. Immediately afterwards, I suggest how information designers might apply these lessons when working on technical communication products. I close with some broader thoughts about these lessons.

"What Separates a Museum Worth Suffering for from One You Wouldn't Stoop to Be Sick In?"

So wondered Judith Stone, writing in a special 1993 issue of *Discover* that focused on the emotional and educational impact of science museums on scientists and science writers.

I asked myself the same question.

Museums have always fascinated me because they are some of the most complex and successful forms of scientific and technical communication. To answer the same question as Judith Stone, then transfer the lessons learned back to the professional communities of instructional and information designers, I undertook a qualitative study of the design for three permanent exhibitions in history and technology museums, and related background and follow-up research.

The primary purpose of the study was to understand how members of the design team addressed instructional issues and to see which design practices for formal learning transferred to the design for informal learning in museum exhibits. The exhibits were purposely selected and included exhibits on

- the history of a major city in the United States at an urban history museum;
- the history of the canning industry in the late 19th century at an industrial history museum;
- computer and telecommunications networks at a high technology museum in the United States.

In the main study, each member of the "core" design team was interviewed three times. Core team members are those who play a primary role in designing and developing the exhibit. These team members include

- An idea generator, who devises the concept for the exhibit, chooses the content, and writes the "storyline" (a detailed description of the exhibit and the preliminary draft of copy for the labels that appear in the exhibit).
- The exhibit designer, who prepares the physical design of the exhibit, including its floorplan and graphic identity; chooses wall and floor coverings; designs display cases; and prepares blueprints.
- An idea implementer, who acts as a general contractor of sorts, securing objects for the exhibit that are not in the museum collection, overseeing the work of the peripheral team (specialists who implement the plans), ensuring conservation of items to be displayed, and making sure that the design is implemented according to plan.

For each exhibit, members of the peripheral team were also interviewed when feasible. These team members provide the specialized skills needed to develop a part of the exhibition. Skills needed on the peripheral team vary among exhibits. Typically, this team includes a museum educator (whose job is to develop programs geared toward school groups that are related to the exhibit content), public programs coordinator (whose job is to develop programs geared toward adults and the general public), registrar (whose job is to oversee the documentation and protection of objects in exhibits), media specialists (including video and interactive specialists), and editor (whose job is to edit the copy for all labels and gallery guides associated with an exhibition). In addition to the interviews, I observed team meetings and reviewed project plans when feasible.

The study followed the grounded theory methodology. A central feature of this methodology is constant comparative analysis. That is, data are constantly analyzed throughout the data-collection process to devise theories; collected data are later compared with the evolving theory to determine whether they support the theory (Strauss and Corbin 1994, 273). Strauss and Corbin suggest a three-phase process for analyzing data. The first phase is open coding, which they define as "the process of breaking down, examining, comparing, conceptualizing, and categorizing data." The next phase is axial coding, "a set of procedures whereby data [are] put back together in new ways after open coding, by making connections between categories" (Strauss and Corbin 1990, 96). The last phase is selective coding, "the process of selecting the core category, systematically relating it to other categories, validating those relationships, and filling in categories that need further refinement and development" (1990, 116).

Whenever they are coding, researchers look for dominant patterns—patterns that appear in all sites studied. Researchers also look for weak patterns: ones that occur in at least two sites. Researchers try to explain why a weak pattern might not be observed at the other sites.

Besides the core research for this study, I conducted preliminary and follow-up research. This research consisted of a literature review; observations of visitor behavior in a science center in a large city in the United States; visits to over 200 museums in the United States, Canada, Europe, and Asia; and participation in two conferences and other events for museum exhibit designers.

1. "Did Anyone Target an Age Group?"

What I Observed in Museums

Because visits to museums are voluntary in nature, museum staffs must motivate people to visit (Csikzentmihalyi and Hermanson 1995). First, museum staffs must motivate visitors to enter the building. To do that, they must work past an impression among the public that museums are primarily intended for people from upper economic classes and majority religious and racial groups (Zolberg 1994). Such impressions have, in the past, made people from outside of those groups feel unwelcome in museums. This challenge is similar to that faced by businesses that want to sell products and services outside of their countries or to historically marginalized groups like women, African Americans, Latinos, and gays and lesbians.

To address this concern, museums have attempted to broaden their constituencies. This is a policy of the museum profession backed by practices in specific museums. Believing that diversity behind the scenes is essential to representing diversity elsewhere in museums (including in exhibits), museums have established formal relationships with constituency groups. For example, the high technology museum in this study has an advisory board of low-income children, and the Brooklyn Museum has an outreach project with the surrounding neighborhood. Museums have also made a concerted effort to broaden the socioeconomic, gender, and ethnic backgrounds of their staffs and boards, and continue to do so (Hirzy 1992).

These behind-the-scenes changes are reflected in exhibits that have a different type of appeal than in the past. In some instances, exhibits are designed to appeal to the general public; called "blockbusters," they are temporary exhibits (running from a few months to a year), focusing on well-known topics with broad public interest; they are primarily intended to lure large numbers of visitors (Lee 1994). One of the first was the 1979 King Tut exhibit that visited major art museums, and it has been followed by blockbusters such as the Monet exhibit that visited the Art Institute of Chicago in 1995, and the Titanic exhibit that visited the Museum of Science and Industry in Chicago in 2000. Museums can see attendance surge by as much as 33 to 50 percent during a blockbuster.

Other exhibits are designed to appeal to targeted constituencies, ones whom museums typically ignored in the past. Some of these exhibits are temporary, like the retrospective of African American artist Jacob Lawrence at the High Museum of Art in Atlanta and an exhibit on the contributions of women engineers at the Franklin Institute in Philadelphia.

Some exhibits for targeted audiences are permanent, like the First People's galleries in the Canadian Museum of Civilization in Ottawa. Following its most recent renovation, the Minneapolis Institute of Arts devoted half of its permanent gallery space to non-Western art; previously, such art had occupied less than a third of the gallery space.

Within these exhibits, staffs design interpretive materials, like labels (signs within the exhibit that contain explanatory text) and media presentations. When developing these materials, staffs take into account the diversity of experiences that affect interpretation of an object because staff members want to avoid foisting their own interpretations on the public. Many interpretive materials now describe the outside factors that shape the meaning of objects and topics on display.

In addition to exhibits, museums also provide related public programs that are targeted to particular communities. Some programs focus on single people, such as the High Museum's Young Professionals, which is geared toward people under the age of 40. Other programs focus on underprivileged youth, such as an after-school program sponsored by the Computer Museum.

Although some exhibits and activities are intended to draw targeted audiences, exhibit designers know that the museum is a public place and the entire public must feel welcome in each exhibit. So ultimately, these designers lack a clearly defined audience. In fact, at a meeting of exhibit designers at the 2000 American Association of Museums Annual Meeting, one designer asked, "Did anyone target an age group?"

Still, efforts to broaden the appeal of museums have changed public attitudes toward them over time; they're places that people increasingly choose to visit. In the United States, for example, more people visit museums in a given year than attend professional sports events (Ivey 2000).

Lessons for Web Design

Like museum exhibit designers, designers of Web sites need to appeal to a variety of demographic groups. As businesses increasingly market globally, the literature on technical communication provides substantial guidance in addressing geographically distinct markets for whom information will be translated and localized (Hoft 1995).

The community of Web site designers and technical communicators pays less attention to other aspects of cultural difference. For example, little has been written about the impact of occupational culture and socioeconomic class on technical documents. More significantly, because many believe that technical communication is objective (that is, free from bias), technical communicators are rarely encouraged to identify their own cultural biases and explore how they might affect the communication products that they develop.

2. "Keep the Collection from Klummeting Your Guests"

What I Observed in Museums

The industrial history museum that I studied does not have enough exhibit space to physically display the tens of thousands of hand tools in its collection, much less the other artifacts, like machinery and manufactured goods. At the time of the study, the museum did not even have enough storage space on the premises to store objects it could not display. The staff stored them in a rented storage space several miles from the museum. The idea implementer explained that museums typically display only 10 percent of their collections at a given time.

Objects form the centerpiece of most museum exhibits. Because of that, and because the primary purpose of museums is educational, museum professionals often refer to their work as object-based learning. One of the most significant choices a museum exhibit design team makes, therefore, is what to display. Choices are carefully considered. As exhibit designers learned when they would cram entire collections into a series of glass cases that visitors would ignore, "You have to keep the collection from klummeting [overwhelming] your guests." But in choosing which objects are displayed, exhibit design teams also choose which objects remain in storage.

This choice is made in the early stages of design. Because museum exhibits effectively involve a major renovation of a building and therefore require budgets that exceed the costs of most homes, they are funded in two phases. The first is the less expensive planning phase, which is similar to the needs analysis and requirements phase of a technical communication project. If funders have concerns about the plans, those concerns can be resolved before spending the large sums of money needed to actually build the exhibit.

In the planning phase, the idea generator works with a team of content experts and educational specialists to devise a focus for a proposed exhibit. Then the idea generator and idea implementer work together to develop the detailed plans for the exhibit, called the "storyline." The storyline is:

> a written document that presents the key elements of the visitor experience. The storyline refines the subject of the exhibition, identifies key topics to be addressed in the exhibition, and discusses possibilities for presentation, including how content in the exhibition might flow and be presented, and the types of objects to be included. Members of the staff who are going to work on the exhibit design team are identified at this time, although only the idea generator and [idea implementer] take the most active roles during this phase. The staff often reviews the museum collection at this point to determine what objects it already has and the objects it might need to collect to effectively realize the exhibition.

(Carliner 1998, 84)

In other words, only after the content is chosen do exhibit design teams choose objects. In some cases, several objects might meet the needs of the content, so design teams choose objects based on their anticipated appeal to visitors and condition. In some cases, because funds for conserving objects are more plentiful when associated with an exhibit, the design team might choose an object that needs conservation. In other cases, the design team might intentionally choose a "touch object"—that is, one that visitors will be encouraged to handle. Touch objects must be physically durable.

Some museums have addressed the problem of large collections at an institutional level. Those museums that have comprehensive collections in each topic area addressed by their missions need buildings of immense physical size merely to display and house these collections. Within a given topic area, some collections are sufficiently large that they could comprise museums themselves.

Museums have tried many approaches to shield visitors from this enormity. Some have spawned other museums. For example, the Washington, DC-based Smithsonian Institution has several museums, each focusing on a particular subject area. The London-based Tate Gallery has opened another London site in which to display its international modern and contemporary art collection. The New York-based Guggenheim Museum has opened satellites in Europe, including one at Bilbao, Spain.

Although museums usually have more objects than they can display, many still find themselves short of objects when planning new exhibits. For example, each of the museums that I studied lacked objects in their collections needed for the exhibits studied. In two of the exhibits, new acquisitions represented over 50 percent of the objects ultimately displayed. In each exhibit, too, designers used fabricated objects (that is, ones that had been built for the exhibit rather than true historical artifacts). Some objects were fabricated because the designers wanted visitors to be able to touch them, and real objects would fall apart under such wear. Other objects were fabricated because real ones did not exist.

On the other hand, entire museums have opened with signature buildings and without extensive collections to support them. Building collections is proving difficult for these museums. For example, the core collections for many natural history museums opened at the beginning of the 20th century are specimens of large animals collected on hunts in wilderness areas. Killing endangered species of animals for display in museums is no longer an acceptable practice. Similarly, as prices for art skyrocketed in the 1980s and 1990s, many art museums that have seemingly large acquisitions budgets still do not have enough money to purchase pieces for their collections.

Lessons for Web Design

As museums have learned to focus exhibits and limit the amount of information to which they expose visitors, so designers of Web sites must learn to focus their content and limit the amount of information to which they expose users.

With easily available computer storage and increasingly sophisticated search mechanisms, communicators have little technical incentive to limit information. Furthermore, with the promise of ready access to all the knowledge in the world through the World Wide Web, some communicators understandably feel an ethical commitment to provide full access to information that the user needs to know. That technical communicators have always been committed to completeness only strengthens this commitment.

But our values and technology conflict with users' needs and experiences. Consider the following:

- According to studies by User Interface Engineering, using a search mechanism leads users to information of interest less frequently than links (1997). That fact places an ongoing premium on the ability to carefully structure and chunk information for users.
- The growth of profiling software and intelligent agents provides communicators with both the incentive and tools to tailor each online experience as much as possible to the unique needs of a user. The effectiveness of the rules that operate this software directly emerges from communicators' ability to identify users' bottom-line goals and scenarios of use, as well as to develop lists of relevant characteristics that affect a profile.

Technology, alone, then does not solve the problem of "klummeting users" with information; only design practice does. For example, one tool in controlling information is behavioral objectives (also called learning objectives). Objectives state what users should be able to do after completing a tutorial. Instructional designers develop objectives before starting work on a tutorial and use them to focus their work. They include only content that directly supports the objectives. Other content is discarded or, if it must be incorporated, changes the scope for the project (Mager 1997).

3. "An Exhibit is Not a Book on a Wall"

What I Observed in Museums

When the design was driven by subject-matter experts called curators, the heart of most exhibits was a series of cases crammed with artifacts (such as paintings, furniture, textiles, photographs, and documents) and accompanied by detailed documentation on each object (usually typewritten). This dense documentation was primarily prepared by one scholar for use by other scholars.

This reference-like approach to displaying objects created a barrier between museums and the public. The public was overwhelmed by the quantity of objects and the technical language and detail of the documentation. In fact, studies indicated that few visitors actually read labels, and, of those who did, most spent less than half a minute doing so. When museums started broadening their audiences two decades ago, they realized that "[the] museums of the past [would have to] be set aside, reconstructed, and transformed from a cemetery of bric-a-brac into a nursery of living thoughts" (La Follette 1983, 41).

In response, exhibit designers transformed their approach to design, using four concepts to guide them in their efforts.

Guiding Design Concept A: Immersion

The first guiding concept is immersion. According to the idea generator at the urban history museum I studied, a museum exhibition should immerse visitors in its story. She noted that a nearby zoo uses this immersion theory of exhibit design. The zoo's designers "put people where the animals are and let [visitors] become a part of the experiment." She applies these beliefs and theories to all the exhibits at her museum. "It's theater," she noted, "yet the objects are real, just as animals are real [in the zoo]." In her exhibit, visitors are immersed in the city at four periods in time: an open field from the time preceding settlement, a city street from the late 19th century, another city street from the early 20th century, and a highway scene from the late 20th century. The designers of the two other exhibits studied also used immersion.

Guiding Design Concept B: Themes

The second guiding concept is dividing complex topics into a limited number of key themes. A designer participating in the exhibit brainstorming session at the 2000 American Association of Museums Annual Meeting called this "modularity." Because topics for exhibitions are often broad and the number of facts presented is greater than a visitor can process during the short period of a typical visit, designers try to identify a limited number of broad points on which to focus, and build exhibits around them. Each of the exhibits that I studied had fewer than five themes. By limiting the number of themes, designers hope to increase the likelihood that visitors will better recall the insights from exhibits.

For example, designers focused on four key themes in the development of the city featured in the exhibit studied at the urban history museum rather than present a timeline of development. These four themes corresponded to four distinct phases of the city's development, and the design team built four galleries, each immersing visitors in a phase of the city's development.

Guiding Concept C: Layering

The third guiding concept is that of layering content. The idea generator at the urban history museum explained it best. She insisted that an exhibit is not "a book on a wall." In other words, visitors should not have to read all the labels to learn about the topic of the exhibit. Instead, they should be able to explore in as much detail as they like and leave feeling as if they learned a complete topic.

She designed her exhibit so that labels—text signs on the wall that provide explanatory information—are presented in three levels of depth. Visitors can look at the label and identify its tier, and read all the labels in a chosen tier to see a complete story. These tiers include:

1. Introduction to the gallery: These labels provide the title of the gallery and an orienting quote. The orienting quotes originated during the time period depicted in the gallery. These labels are the largest, so visitors can easily identify them several feet away.
2. Theme labels: These labels introduce key themes in the exhibit. The labels consist of a heading, a limited amount of text (no more than 12 lines), and, occasionally, a drawing or reproduced photograph. The text on these labels is large enough to be seen a few feet away.
3. Object labels: These labels, the most numerous in the exhibition, describe characteristics of individual objects, such as their significance or the materials used to make them. Not every object has a label. The text on these labels is the longest, but rarely longer than 12 lines. The type on the labels is small; visitors must stand close to read it. Some of the object labels also have pictures to further amplify points.

Guiding Concept D: Skimmability

The fourth guiding concept is "skimmability." Because visitors come from all ages and educational and professional backgrounds, designers cannot assume they know the technical language associated with the subject matter of the exhibit. In addition, because visitors are usually standing on their feet, reading labels can quickly become an uncomfortable experience. Finally, most visitors usually have a limited amount of time, either because they have other activities scheduled, want to leave time to see other parts of the museum, or are visiting with an impatient friend or relative. Therefore, designers must write the labels so that they can be skimmed while standing, rather than studied while sitting.

Lessons for Web Design

Just as the designers of early television quickly realized that a television show was not a radio show with pictures, so designers of Web sites are learning that readers do not prefer to read long passages of text on a computer screen, electronically distributed books not withstanding (Marsh 1997). In fact, some studies show that users do not read online; they skim. Users don't skim everything, merely the first few lines on a screen. In those instances where they do read word for word, users typically read more slowly online than they do in a book (Horton 1995).

As objects distinguish museum exhibits from books, and pictures distinguish television from radio, so the ability to interact and the ability to integrate several media distinguish computers from books and other types of media. Many of the design techniques used to control the flow of data in a museum exhibit may also work online:

- As exhibit designers use immersion to recreate environments for visitors, so Web site designers can use simulation to recreate environments for users.
- As exhibit designers layer content so visitors can choose a desired level of complexity, so interface designers can create layered interfaces to match users' experience levels and layered help systems to match users' appetite for information (Wilson 1994).
- As exhibit designers can design skimmable exhibits, so Web site designers can present content in a scannable mode, using such devices as navigational tools, headings, lists, charts, and graphics to promote scanning (Carliner 2000).

4. "Even the Best Signage Can't Fix a Poorly Designed Museum"

What I Observed in Museums

The designer of the exhibit on computer and telecommunications networks at the high technology museum I studied commented that visitors should have the "realization that what [they]'re experiencing is unique, powerful, and challenging." A good exhibition "keeps [visitors] coming around the corner" and "makes [them] want to explore."

Because the physical location of objects within an exhibit has a significant impact on visitors' experiences, exhibit designers try consciously to use space.

Conscious use starts with the general layout of the exhibit. Some designers like to create a hub of activity, such as the designer of the exhibit on networks:

> I wanted a big circle in the center, as if the exhibit radiated from a hub. I like to start with a larger metaphor. . . . Even if people don't realize it, the exhibit has strength of that organization. It makes everything flow naturally, according to a plan. Otherwise, it's just a space layout. . . . Whether people understand or not, they know something's there for a reason. . . . [Visitors] should always see the hub. That's how it is on the network.

> The concept evolved from my work in retail. The [bookstores I designed] have a book layout, with "pages" on either side [of a central aisle]. Nobody thinks about it but it's an organization method that, at the least, makes sense.

Others prefer a layout that lets visitors enter from any point. That's what the idea generator at the urban history museum I studied prefers. Rather than following a timeline, she wanted to make it possible for visitors to enter the exhibit at any point in time and coherently follow the story forward or backward from that point.

Sometimes a controlled approach is necessary. Because sequence is integral to telling the story of the canning factory, designers planned for it to be followed in a specific sequence, with definite starting, middle, and ending points. Some museums use the sequential approach as a means of controlling crowds. For example, the temporary galleries in the Boston Museum of Fine Arts and Minneapolis Institute of Arts are intended to be followed in sequence because it is the only way to manage the large crowds in blockbuster exhibits and because these exhibits are separately ticketed, requiring a single entrance.

Floors and walls also become design elements. For example, the designers at the urban history museum in my study used flooring that would simulate a sidewalk in one gallery and a highway in another. The designer of the exhibit on networks that I studied chose a mesh wall covering to enhance the high-tech mood and image of the exhibit.

Raising or lowering the level of light in a gallery also helps create the mood of an exhibit. For example, lighting in the galleries of street scenes in the urban history museum I studied had a high lighting level to simulate daylight. Sometimes lighting levels are dictated by practical considerations. Because fragile textiles, books, sketches, and paintings fade in bright light, exhibit design teams must often lower light levels to preserve the objects.

Idea generators and idea implementers also become involved in the design of floor space. They choose signature objects to catch visitors' attention and beckon them forward in an exhibit. Placed in one section of an exhibit, signature objects are large objects that can be seen from another part of the exhibit. For example, a fire engine in the urban history museum in my study and an Egyptian temple (complete, and inside the gallery) in the Metropolitan Museum of Art are examples of signature objects. Similarly, museum educators become involved in the design of floor space. The educator at the urban history museum noted that she always has to remind the design team to leave a "gathering space" in exhibits so she has a place where she can speak to a group of 20 to 40 students at a time.

Laws in some jurisdictions require that exhibits be accessible to all visitors, regardless of their physical disabilities. For example, exhibit designers typically add ramps to exhibits that have sunken or raised areas, to ensure that visitors in wheelchairs have sufficient clearance between objects, and labels are readable from sitting positions. Although not required by law, many exhibit designers also include seating areas in exhibits because older adults and young children need a place to rest in the middle of an exhibit. The design team at the urban history museum I studied also tested its exhibit with people in wheelchairs to make sure that the accommodations met the needs of these visitors.

Fixed architectural elements also affect the design of the floor space. For example, one of the obstacles facing the design team at the high technology museum in my study was a stairwell in the middle of the exhibit (not part of the exhibit). It could not be moved, so designers had to figure a way of incorporating it.

In addition to considering the floor space of the exhibit, designers also consider traffic patterns in the museum building. Some staffs place popular temporary exhibits at the end of a hallway, subtly requiring that visitors walk by permanent exhibits they might otherwise miss. Architect Richard Meier designed the High Museum of Art in Atlanta so that visitors could see nearly all the exhibits from the atrium at the entrance. Based on this initial scan, visitors can decide where to begin exploring.

Despite research into the traffic patterns of visitors in museums, not all museum buildings are easily traversed. Some museums try to compensate for a non-intuitive floor plan with extra signage. But as one exhibit designer noted in the brainstorming session of museum exhibit designers, "even the best signage can't fix a poorly designed museum." Another commented that way-finding within a museum has "little to with signs and maps. [It] has to do with the layout of the building."

Lessons for Web Design

As museum exhibit designers have learned that physical space is a key communication resource, so Web site designers have learned that screen real estate is a key communication resource. Consider:

- Because of the consistency of the Windows and Macintosh interfaces, users expect to find certain types of information at certain locations on the screen, like the menus and button bars.
- Similarly, because of the patterns of eye movement, users are more likely to see information placed in certain areas of the screen than in others (Horton 1995).
- Because many users typically do not scroll down, communicators have learned that they need to include mechanisms for encouraging users to scroll down and move forward to related pages (User Interface Engineering 1998).
- As exhibit designers have learned that the physical layout of a building constrains their ability to help visitors effectively find their way through the museum, so interface designers have learned that the structure of the underlying code constrains their ability to effectively design an interface. For example, a poorly structured program often results in a confusing menu. One software developer commented, "I can usually look at an interface and tell you the underlying structure of the data."

Based on these observations and experiences, Web designers might do the following:

- *Consciously place information on the screen*, making sure that key information appears in places where users are most likely to see it. Commercial sites have already learned to place advertisements at the top of a page and along the right margin to increase attention to them. We haven't developed similar conventions for technical information. Perhaps we could follow the example of cnn.com on its long stories, and place a table of contents at the beginning of the page. Or perhaps we can place summaries of key points along the right margin.

- *Create "signature objects."* The most likely signature object for a Web site is exclusive content. The challenge is most acute on commercial Web sites (whether business to consumer or business to business), because so many Web sites license content from third parties who, in turn, license the same content to other parties. Consider news. Many sources suggest that a news feed brings visitors back to a site. But if the news feed to one site comes from the same source feeding a competitive Web site, that news is not a signature object. As museums have learned, a copy of the *Mona Lisa* does not have the same signature value as the original.

- *Design for accessibility by people with disabilities.* The technical term for this type of design is universal design because it is a strategy to provide access to all. Many designers assume that adaptive equipment and specialized software can handle many of the challenges faced by persons with disabilities. For example, large screens and specialized software can increase the size of a display for people with visual impairments. But such hardware and software do not solve the problem of an inconsiderate design. For example, consider the problems encountered by a user of a Web site that relies heavily on audio cues, and does not provide alternate presentations of those data, such as transcriptions.

- *Consider traffic patterns.* On the one hand, designers want to make sure that visitors notice the most important or sought-after information on the Web site. However, as museum designers place less-known exhibits in the path of the sought-after ones to give those less-known parts more exposure, so Web site designers might place less-known content ahead of the better known material as a means of introducing visitors to other content.

5. "Museum Exhibits Must Capture the Visitor's Curiosity"

What I Observed in Museums

The idea generators at each museum all agreed: at the heart of a good museum exhibit is a good story. Like stories in books or film,

> museum exhibits must capture the visitor's curiosity. . . . Our attention is attracted by novel or unexplained stimuli—a loud noise, a sudden bustling activity, a strange animal, or a mysterious object. It is by appealing to this universal propensity that museums can attract the psychic energy of a visitor long enough so that a more extensive interaction, perhaps leaning to learning, can later take place.
>
> (Csikzentmihalyi and Hermanson 1995, 36–37)

The recipe for successful storytelling in exhibits is the same as that in literature: riveting plots and engaging characters. To create riveting plots, museum exhibit designers employ a number of standard storytelling techniques. One of the most basic is making sure the exhibit has a distinct beginning, middle, and ending. For example, the exhibit on networks that I studied begins with a two-part opening: a video overview, followed by a room where visitors received an "identity card." Visitors use the card to choose one of four virtual tour guides to lead them through the exhibit (seen by visitors on interactive display terminals); the computer records the choice on the identity card so visitors see related material at each guide station in the exhibit. The middle of the exhibit is a sequence of galleries, each of which describes a different type of network. The exhibit ends with another two-part sequence: a gallery presenting the negative side of networks, followed by a room where users can connect to the Internet.

Within the exhibit, exhibit designers use common storytelling techniques such as immersion, juxtaposition, repetition, and subliminal messages to engage the visitor. In addition to serving as a guiding principle of content development described earlier, immersion also serves as a storytelling technique, much like establishing shots in film and description in novels. It physically places visitors in the environment of the objects.

For example, the Scandinavian Heritage Museum in Seattle tells the story of immigration from Scandinavia to the United States by literally guiding visitors through a sequence of scenes depicting the journey. Visitors see such scenes as rural poverty in the old country, a crowded ship carrying immigrants, and homes in the new country. The Minneapolis Institute of Arts recreates period rooms from Charleston, Paris, and London to showcase furniture styles of the past. Exhibit designers believe that experiencing a subject through immersion is so essential to the success of an exhibit that they include it in grant proposals to persuade funders to support the exhibit.

Even a seemingly minor detail contributes to the authenticity of the immersion environment. For example, the walls in each gallery of "Without Boundaries" were painted specific colors to enhance the authenticity. Green walls in the first gallery offered a pastoral feeling, typical of a newly settled rural area, while gray walls in the gallery depicting the commercial growth of the city evoked a business like mood. Sometimes the building itself creates authenticity. The industrial history museum that I studied is housed in a former canning factory. In addition to adding authenticity, this history actually inspired the subject of the exhibition.

Another storytelling technique is juxtaposition, in which two opposing images or concepts are positioned near one another so visitors can make the contrast. The designers of the exhibit on the history of the city in my study juxtaposed the clothing of early European settlers with that of Native Americans, so viewers would sense the culture clash that would define the early history of the region. Later in the exhibit, the designers recreated a street with scenes from white culture on one side and scenes from African American culture on the other, to show their separate histories in the community. An activity that takes place within the exhibit on the canning factory juxtaposes managers and workers in the same work environment.

Another technique is repetition, where an image or concept appears more than once in an exhibit to reinforce a point. Exhibit design teams intentionally repeat points to increase the likelihood that visitors will remember them. For example, clothing typical of an era was included in each gallery of the exhibit on the history of the city to emphasize its importance as a cultural statement in each period of the city's development.

Exhibit designers also include subliminal messages; ones they hope will make an unconscious impact on visitors. Three stones in the first gallery of the exhibit on the history of the city in the urban history museum each represented a different phase in the early growth of the city. Designers did not expect most visitors to recognize the significance of these stones. In fact, designers at each museum I studied did not expect visitors to understand their subliminal messages, but the idea generator at the urban history museum said that some visitors tell her that they do get these messages.

In addition to a tightly crafted plot, a good story must be populated by engaging characters. Exhibit designers address this issue, too. Each of the three exhibitions I studied included key characters. In two of the museums studied, the characters were fictional but emerged from extensive research and were composites of real people. For example, the virtual guides through the exhibit on networks were intended to represent different segments of the local population. One was a homeless person. Research with the homeless population helped exhibit designers flesh out this character. Similarly, the designers of the exhibit on the canning factory included descriptions of workers and managers. Although the names were fictional, their life stories were based on information in the museum archives.

As stories are about people, so they must appeal to people. Therefore, the gauge for assessing planned storytelling techniques is their anticipated appeal to visitors.

> The link between the museum and the visitor's life needs to be made clear . . . the objects one finds and the experiences one enjoys, while possibly inspiring awe and a sense of discovery, should not feel disconnected from the visitor's experience.
>
> (Csikzentmihalyi and Hermanson 1995, 37)

At the most basic, museum exhibit design teams try to appeal to the everyday and to today. For example, a living room in the exhibit on networks showed how networks affect modern home life. The last activity in the exhibit on the canning factory gives visitors an opportunity to relate work of the late 19th century to work today.

Although exhibit designers make liberal use of storytelling techniques, they sometimes have difficulty finding the human story in the otherwise academic topic of a proposed exhibit. "What's your story? [Sometimes, it's] really hard to get it out to visitors," commented an exhibit designer attending a meeting of her colleagues at the 2000 American Association of Museums Annual Meeting.

In contrast, another designer commented that she is "more interested in the voices and stories than the technical aspects of the exhibit." She admitted, though, that the technical aspects are essential in the practical challenge of bringing the story to the public.

Lessons for Web Design

As exhibit designers rely on storytelling techniques to engage visitors in the content of exhibits, so some Web site designers are relying on storytelling techniques both to engage visitors and to use as a planning tool.

Here are some of the ways that Web site designers use storytelling techniques as a planning tool.

- *Describe the background story.* One common writing technique in storytelling is sketching out a character's backstory: the experiences that preceded those told in the piece being written. Web site designers use a similar technique called scenarios or user cases (Nurminen and Karppinen 2000). A scenario describes the real-world situation (or backstory) that drove a user to consult a particular Web site.
- *Describe users as real people rather than demographics.* Author Alan Cooper (1999) recommends that Web site designers also prepare descriptions of archetypes—that is, provide character descriptions of typical users. As the many characters of a good story often represent a diversity of experience or perspectives, so do archetypes. Cooper recommends that, at the least, the archetypes represent a user who will easily adapt to the changes, one who will have difficulty, and one who represents the middle-of-the-road user. By defining this spectrum, communicators are more likely to address everyone rather than a single type of user who is represented by the demographics of the intended users.

In addition, designers can employ many of the same storytelling techniques used in museum exhibits in Web sites. For example, as exhibit designers "immerse" visitors in a setting, so Web site designers simulate experiences. The technique is widely touted in games and online learning. For example, the game SimCity (although not yet on the Web) immerses visitors in the development of a city. A Web-based simulation developed for internal use by marketing representatives at Dell Computer mimics a virtual pet, but instead of participants following the life of an animal, they follow the day of a marketing representative (Hartley 2000).

Similarly, as exhibit designers try to create a mood for their exhibits, so can Web designers. For example, a graduate student who was visiting a cybercafé in Manhattan (New York City) commented on the way that the home pages used in this café recreated the Gotham mood online:

> I sat in the cool air [as] I waited for the default homepage to load, I noticed that a designer did a wicked cool thing with the interface. As the gray letters emerged from the black background, the designer played a movie in the background. Cars, people, and trucks passed by. The sound was cool, too, and when a horn sounded I jumped! The sound didn't come from the speakers! I watched the reflection of real life—a busy Manhattan street—in my screen!

Just as it works as a storytelling technique in exhibits, so juxtaposition is an effective storytelling technique online. On some Web sites, designers visually juxtapose contrasting content. For example, on vote.com, designers present a series of issues. Beneath a value statement, designers place the description of the "pro" position on the left and the "con" position on the right. Similarly, following a news story, CNN lists Web sites with related content, letting visitors surf to sites representing opposite points of view.

Subliminal and subtle messages are also an important part of Web sites. They tend to show up more in design efforts helmed by graphic designers and artists than by those led by usability experts, who tend to take a more utilitarian approach to design (Every 1999).

6. Work Toward "Wow!"

What I Observed in Museums

"The first thing we're looking for is for people to say 'Wow!'" commented the director of the then-new Futures Center at the Franklin Institute in Philadelphia (Behr 1989). He's not alone. When reviewing the designs for a proposed exhibit at the high technology museum I studied, the museum educator asked her colleagues, "Where's the fun factor?" Almost universally, the designers of museum exhibits hope their visitors have a pleasant experience.

Part of this interest stems from a genuine desire of the design team to share their passion for a subject with visitors. For example, the idea generator for the industrial history museum wants to help visitors understand their ancestors' experience at work. "[They] spent more than a third of their lives at work; the museum fills a large void in people's understanding of the past." The director of public programs at the urban history museum wants her visitors to "enjoy the experience" and leave exhibits "knowing, thinking, and feeling."

In some cases, the need to "wow" visitors emerges from more practical considerations. According to one exhibit designer, museums must compete for visitors with other "cool stuff," including other museums, movies, theme parks, performing arts, and sporting events (Mintz 1994; Zolberg 1994). Some of these competitors are becoming more like museums. For example, theme parks such as EPCOT in Orlando, Florida, and the Luxor Casino in Las Vegas, Nevada, are displaying and interpreting objects as museums do, but with larger budgets and more lavish presentations. This competition raises visitor expectations for effective exhibits (Mintz 1994, 33). In other instances, museums compete with other types of entertainment, such as movies, theatrical and musical performances, and sports.

Exhibit designers choose topics with strong popular interest not only to broaden their audiences, but also to attract visitors. For example, because many young children have a fascination with dinosaurs, most science and natural history museums regularly schedule dinosaur-themed exhibits. When possible, they have dinosaur skeletons and eggs in their permanent collections, as do the American Museum of Natural History and the Los Angeles County Museum of Natural History. Well-known artists (especially Impressionists) are similarly popular attractions for art museums.

Well-known objects can also attract visitors. People visit the Art Institute of Chicago to see the painting *American Gothic*, the British Library in London to see the original draft of the Magna Carta, the Israel Museum to see the Dead Sea Scrolls, and the Smithsonian's American History Museum to see the collection of gowns worn by American first ladies to the balls celebrating the inaugurations of their husbands as United States presidents. The idea generator at the urban history museum noted that objects are powerful teachers because "[they] hold their own experiences. People ask 'Is this real?' If it weren't, it wouldn't be here."

An object in a temporary exhibit can have a similar drawing power. A pre-opening furore sparked by comments made by the mayor of New York City over a painting of the Virgin Mary, composed, in part, of elephant dung, lured visitors to "Sensation: Young British Artists from the Saatchi Collection," an exhibit at the Brooklyn Museum. The need to wow visitors continues after they arrive in the exhibit; designers must maintain visitors' interest. The signature objects mentioned earlier serve such a purpose. Sensory experiences, like the simulated earthquake at the California ScienCenter in Los Angeles, are intended to engage senses other than the sight. Although admittedly more sedate, some museums create a multisensory experience through music or continuous playing of recorded environmental sounds in the exhibit area. For example, the urban history museum I studied plays recordings of ambient sounds in the exhibit. Some exhibit designers try to create an emotional reaction among visitors. For example, the urban history museum displayed a robe from a Ku Klux Klan member. A dark gallery with metallic accents in the exhibit on computer networks at the high technology museum was intended to create a "big brother is watching you" feeling.

Exhibit designers try to transport visitors to other times and places. The exhibit on the canning factory in the industrial history museum recreates the world of work in the late 19th century. The National Maritime Museum in Greenwich, UK, recreates scenes from the journeys of British explorers in the 16th and 17th centuries. In its "Traveling the Pacific" exhibit, the Field Museum of Natural History recreates a market in the Philippines.

But the exhibits that seem to create the strongest feeling of "wow" among visitors are interactive ones. The Exploratorium, a science center in San Francisco, pioneered the interactive exhibit. At that museum, visitors perform mini-experiments to discover scientific principles; then, if they want, they read the explanatory material to learn more about the principles (Hein 1990).

Another interactive technique is the use of touch objects that was mentioned earlier. Museum exhibit designers believe that one of the most powerful learning experiences in museums occurs when a visitor can touch real objects, so they try to provide this experience whenever possible. In some instances, however, it is not. Contact with oil from human hands, for example, can damage fragile artwork. Climatic conditions can destroy documents. Light fades fabrics. Visitors sometimes damage objects, though not always intentionally. A visitor to the Minneapolis Institute of Arts thought a chair in one of the galleries was intended for weary visitors. When it broke after he sat on it, he learned that the chair was actually a delicate Chinese antique. But in cases where the potential for damage is slight, or the museum has a duplicate of the object, designers like to place it on display as a touch object. The public programs and education staffs can enhance

> the sense of "wow" in an exhibit. Public programs are those aimed at the general public. Sometimes the programs involve craftspeople demonstrating implements on display. For example, the Fruitlands Museum in Harvard, MA, scheduled demonstrations by carpenters and blacksmiths to complement its exhibit of tools. Science museums often schedule demonstrations. For example, SciTrek, the Science and Technology Museum of Atlanta, schedules several demonstrations each day.

The education staff focuses almost exclusively on school groups visiting the museum. According to the museum educator at the urban history museum I studied, her colleagues at other museums typically develop scavenger hunts and activity baskets as tools to help young visitors notice all parts of an exhibit or focus on parts of special interest to their teachers (if the students attend with a school group). The idea implementer at the industrial history museum in my study added that she also develops materials that classroom teachers can use to prepare students for an upcoming visit and debrief the visit afterwards.

Lessons for Web Design

As exhibit designers try to "wow" visitors with provocative subjects, interactivity, and similar techniques, so must Web site designers. One particular area of interest to Web site designers is the design of the interaction between users and the computer. Web site designers try to "wow" users in a number of ways.

- "Splash" screens, which display a brief animated sequence, are intended to capture and hold user interest. Web sites for commercial films, for example, usually start with an elaborate splash screen intended to generate excitement about the film. But the scene must splash quickly, or visitors will surf elsewhere.

- Profiling—the act of capturing information about a given user and using that information to tailor the Web site to that user's interests—attempts to "wow" visitors through personalization.
- Online communities and scheduled chats can foster a sense of loyalty among users and increase the number of visits to the site, just as public programs and education are also intended to help visitors discover parts of museum exhibits.

Two challenges face designers in bringing "wow" to their Web sites. The first pertains to technology. With each technical development often comes a new means of "wowing" users. But the challenge to Web site designers is finding techniques that engage users within the context of the Web site content, rather than as merely demonstrating the technology.

Furthermore, the same technologies that let Web sites develop and enhance profiles of users also involve an invasion of user privacy. Web site designers must determine at what point the value of better knowing users exceeds the risk of offending users by collecting and using information that users might not want anyone to be collecting. European law severely limits such practices. In contrast, American Internet users have shown a surprisingly high tolerance to tracking.

The second challenge facing designers in bringing "wow" to their Web sites comes from the almost religious battle between usability experts and graphic designers on ideal approaches to Web design. Usability experts, led by the likes of Jakob Nielsen, tend to focus on observable, measurable patterns of effectiveness that can be independently verified through usability research. But measuring affective responses like "wow" will tax even the best-refined research methodologies, and graphic designers and others with backgrounds in the arts and humanities are often hard pressed to produce data from universal research that would support the use of nonstandard approaches, like those of storytelling (Cloninger 2000).

7. Avoid "Sound Bleed" and Other Media Nightmares

What I Observed in Museums

Two-thirds of the way up the back wall of the entrance lobby to the Walker Art Center in Minneapolis is a horizontal line of lights that lead around a curve and beckon visitors through a hidden doorway. Beyond the doorway is a long, low, dark theater with built-in benches. On the three oversized screens at the front of this theater, a slide and sound show continuously plays. It introduces visitors to the primary temporary exhibit. When visitors leave the theater, they walk up a half-flight of stairs and enter that exhibit.

Visitors need no beckoning lights to see the "[city] in your face" video in the city history exhibit at the urban history museum I studied. It simultaneously plays on 12 monitors of various sizes hanging from the ceiling at the entrance to the exhibit. A glass wall behind the monitors gives visitors a glimpse into each of the four galleries; visitors can enter any gallery they choose.

At the end of a visit to the exhibit on the canning factory in my study, visitors participate in a computer-based survey that asks them about the types of jobs they saw in the exhibit and helps them relate them to jobs in today's economy that might interest them.

As exhibit design teams at the Walker Art Center, the urban history museum, and the industrial history museum have done, so exhibit design teams at many museums are integrating media into their exhibits. In some instances, media presentations are as central to an exhibit as the objects. For example, in addition to the "in your face" video, the exhibit at the urban history museum includes three video theaters. The theaters are placed between pairs of galleries and are used to explain the transition from the time covered by the first to the time covered by the second. The videos playing there were created from photos and film footage in the museum archives. The idea generator explained that these videos provided an efficient means of telling the stories of these transitions; stories that the museum had neither the objects nor the gallery space to tell.

Other museums use computers in the exhibit area to provide visitors with access to additional information. For example, the Minnesota Historical Society has included some of the oral histories in its collection on a computer in its exhibit "Minnesota A to Z" so visitors have access to life stories of local citizens while learning about Minnesota culture. On interactive stations placed in education rooms near the galleries, the Seattle Art Museum provides an additional level of documentation about objects from its permanent collection and links users to background and related material.

The Internet is also becoming increasingly important to exhibits. Some Web sites serve as online brochures for exhibits, as at the Wing Keye Museum in Seattle. Some Web sites extend the visit by providing information that visitors might explore in advance and other information they might explore afterwards, such as the information accompanying the Field Museum of Natural History's permanent collection. Some Web sites serve as exhibits in their own right, either displaying digital versions of materials that are no longer on exhibit or separate displays that are available only online, such as the Museum of Modern Art's "Art Safari."

Although videos and computer displays can extend an exhibit, each of the exhibit design teams I studied expressed frustration in working with media. New technology, inexperience, and significant under-budgeting affected the development of the virtual tour guides for the exhibit on networks at the high technology museum. Programming bugs plagued some of the computer displays in the exhibit on the canning factory. What frustrated exhibit designers most, however, was that the program was written with proprietary software and was not documented. So when the company that wrote the program went bankrupt and the programmers literally left the country, the US$10,000 station (about 5 percent of the exhibit budget) was unusable. Information on another computer was still usable, but the content was out of date and the staff did not have funds to revise the content.

Other than at the high technology museum, video production went smoothly for the museums studied. But exhibit designers wondered whether visitors actually watched the videos. My observation of visitors to a science center suggests not. The center had several video stations within its exhibit space. Each played video on demand; that is, a video would play after a visitor pressed a start button. Few visitors stopped at the videos.

Perhaps they were concerned about the noise from the video calling attention to them in the otherwise quiet space. Such sound bleed (that is, sound that can be heard outside its display area) is a practical issue in using video and other audio tools. Because many visitors like to read labels or think as they ponder an exhibit, museums are typically quiet places. Loud sounds from a video within the exhibit could break their concentration. Worse, should sounds in one gallery "bleed" (that is, be heard) in the next, the sound seems illogical and out of place, and reflects poorly on the designers.

Lessons for Web Design

As in museum exhibits, video, audio, and specialized software can provide significant value to a Web site. But using them can also create substantial practical challenges.

- Although narration is often helpful for people with reading difficulties and sound effects demonstrate audio content, the noise created by one user's computer can "bleed" throughout a workplace and distract other workers in the area. Furthermore, users can typically read a passage to themselves faster than a narrator can, and as a result, they may find narration more of a roadblock than a benefit.
- Slower Internet connections can slow the display of some Web content (large graphics, animation, and video, for example).
- In some implementations, users need special software called plug-ins to play video and audio. Some users do not have access to plug-ins, making reliance on them impractical. Even when users have access to plug-ins, some experts warn against using them, in case users have difficulty or the image becomes garbled. For example, students of one online university had difficulty viewing lectures because the intranet from which they worked did not allow plug-ins. For these reasons, some Web gurus like Jakob Nielsen recommend against using plug-in technology.
- Some Web site designers like to take advantage of the latest technical improvements to Web technology. Because users are often slow to upgrade their browsers, they might not have access to that technology. For example, when frames were first introduced, designers who wanted to use them

immediately had to implement frame and non-frame versions of their Web sites.

Finally, as computing increasingly moves off of desktop and laptop computers and onto other types of devices like mobile phones and personal digital assistants, designs for one type of display will increasingly fail to display effectively on other equipment.

8. "Attendance Figures Measure Marketing Strategies, Not Exhibition Strengths"

What I Observed in Museums

One of the most challenging aspects of exhibit design is assessing its effect on visitors. One common measure used by museums is attendance. In most instances, because most museum admissions let visitors see any exhibit, attendance figures pertain primarily to the museum in its entirety. In these cases, attendance spikes (that is, sudden increases in attendance) are usually attributed to changes in the make-up of exhibits. For example, a spike that follows the opening of a new permanent exhibit is attributed to that exhibit. In some instances, however, museums charge separately for blockbuster temporary exhibits or exhibits located in another facility. In those cases, attendance figures pertain to the separately charged exhibit.

"Wouldn't it be better to judge an exhibition's success—or failure—by attendance figures?" asks Chambers (1999). No, she determines, observing that "attendance figures measure marketing strategies, not exhibition strengths" (1999, 31).

According to the American Association of Museums' Standards for Museum Exhibitions, an exhibit is successful if it is physically, intellectually, and emotionally satisfying to visitors. Visitor research is a discipline within the field of museum studies that assesses the impact of exhibitions and their components. Visitor research explores a variety of issues, such as the demographics of visitors to particular types of museums (like science museums), the amount of time visitors spend reading labels, the objects that visitors focus on, the themes that visitors recall from exhibits, and the exhibits that visitors actually go through and the ones that they ignore. Chambers notes that it is

> significant that museum exhibitions began to be a topic for professional discussion just when American advertising was developing into a science. Research into the power of advertising design to attract and hold attention (to sell a product) soon spilled over into the museum world and created new criteria for visual presentations and their power of persuasion. Early visitor studies of the 1930s took their cue from psychological studies about the manipulative techniques of advertising, as many of them still do.
>
> (1999, 33)

Most of these studies are quantitative and results are used as much to generate design guidelines as to assess effectiveness.

Although they value it in theory, few museums actually have the resources to perform their own visitor research. Certainly the museums in this study did not. Other than attendance figures and evaluations from public programs, the design teams in my study relied almost exclusively on anecdotal evidence to assess the effectiveness of their work. The designers at the high technology museum were the most rigorous, using a form of usability test to assess the effectiveness of proposed exhibits. They would place prototypes of interactive displays in a gallery and observe visitors' interactions with them. In some cases, staff members would also interview visitors to get more specific feedback. They did not apply such rigor to assess the effectiveness of a completed exhibit, relying primarily on comments from feedback forms placed at the end of the exhibit and comments relayed by docents working in the exhibit.

The staff at the urban history museum placed a comment book at the end of its exhibit on the history of the city. Once a month, the idea implementer would record comments from the book and share them with the rest of the exhibit design team. Sometimes visitors to that museum would contact the staff. The exhibit design team generally considered themselves to be successful when they received requests from visitors for more information or the opportunity to visit the museum library.

When feasible, museum staffs try more rigorous approaches. The National Aquarium in Baltimore, Maryland, commissioned a study to assess the long-term impact of an exhibition on visitors. According to Adelman (2000), they wanted to see whether or not the exhibit had a transformative effect on visitors. Paris (2000) noted that transformation results from a combination of process and outcomes that are neither well understood nor documented. The researchers noted that because of the long time frame and high cost of this type of study, few museums can maintain such evaluation programs on an ongoing basis.

Lessons for Web Design

Because we can easily do so, it is tempting to report the number of visitors as the measure of effectiveness of a Web site. We even have technology that tells us where visitors come from and where they go when they leave our sites. But as counting attendance at museum exhibits may only measure marketing strategies, so counting the number of visitors to a site may only measure marketing strategies.

Another tempting measure might be measures of system performance (such as the speed of loading) or checklists of usability items (such as the number of links per page or the extent of use of passive voice). But these characteristics only correlate with usability; they do not guarantee either usability or users' ability to perform the tasks for which the Web site was designed.

Only when users can perform the tasks for which a Web site is intended is the Web site successful. If the Web site was designed with clear objectives, then one key measurement of effectiveness is whether users can achieve those objectives.

To ensure the long-term success of the Web site, however, it is important to also gauge user satisfaction with the site. If users are not satisfied with the experience of using the Web site, they are not likely to use it in the future if given an alternative.

Closing Thoughts

Even with the creation of immersive environments, exhibit designers recognize that exhibits are, at best, artificial environments. A room that has been rebuilt inside a museum exhibit is no longer part of a real house. Fire alarms from the 1890s that sit on walls in the exhibits on urban history and networks are no longer working. Exhibit designers acknowledge that one of their main tasks is to give visitors tools to better understand the outside world—not replace it.

Although we recognize that we are creating online communities with our Web sites, and especially when we provide opportunities for users to interact with one another online, we too must recognize that online worlds are ultimately artificial ones and that people still need direct, ongoing contact with one another to learn and work.

I learned this in my work also. Although my primary interest was design techniques, what struck me most was the cohesiveness of design teams in this study. In each museum, I observed that the idea generator served as more than the nexus of ideas; this person also served as an informal educator of the team. The idea generator and, in some instances, the idea implementer were the only ones who had personally learned exhibit design for museums. Other team members relied on the idea generator for guiding concepts of design, and terms introduced by the idea generator were used by all members of the design team. For example, the idea generators at the urban and industrial history museums used the term "immersion," as did their staffs. The idea generator at the high technology museum used the term "immersive," as did his staff.

Similarly, rather than learn about museum studies, exhibit designers in my study often look to other design disciplines for ideas. For example, the designer at the high technology museum relies on his retail experience for ideas.

The enduring lesson that Web site designers can learn from this study is that we must take responsibility for our roles. Not only do we design Web sites for users, but we also provide intellectual and emotional leadership for our entire design teams.

Acknowledgment

*This article was originally published in *Technical Communication* (2001), 48: 66–81.

References

Adelman, L. 2000. "Results of a Study for Transforming Visitors' Attitudes and Behaviors." American Association of Museums Annual Meeting. Baltimore, MD.

Behr, D. 1989. "Museum Welcomes the 21st Century." *Los Angeles Times*, October 24: S-1, S-7.

Bloom, J., and E. A. Powell. 1984. *Museums for a New Century: A Report of the Commission on Museums for a New Century*. Washington, DC: American Association of Museums.

Carliner, S. 1998. "How Designers Make Decisions: A Descriptive Model of Instructional Design for Informal Learning in Museums." *Performance Improvement Quarterly* 11 (2): 72–92.

——. 2000. "Twenty-Five Tips for Communicating Online." Online learning 2000. Lakewood Conferences/Bill Communications. Denver, CO.

Chambers, M. 1999. "Critiquing Exhibition Criticism." *Museum News* 78 (5): 31–37, 65.

Cloninger, C. 2000. "Usability Experts are from Mars, Graphic Designers are from Venus." *A List Apart*. http://www.alistapart.com/marsvenus/index.html.

Cooper, A. 1999. *The Inmates Are Running the Asylum*. Indianapolis, IN: Sam's.

Csikzentmihalyi, M., and K. Hermanson. 1995. "Intrinsic Motivation in Museums: What Makes Visitors Want to Learn?" *Museum News* 74 (3): 35–37, 59–62.

Every, D. 1999. "Real Interfaces: UI Religious Wars." http://www.mackido.com/Interface/RealInterfaces.html.

Hartley, D. 2000. *On-Demand Learning: Training in the New Millennium*. Amherst, MA: HRD Press.

Hein, H. 1990. *The Exploratorium: The Museum as Laboratory*. Washington, DC: Smithsonian Institution.

Hirzy, E. C., ed. 1992. *Excellence and Equity: Education and the Public Dimension of Museums*. Washington, DC: American Association of Museums.

Hoft, N. 1995. *International Technical Communication: How to Export Information About High Technology*. New York, NY: John Wiley & Sons.

Horton, W. 1995. *Designing and Writing Online Information*. 2nd edn. New York, NY: John Wiley & Sons.

Ivey, W. 2000. Keynote Address to the American Association of Museums Annual Meeting. Baltimore, MD, May 15.

La Follette, M. 1983. Introduction to Special Section on Science and Technology Museums. *Science, Technology and Human Values* 8 (3): 41–46.

Lee, E. W. 1994. "Beyond the Blockbuster: Good Exhibitions in Small Packages." *Curator* 37 (3): 172–185.

Lynch, P., and S. Horton. 1995. "Yale C/AIM Web style guide." http://info.med.yale.edu/caim/manual/contents.html.

Mager, R. 1997. *Measuring Instructional Results*. 3rd edn. Atlanta, GA: Center for Effective Performance.

Marsh, J. 1997. New Media Instructional Design Symposium. Chicago, IL, July.

Mintz, A. 1994. "That's Edutainment!" *Museum News* 73 (6): 32–36.

Nurminen, M., and A. Karppinen. 2000. "Use Cases as a Backbone for Document Development." 47th STC Annual Conference. Orlando, FL.

Paris, S. 2000. "Motivation Theory: Transforming Visitors' Attitudes and Behaviors." American Association of Museums Annual Meeting. Baltimore, MD.

Stone, J. 1993. "Ten Great Science Museums." *Discover* 14 (11): 105.

Strauss, A., and J. Corbin. 1990. *Basics of Qualitative Research: Grounded Theory Procedures and Techniques*. Newbury Park, CA: Sage Publications.

—— and ——. 1994. "Grounded Theory Methodology: An Overview." In *Handbook of Qualitative Research*, ed. N. K. Denzin and Y. S. Lincoln. Newbury Park, CA: Sage Publications.

User Interface Engineering. 1997. "Why On-Site Searching Stinks." *Eye for Design* September/October. http://world.std.com/uieweb/searchar.htm.

——. 1998. "As the Page Scrolls." *Eye for Design* July/August. http://world.std.com/uieweb/scrollin.htm.

Wilson, C. 1994. Presentation on Usability to the Board of Directors, Society for Technical Communication. Boston, MA, September 28.

Zolberg, V. L. 1994. "'An Elite Experience for Everyone': Art Museums, the Public, and Cultural Literacy." In *Museum Culture: Histories, Discourses, Spectacle*, ed. D. J. Sherman and I. Rogott. Minneapolis, MN: University of Minnesota Press.

Chapter 5

Toward a Meaningful Model of Technical Communication*

Hillary Hart and James Conklin

Editors' Introduction

"Toward a Meaningful Model of Technical Communication" by Hillary Hart and James Conklin combines quantitative and qualitative methods to examine the typical practice of current technical communication, and to explore whether it is possible to construct a model or suggest a metaphor that accurately describes the profession as it exists today.

Most of the data that the authors gathered were qualitative. They planned and conducted three geographically dispersed focus groups across the United States, with a total of 47 participants. As they themselves note, Creswell (2002) and Krueger (1994) have established the focus group as the best means of examining a subject (in this case, the work being done today in our profession) from the inside out. As with many of the articles selected for this anthology, their methodology section is admirably detailed, explaining not only what they did, but why they chose the techniques they used.

In addition to the focus group activities (word association, structured conversation, and participant dialog), the participants were asked to complete a survey that requested quantitative information about demographics, level of job satisfaction, training and preparation for their current positions, their roles in the workplace, tools used, teamwork, and time spent on various types of activities. The focus groups probed essentially the same topics and also asked participants to describe in metaphor or illustration their view of the current practice of technical communication. The sessions were transcribed and coded, and the results were clustered around a number of common themes.

The result is an eminently readable report that summarizes what kinds of work the participants do on the job, how they are regarded by peers and managers in their organizations, and how their work is valued. Although we cannot generalize from the results of this study of 47 technical communicators, we can compare the trends reported here to reports by other technical communicators and find many similarities.

Not surprisingly, Hart and Conklin address many of the same themes explored by Giammona in "The Future of Technical Communication" (Chapter 3 in this volume). Although not all of them are investigated here in the same depth as in Giammona's article and some are only suggested, the following themes play at least a limited role in Hart and Conklin's article:

- What is a technical communicator today?
- What is our future role in organizations?
- How do we contribute to innovation?
- How should we be educating future practitioners?
- What technologies are impacting on us?
- Does technical communication matter?

We selected this article for several reasons (and the fact that one of us was a co-author of the report in question played no role in our decision). First, focus groups are a particularly rich technique for qualitative research because of the interaction of the participants and the multiplicity of viewpoints that they provide. As a result, focus groups tend to provide a richer depiction of the population of interest than surveys. However, they are not frequently used in technical communication research because they are relatively expensive (they almost invariably require travel), and research in our field is seldom funded externally. Moreover, recruiting a sufficiently large group of participants can be difficult, especially when multiple sessions are held in multiple geographical locations.

Hart and Conklin's approach resonates not only with the article by Giammona earlier in this volume, but also with another influential article by Rainey, Turner, and Dayton (2005) that similarly collected both quantitative and qualitative data (surveys and a small number of interviews).

References

Creswell, J. W. 2002. *Educational Research: Planning, Conducting, and Evaluating Quantitative and Qualitative Research.* Upper Saddle River, NJ: Prentice Hall.

Krueger, R. A. 1994. *Focus Groups: A Practical Guide for Applied Research.* 2nd edn. Thousand Oaks, CA: Sage Publications.

Rainey, K. T., R. K. Turner, and D. Dayton. 2005. "Do Curricula Correspond to Managerial Expectations? Core Competencies for Technical Communicators." *Technical Communication* 52: 323–52.

Introduction

Technical communicators emerging from education and training programs in the discipline are experiencing a clash between expectations and the reality of the workplace (Wilson and Ford 2003). They aren't prepared for the multiple roles required of them in today's business environments, and are surprised to find that employers tend to undervalue and "misuse" technical communication professionals. Educators have begun to realize that technical communicators working on cross-functional teams are focusing more on people and less on texts, and that future curricula need to focus more on human interaction and knowledge creation (Wojahn and colleagues 2001; Rainey, Turner, and Dayton 2005). Reporting on the results of a survey of managers that sought to identify the most important technical communication competencies, Rainey, Turner, and Dayton argue that "collaborative competencies" and "people skills" are vital to technical communication success, while communication skills are generally seen as "a given" (332–33).

In his introduction to the same special issue of *Technical Communication*, guest editor Michael Albers begins with the observation that the practice of technical communication has moved from a focus on "writing documents" to an increased need "for both collaboration and project management" (2005, 267). These conclusions echo slightly earlier statements from researchers who have declared that technical communication is shifting "from a product focus ... to a performance focus" (Hughes 2004) and that recent advertisements for technical communication jobs increasingly stress the importance of "interpersonal skills" (North and Worth 2000). In their chapter "Moving from the Periphery . . .," from *Power and Legitimacy in Technical Communication*, Sullivan, Martin, and Anderson call for technical communicators to acquire "social knowledge" to become successfully integrated into "the life" of their organizations (2003).

These observations point to the need for a redefinition of technical communication roles. The models and descriptors (for example, "writer") that explain the purpose and role of technical communication as having to do mainly with written communication are obviously outdated and inaccurate. The contemporary workplace is experiencing the emergence of more inclusive management practices and more collaborative and empowered work practices (see Chemers 1997; Kouzes and Posner 2005; Lawler 2001; Lipman-Blumen 2001). As Lawler writes:

> Hierarchical organizations are simply too inflexible and rigid to compete effectively in today's business environment. They fail to attract the right human capital and to produce the right core competencies and organizational capabilities. As a result, they need to be replaced by lateral forms of organization that rely heavily on teams, information technology, networks, shared leadership, and involved employees.
>
> (Lawler 2001, 17)

An empowered workforce requires effective relationships, clear communication, a spirit of initiative, and a willingness to engage in respectful conflict. These changes are having a profound effect on the ways in which people communicate with each other within and across organizational boundaries, as well as on the ways in which organizations create and use knowledge. One researcher recently claimed that traditional technical communication belongs to an earlier style of scientific management that attempts to control and direct human behavior, while a research team in the UK associates traditional documentation approaches with an epistemology of transfer whereby experts write prescriptive instructions that are to be followed by the workers (Orr 1996; Boreham and Morgan 2004).

If practicing technical communicators feel caught in these shifting workplace practices, how much more difficult is it for teachers of technical communication to keep up with the shifts, design appropriate exercises and projects, and decide what skills are the right ones to be teaching? Recent literature is full of advice on how to improve college and university teaching (for a sample, see Hayhoe 2003; Faber and Johnson-Eilola 2003; Davis 2004; Rehling 2004; Kain and Wardle 2005; Tebeaux 2004), but the dizzying number of strategies offered can be contradictory (see Bushnell (1999) for a "subversive" strategy). And there is still a need for more narratives, metaphors, and images that capture the reality of contemporary technical communication work and can inform pedagogical frameworks for educating students.

It is time that a new, more accurate, and helpful model of technical communication was developed and accepted by the entire technical communication community (including employers, employees, and academics), to help technical communication professionals better understand their emerging role and contribute more fully to the contemporary workplace.

In response to this need, we have undertaken a qualitative research inquiry into emerging models that illustrate the purpose and direction of technical communication. By "models," we mean conceptual frameworks, images, and metaphors that are able to communicate rich insights and deep meanings about their referents. For example, at one time many technical communicators referred to themselves as a "bridge" between technology experts and technology users. "Bridge" was a model for technical communication, and many found it to be a powerful way of communicating the intermediary or advocacy role played by technical communicators.

In conversations with technical communication professionals and academics, however, we have noticed that many within the profession have come to see this image as misleading or even degrading. Bridges are passive conduits for traffic, rather than active participants and contributors to complex human processes. Moreover, bridges are "walked on"—a complaint we heard more than once by members of the profession. Today, the Society for Technical Communication mission statement makes no reference to a bridge metaphor (or to any other metaphor), and says that a technical communicator is "anyone whose work makes technical information available to those who need it" (http://www.stc.org/about/).

Through our consulting work, our interactions with others in the profession, and our review of the literature on technical communication practice, we strongly suspect that technical communication is now placing less emphasis on one-way communication through texts, and more emphasis on the creation of opportunities for two-way communication between those who create technology and those who implement and use technical innovations in specific workplace or social environments. It occurred to us that one way of exploring this idea would be to invite technical communicators to speculate on the sorts of images and metaphors that best illuminate the type of work that they do today. We had heard practitioners and academics complain that the bridge metaphor is inadequate because it fails to recognize that technical communication is a creative or constructive activity. It thus seemed to us that it would be appropriate and interesting to invite technical communicators to work with us to construct a new metaphor for the profession. What is the appropriate model and metaphor for the contemporary technical communicator? If the "bridge" no longer works as a representation of the mediating role played by technical communicators between experts and users, what image does work?

The purpose of our project is to explore the state of current practice in the technical communication field and to develop visual and verbal models that more accurately describe that practice. Our research explores the possibility that the field of technical communication is moving from a linear model of one-way communication toward a new model of communication as a two-way collaborative process. To investigate this possibility, we gathered information from groups of technical communicators in three North American centers. We describe the research methodology and setting below.

Methodology and Setting

To gain an understanding of how technical communicators make sense of their day-to-day work experiences, we set out to gather information about how experienced communicators describe their work: what sorts of tools they are using to produce what sorts of products, what types of processes they participate in, and what personal and professional objectives they are pursuing. Primarily, qualitative data-gathering methods seemed best suited to our project, although we did use a brief written questionnaire to collect demographic information about our sample population. Because we were interested in constructing metaphors and models and not in identifying correlations or determining quantities of particular tools or techniques used in the workplace, we decided that focus groups composed of experienced communicators would provide the in-depth analyses we were seeking.

We determined that experienced communicators would be the appropriate group to work with since we wanted to gain a sense of the changes that have occurred in the profession over the past several years. If we hoped to construct a meaningful model we needed to understand technical communication work today from the inside out, and focus groups are an established social science method for teasing out such information (see Creswell 2002; Krueger 1994). So we asked three sets of communication professionals to share an afternoon with us, recounting their day-to-day experiences, achievements, and challenges over the last 5 to 30 years.

Beginning in 2004, with the help of several Society for Technical Communication (STC) leaders, we set up and conducted three focus groups of 10 to 14 technical communicators each (Washington, DC; Houston, Texas; and

Seattle, Washington). The STC leaders who assisted us invited experienced technical communicators (with at least five years' experience in the profession) to participate in our sessions. At each focus group meeting we used various methods to capture information: one individual written questionnaire and several kinds of interaction. Participants were asked to fill out the questionnaire as they arrived. We wanted to gather demographic information as well as some hard data about tools used and processes, teamwork, and activities engaged in by participants.

Two of the focus groups took place in homes, and one took place in a corporate boardroom. In Washington, DC, we were invited into the residence of a local technical communicator. We held the session in the dining room, and the 12 participants overflowed into the adjacent living room. The conversation was friendly and intense, and one participant took numerous digital photographs that were later shared with us.

In Houston, we were invited into a private residence once again. We held the session on a warm spring day. The 11 participants arranged themselves in the living room, and we stood in front of open French doors leading onto a terrace. A breeze blew the curtains and ruffled our flipchart sheets throughout the afternoon. Once again, the conversation was both intense and collegial, and at times participants engaged directly with each other to find better ways to articulate the unique challenges that currently face the profession.

Our third focus group took place in the boardroom of a technical communication firm in Seattle. Fourteen people sat around the table and were delighted to find that our corporate sponsors had assembled "goodie bags" for participants to take home. Once again we noticed the collegiality of the group as participants presented individual views and also developed or disagreed with points of view expressed by others.

In all cases, we were struck by the commitment and generosity of participants. Most had given up valuable free time to participate in our sessions, and all were good-natured and serious in their manner of engagement. We noted the creativity and humor of the participants. Most seemed deeply interested in making a worthy contribution to their organizations and to their audiences. Most described work situations that are diverse and complex, and many talked about a need to find ways to gain more influence and have a greater impact on decision making.

Once the session actually started, all information gathering was interactive: word-association exercises, structured conversations, and participative dialog. The two latter activities were based on a facilitation technique designed to help groups engage in conversations that move from objective data to deeper interpretation of meaning (Stanfield 2000).

To loosen up people's thinking, we began the session by tossing out words and asking for immediate responses, unmediated by thoughts of appropriateness or tact. During the structured conversations, we asked specific, work-related questions and captured all responses on flipcharts and in our notes. Finally, the participant dialogs engaged people in helping us create models for the work they do. A typical session posed the following questions:

- *Overall focus question for the session:* Taking into consideration the ways in which you have seen technical communication change over the past decade, what models or metaphors would you use to describe technical communication?
- *Structured conversations:* What types of work products are you responsible for? What types of processes do you lead or participate in? What are your professional objectives today, and how have they changed over time? What are your major accomplishments in the field of technical communication over the past two years?
- *Participant dialog:* Looking at the data that we have created so far, can you suggest models or images or metaphors to describe this rich, changing field of technical communication work? Can you suggest aspects of the future of technical communication based on the dialog so far?

Toward the end of each session, as participants discussed possible models, metaphors, and images for technical communication, we showed an illustration of the evolving role of technical communicators that we had adapted from Craig Waddell's "Four Models for Public Participation" (1995). Waddell's model describes the evolution of public participation in decision making about scientific and environmental projects, especially those with some risk attached to them. In recent publications, federal and state environmental agencies in the United States promulgate a participative process in which the "non-expert" public is empowered to communicate concerns, values, and emotions about a project (such as expanding a landfill) to the regulatory and scientific "experts."

In fact, the most influential risk-communication theories now recognize that experts and the public each have their own concerns, values, and even technical knowledge, and both should be equally empowered in exchanging information (Morgan and colleagues 2002). The literature we had reviewed and our own experiences as technical communicators made us curious about the extent to which new relationships and interactions between subject matter experts (such as engineers, software developers, business consultants) and frontline workers might be affecting communication flows between them, and thus impacting the field of technical communication. It seemed possible that technical communicators could be playing new roles in mediating the relationships between experts and users, as suggested in Figure 5.1. The focus groups were our attempt to discover those roles and to elaborate on the model adapted from Craig Waddell.

After the data-gathering process was complete, we organized and analyzed the data. We wanted to answer three questions:

1. Who are the people we are studying?
2. What sorts of things are they working on, and what processes and tools are they using?
3. How do they feel about the work they are doing today?

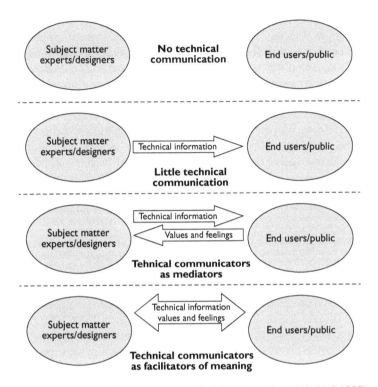

Figure 5.1 Technical Communication Model (adapted from Waddell 1995)

The two-page questionnaire was administered before the start of each focus group meeting: participants were asked to answer the questions in writing as everyone gathered and before the interactive dialog began. Aside from demographic information, nine questions were asked. The first four used a five-point Likert scale to measure levels of agreement (from strongly agree to strongly disagree); the remainder used categorical lists of possible answers. See Appendix 5.1 for the survey instrument.

Thirty-seven participants completed this questionnaire; 12 in Washington, DC, 11 in Houston, and 14 in Seattle. Responses to the Likert-scaled questions therefore usually totaled 37, although not everyone answered every question. For most of the non-Likert questions, participants wrote many answers, except for questions 8 and 9, where people generally did as instructed and calculated one percentage per activity.

We began our analysis of the questionnaire results by counting responses for each level of the Likert scale and for each of our other categories (types of tools, and so on). Using this simple method, we could deduce certain general trends (whether most people were satisfied with their current job, for example). To gain a more detailed picture of our sample group and their experience with their current jobs, we plotted responses against each other for questions 1 and 3, using a 2 × 2 table that created composite scores by collapsing together "strongly agree" with "agree," and "strongly disagree" with "disagree" (this technique of grouping frequency distributions in statistical data gathered through survey techniques such as Likert scales is a common way to reduce the volume of data and facilitate the process of interpretation). Analyzing questions 1 and 3 this way allowed us to correlate job satisfaction with role definition. For the other questions, we charted the number of responses on the y axis for each category on the x axis.

For the qualitative data, we performed a variety of categorization and theming analyses. Clustering the word-association data by key words became our first task (for example, we clustered all of the data related to the word "writer," and then we identified common words or meanings that occurred in more than one focus group). We also reviewed all of the words that participants used, and clustered them together into themes (for example, we clustered all words related to the theme "self and technical communication").

For the conversational data, we categorized or grouped the data according to themes that emerged through a careful review of the transcripts. We then regrouped the data in tables that allowed us to note commonalities and differences in how people responded to our questions. We also identified the words that were used most often. Finally, we created a consolidation table that related the questions we asked with the themed responses (shown in a later section as Table 5.3).

The dialog data resulted in lists of metaphors and verbal descriptions of technical communication practice, along with several drawings that attempted to depict visually the essence of technical communication practice. We recreated the visuals that participants had drawn on flipcharts or whiteboards and identified the major characteristics of each drawing. We then looked at the repeated occurrence of certain characteristics and listed those attributes that seemed to occur repeatedly when technical communicators attempted to describe their profession with an illustration. We also looked at all of the metaphors and verbal descriptions that participants provided, and we grouped them into types. We then listed the number of items that were present under each type.

After carrying out these organizing and analytical procedures on the data, we examined the results of our analyses and looked for patterns and anomalies. We present our findings below.

Research Findings

Given the goal of this research—to explore with practicing technical communicators the current state of the profession in order to develop appropriate models—our primary methodological approach was qualitative. We did want, however, to capture written demographic information about those participants and see whether such a small sample could tell us anything statistically relevant about the way they spend their work time. We present our findings from an analysis of the quantitative data gathered through the questionnaire, followed by our findings from an analysis of the qualitative transcript data.

Quantitative Findings

The technical communicators we surveyed and worked with were an experienced, well-educated group with relatively stable job situations. Out of 37 respondents to the demographic questions, 25 percent were self-employed and 75 percent were working on contracts or in regular full-time positions. The average length of time in their current position was four years and the average length of time working in the field of technical communication was almost 18 years (minimum was four years and maximum was 33). As shown in Figure 5.2, the most common terminal degree for these people was the bachelor's, but 46 percent had a higher degree of some sort.

Many Professional Roles

The close relationship between job satisfaction and well-defined professional roles was established by creating a density plot of the responses to Q1 and Q3; see Figure 5.3. The figure shows that 23 out of 37 participants strongly agreed or agreed to both questions. This quantitative finding seems to support the commonsense notion that technical communicators like to know what they are supposed to be doing on the job.

Overall, people in our focus groups felt quite satisfied with their current positions—more satisfied than with previous positions—and felt prepared for the roles they were filling. Figure 5.4 provides an overview of responses to questions 1–4; the gray and cross-hatched bars represent those who agree or strongly agree with the positively worded questions. Clearly, a large majority of respondents feel confident and satisfied with their jobs.

The previous two figures, however, leave two issues hanging. The first issue is this: To what extent did people in our focus groups define their roles themselves? This question is partly answered in the qualitative discussion of this paper, which

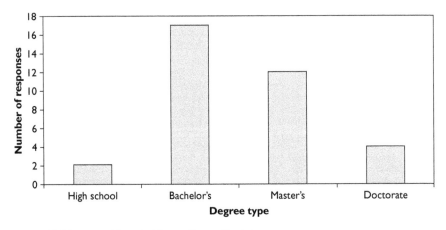

Figure 5.2 Educational Level of Focus Group Participants

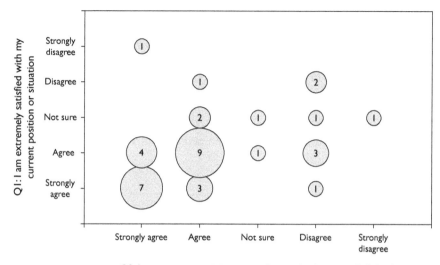

Figure 5.3 Relationship between Current Job Satisfaction and Definition of Professional Roles (n = 37)

reveals that these communicators were a relatively skillful, experienced, and confident group. However, while these people felt prepared for the roles required of them (Q4: 35 out of 37 agree or strongly agree), they did not feel as confident that their roles were well defined (Q3: 27 out of 37).

That fact leads to the second issue left hanging by our statistical correlation of Q1 and Q3 (Figure 5.3). While we can say that satisfaction with position correlates to having well-defined roles, we cannot say that the opposite is true. Half (four) of those who strongly disagreed or disagreed that their roles were well defined were

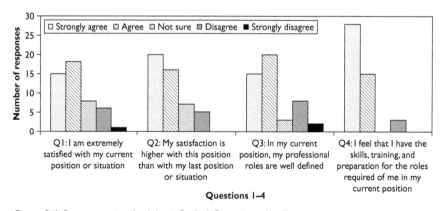

Figure 5.4 Responses to the Likert-Scaled Questions (1–4)

nonetheless satisfied with their jobs, and two were simply unsure (see the two right-side columns of bubbles in Figure 5.3). Only two (25%) of the people with undefined roles were definitely dissatisfied with their jobs. This data suggests that some technical communicators may be getting used to the idea that rewarding work may involve greater creativity and less definition than was previously the case.

This finding is also reflected in the responses to Q5, in which participants described their current role. "Writer" and "editor" appeared most frequently, but only a minority of respondents used those terms exclusively. Notice in Figure 5.5 that most people had multiple roles and the number of listings in the "other" category is the same as the number of people calling themselves "writer." Only 8 responses out of more than 75 listed "writer" as a unique identifier.

Teams and Workflow Activities

Answers to Q8 provide an interesting portrait of how these technical communicators spend their time at work: they are likely to spend a lot of time working on teams. In fact, 83 percent of respondents spend at least 20 percent of their work time participating on teams, and 38 percent spend at least 80 percent of their time participating on teams. Even more interesting, many of those people seem to be communicators who do not feel that their roles are well defined, pointing to an inverse relationship between definition of professional roles and time spent on teams. Figure 5.6 shows the team experiences of people in relation to role definition. The bars on the far left show how people who spend less than 20 percent of their time on teams feel about their role definition, and the far right shows how people who spend 80 percent of their time on teams feel about their role definition. As the graph shows, of the eight people who disagreed or disagreed strongly that their roles were well defined, six of them (75 percent) work at least 80 percent of the time on teams.

Because the sample numbers here are small, we cannot claim statistical significance for this observation, but it seems striking that the distribution across the percentage categories is relatively even for those who feel their roles are well defined, in contrast to the distribution of the "disagrees," who are clustered at the far-right side of the chart. As we discovered in correlating questions 1 and 3, some technical communicators may be satisfied with jobs in which roles are more fluid (remember that half the people with undefined roles are happy in their job), and these fluid roles, as we discovered in the discussions, involve more teamwork. By the same token, there certainly looks to be a correlation between ill-defined roles and spending a lot of time on teams.

As for questions 9a–9d (which asked respondents to tell us about the kinds of things they worked on during a typical week), the data may reveal a shift in the occupation of technical communicators (note: responses to Q9d, the "other" category, revealed no discernible pattern, so these data are not included here).

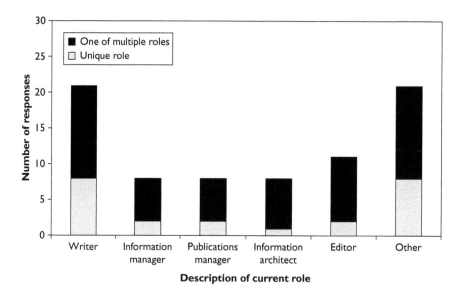

Figure 5.5 Self-Described Roles of Technical Communicators in their Professional Positions (sample size is 37, but respondents could list any number of roles)

Figure 5.7 displays the percentage of the work day that responders spend on various activities.

What struck us first was that a sizeable number of people (nine, or 26 percent of the sample) spend 0 percent of time doing what technical communicators have traditionally spent the bulk of their time doing: creating documents for end-user clients (see the dark-gray bar at extreme left of the figure). By contrast, only three people (9 percent) of this group spend 0 percent of time planning and facilitating communication processes (see the gray bar at extreme left of the figure). When we compare the dark-gray and the gray bars in the remaining percentage-of-time categories in the graph, both of those activities were distributed rather evenly across the categories; in other words, these people are now spending about the same amount of time on communication processes as they are on creating end-user documents or products.

This relatively even distribution between creating documents and facilitating communication processes is reflected in the correlation between how people spend their work time and how defined they feel their roles are within the organization. Table 5.1 shows the distribution of work activities reported by two subsets of our focus group sample: those who agreed or strongly agreed that their roles were well defined, and those who disagreed or strongly disagreed. Of the 27 people who agreed or strongly agreed that their roles were well defined (the middle column of the table), two-thirds (18) spend the same amount of time, at least 20 percent, on

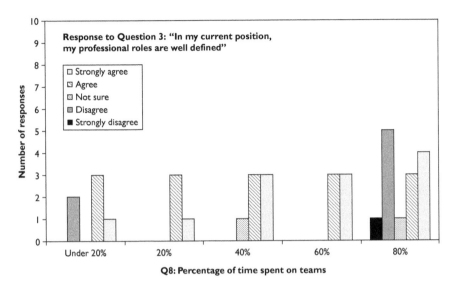

Figure 5.6 Relationship between Percentage of Time Spent on Teams and Definition of Professional Roles

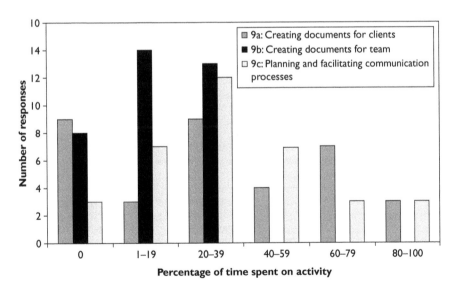

Figure 5.7 Distribution of Work Activities by Percentage of Work Day for Sample Population (n = 35)

each activity: creating documents for user clients and planning and facilitating communication processes.

Of the eight people answering questions 9a–9c who disagreed or strongly disagreed that their roles were well defined (right column of Table 5.1), 63 percent of them spend at least 20 percent of their time creating documents for user clients, while all of them spend at least 20 percent of their time planning and facilitating communication processes. Again, although the statistically small sample numbers make it impossible to generalize these findings, it seems that the technical communicators in our study who are mainly involved in facilitating often-shifting communication processes do not have firmly defined roles within the organization—this fact leads us to wonder whether their value to their organizations relates more to the skills and competencies they possess than to the deliverables they produce.

Conclusions Based on Quantitative Data

Analysis of the questionnaire data reveals that our sample of technical communicators can be satisfied with jobs that do not have well-defined roles (although statistically the correlation is stronger between those who are satisfied and who do have well-defined roles). As other studies have revealed (see Rainey, Turner, and Dayton (2005, 3) for a recent survey), the descriptors for what these communicators do vary considerably: every respondent used more than one term. "Writer" is still the single most popular term, but (since people could list any number) there were as many different terms as there were respondents. Our analysis also shows that these communicators are spending a considerable amount of time working on teams and on planning and facilitating work processes.

We now turn to our analysis of the structured conversations with focus group participants to see whether these findings are corroborated or further elucidated.

Table 5.1 Distribution of Work Activities Reported by Two Sample Subsets: People Who Feel Their Professional Roles are Well Defined, and Those Who Do Not Feel That Way

Self-report on questions 9a–9c	Strongly agree or agree that roles are well defined (N = 27)	Strongly disagree or disagree that roles are well defined (N = 8)
I spend at least 20% of time creating documents for end-user clients	18 (67%)	5 (63%)
I spend at least 20% of time creating documents for my team	7 (26%)	4 (50%)
I spend at least 20% of time planning and facilitating communication processes	18 (67%)	8 (100%)

Qualitative Findings

We grouped our qualitative data into three categories: data gathered from an initial word-association exercise, data gathered from the structured conversations, and the lists of metaphors and the drawings of models that participants created at the end of each session.

Constructing Meaning from Words

The word association exercise was intended to act as a warm-up for the focus group participants before moving into the more substantive conversations. We found, however, that these exercises produced a significant amount of intriguing data, and therefore we experimented with a number of analytical approaches to see whether these data could help us in our effort to reveal how technical communicators are making sense of the current state of technical communication practice.

In the word-association exercise, one facilitator would say a key word, and participants were encouraged to say whatever words came into their minds. The other facilitator would record the answers for subsequent analysis. We asked participants to respond to the following key words, one by one, in this order: *writer, information, communication, deliverable, user, stakeholder, team, participate, collaborate,* and *success.*

After compiling and counting the responses to these key words, we clustered the words by theme. We selected themes by reading through the words and making notes on emerging patterns. Then we assigned names to each of the themes and, finally, put all of the words from the three focus groups into the appropriate row. We did not try to fit every single word into a theme, nor did we worry if one word seemed to support more than one theme. Table 5.2 shows the themes, the number of words that fit into each theme, and a few example words for each.

These results may be suggestive of the emerging perspective of technical communicators. Emphasis is placed not so much on technical communication as a distinct practice within organizations, but rather on the overall organization (the "business" theme) and on the need to engage in interaction in order to have an impact on the wider organization. We are also intrigued by the relatively few words having to do with technology, and suggest that it will be interesting to see whether this balance in emphasis—an emphasis on organizational goals and on interaction and interdependence, rather than on technology—continues to be apparent in the remainder of the data.

The pattern-analysis procedure that we used on the word-association data suggests an emphasis on organizational or business goals and on the people with whom technical communicators interact and to whom they provide services and products. Moreover, there are some indications that technical communicators place more emphasis on processes and interactions than on products and technologies.

Table 5.2 Word-Association Data Clustered by Theme

Theme	Number of words	Description	Example response words
Business	139	Words suggesting an orientation toward the business objectives and vision of management	Management, processes, structure, advertising, contract, schedule
Interaction	116	Words suggesting the views of participants on working with others	Team, people, talking, two-way, meeting, community, exchange, discussion
Self and technical communicaton	54	Words indicating the image that technical communicators hold of themselves and their peers	Researcher, editor, typist, translator, communicator, author, psychologist
Clients and users	41	Words indicating the views of participants on the people to whom they provides services and products	Reader, audience, customer, users, advocate, usability
Organizing	41	Words suggesting how technical communicators make sense of and bring order to work contexts	Detail-oriented, organizer, analyzer, thinker, structure, milestone, project, deadline
Technology	16	Words suggesting an orientation toward the technology that the profession uses and supports	Systems, networks, data, technology, Web site, MS Word

The Structured Conversations

The structured conversations were organized around four key questions:

1. What types of work products are you responsible for?
2. What types of processes do you lead or participate in?
3. What are your professional objectives today?
4. What are your major accomplishments in the field of technical communication over the past two years?

Session participants engaged in lively—and sometimes emotional—conversations as they responded to these questions. Initially, they listed specific items (products, processes, and so on), and then they tended to shift naturally into a conversation that attempted to make sense of their emerging list. We recorded all of these comments in our transcripts, and then carried out a comprehensive thematic analysis that involved clustering, analyzing, and naming the themes. Table 5.3 shows the high-level results of this analysis.

The table lists the questions in the left column and the themes (along with the number of data items that support each theme) in the right column. In their discussion of work products, four major themes emerged. About 46 percent of the work products listed were products intended for users (training materials, quick reference guides, user documentation, online help, newsletters, computer-based training, Web sites, information plans, and so forth). About 29 percent of the products mentioned were products related to the operation of the organization for which the technical communicator works (such as proposals, white papers, marketing materials, project plans, business cases, process maps, news releases, and strategic plans). Only about 15 percent of the products mentioned were highly technical in nature (server-hardening guidelines, technical reference guides, software development kits, technical sales material, functional documents, requirements documents, and so on).

Perhaps most interestingly, in all three focus groups, participants advised us that they are responsible for a wide variety of deliverables. We learned that 43 percent of the participants in one session were responsible for producing a few of the products on their list, while the remaining 57 percent of the group told us that they

Table 5.3 Structured Conversation Data Clustered by Question and Theme

Question	Summary of responses
What types of work products are you responsible for?	• Work products related to users (64 items) • Work products related to the operation of the business (sales docs, business cases, project docs) (40 items) • Work products related to the organization's technology (21 items) • Too much variety to classify (13 items)
What types of processes do you lead or participate in?	• Documentation and training development processes (52 items) • Cross-functional business processes (30 items) • Interpersonal processes (17 items) • Product development processes (16 items) • Multiple membership (7 items)
What are your professional objectives today?	• Find meaning, focus, balance (32 items) • Leverage and develop my expertise (19 items) • Be creative and succeed outside the discipline of technical communication (12 items) • Stay employed and then retire (survive) (9 items)
What are your major accomplishments over the past two years?	• Documentation and training accomplishments (25 items) • Personal development, fulfillment, recognition (23 items) • Leadership accomplishments (14 items) • Promoting the success of others (4 items)

are responsible for producing "a lot" of the products on the list. In another session, 33 percent of the participants told us that they are responsible for a wide variety of products and they rarely know from week to week what they will be working on next. In the third session, most participants said they are responsible for between 10 and 20 different types of products, and 46 percent of the participants told us they are responsible for producing more than 20 different types of products.

These data suggest a continuing emphasis on deliverables intended for end-user audiences; however, this emphasis is now "competing" with other demands that are being placed on technical communicator talents. Technical communicators are being called on to create a broader range of documents intended to support business goals. True to their name, technical communicators still work on highly technical materials, but technology appears to be subservient to broader business concerns. Participants in all of the focus groups indicated that it is common for a technical communicator to produce many different types of deliverables, rather than just a few. All three groups also showed a clear inclination to focus more on processes than on products; we were told, for example, that people work mostly on teams, that they provide services as well as products, and that their value has more to do with the activity of their work than with the products they produce.

In the discussion of processes, approximately 43 percent of responses referred to processes related to the development of documentation or training materials (course design, managing documentation projects, localization, editing, reviews, and so forth). Cross-functional business processes (such as quality assurance (QA) processes, continuous improvement, facilitation, project evaluations, budgeting, and so on) received the second-largest number of responses (25 percent). The remaining responses were divided almost equally between interpersonal processes (networking, conflict resolution) and product development (GUI design, software development, interaction design, testing).

These results point to one of the general findings evident in the study. If the technical communicator's role is evolving and changing, that change does not mean that technical communicators are no longer working on the types of products and processes that are associated with traditional technical communication deliverables. They continue to spend much of their time working on the development of documentation and training materials; however, their role has expanded to include participation in a wider variety of processes and teams.

The discussion of professional objectives revealed a distinct tendency to look holistically at one's life and to look outside the domain of technical communication. Approximately 44 percent of responses had to do with seeking greater meaning, focus, and balance in life (such as becoming more influential, making a contribution, making a difference, acting in accordance with values, empowering people to solve their problems, creating community, helping people, working on enjoyable projects). Another 17 percent had to do with seeking creativity and success outside of technical communication (people talked about writing novels and plays, and about applying their expertise to areas of business outside of technical communication). About 26 percent of responses focused on leveraging or developing expertise as a technical communicator (creating better technical communication materials, achieving a deeper understanding of user needs, becoming more proficient as a technical communicator).

These findings may suggest that the experienced technical communicators who participated in our research are feeling confident of their worth and ability, and do not see a need to focus exclusively on developing their core technical communication skills. We were struck by the conversations about empowering users and achieving a deeper understanding of the information and communication needs of audiences, and about creating communities of users or of like-minded professionals. Participants talked about their values, about thinking strategically, and about finding ways to be more influential on development teams. Our impression was that these technical communicators are looking for ways to contribute, to have an impact, and to make use of their talents.

We were curious to see how these technical communicators would describe their major accomplishments, after telling us about their personal objectives. Were their accomplishments aligned with their objectives, or was there some degree of misalignment between their personal objectives and their actual achievements? When we asked participants about their recent major accomplishments in the field of technical communication, we found that accomplishments related to documentation and training and accomplishments related to personal development, fulfillment and recognition were mentioned at about the same rate (38 percent of responses related to the former, while 35 percent related to the latter). Another 21 percent of responses related to leadership accomplishments, while 6 percent of responses described accomplishments having to do with promoting the success of others.

We compared the statements about objectives and accomplishments, and created Table 5.4 to align the most commonly mentioned themes. The emphasis placed by participants on objectives and the accomplishments related to a desire for meaningful work, personal development, and achieving influence seem largely congruent. However, the bottom row may indicate a slight but noticeable discrepancy between the emphasis placed on objectives and the reported accomplishments related to the traditional work of technical communicators. This discrepancy may indicate that technical communicators continue to receive recognition for performing their traditional roles, while their aspirations have in fact been shifting toward other ways of contributing to organizational success.

Table 5.4 Aligning Objectives with Accomplishments

Objectives	Accomplishments
Creativity, meaning, focus, balance (61% of responses)	Personal development, fulfillment, recognition, and leadership (56% of responses)
Leverage and develop my expertise (26% of responses)	Documentation and training accomplishments (38% of responses)

These conversation data appear to indicate that traditional deliverables remain the major focus of technical communication practice but that technical communicators are also responsible for a wide variety of other deliverables. Variety, in fact, is the main point here: most technical communicators are responsible for producing a wide variety of deliverables and for participating in numerous business processes. Some data suggest that the value of the technical communication discipline is related more to processes than to deliverables—a fact that may indicate that it is through interaction and engagement that technical communicators are making their major contribution to organizational and customer success.

We note that these technical communicators tend to focus on personal objectives related to empowering users, creating communities, achieving a deeper understanding of user needs, thinking strategically, and becoming more influential. It is possible that technical communication accomplishments are "lagging behind" these emerging aspirations: technical communicators continue to be recognized for accomplishments related to successfully completing traditional technical communication tasks (such as developing user manuals), but their objectives are moving in new directions.

Images and Metaphors

After talking about the deliverables they produce, the processes they participate in, their personal objectives, and their recent accomplishments, participants suggested models, images, or metaphors to describe the rich and changing technical communication practice that they had been discussing. Participants tended to brainstorm verbal images to begin with, and then some participants accepted our invitation to sketch an illustration. We captured 11 drawings through this exercise, with some overlap among drawings.

When creating illustrations to depict their current experience of technical communication practice, most participants created non-hierarchical models that showed relationships arranged as a network or web. The models tended to emphasize relationships and communication across disciplines, with technical communication pictured as a hub or center of communication activity. The function of technical communication is not usually pictured as the creation of a product, but rather is seen as creating linkages across a complex, networked organization. The technical communication function is generally not pictured as subservient or marginalized; on the contrary, technical communication is usually pictured as highly interconnected and as an equal among peers. Table 5.5 summarizes the characteristics of all the visual models suggested by participants (the numbers in the right-hand column indicate the number of models that possess that characteristic).

Figure 5.8, a–h, provides eight examples of the visual models that participants provided. These models were hastily sketched on whiteboards or flipchart sheets, with the individuals explaining the main points that they were trying to convey, and with other people occasionally adding their comments and suggestions.

Table 5.5 Shared Characteristics of Models Suggested by Participants

Characteristic	Number of models possessing the characteristic
The model is non-hierarchical.	10
The model includes a focus on relationships.	9
The model includes a focus on communication.	8
Technical communication enables linkages within the organization.	8
Organization is pictured as a web.	7
Technical communication is a core that allows communication to occur within the organization.	6
Technical communication is pictured as a "center" in the model.	5
Technical communication enables communication among teams and functions.	5
Technical communication is an equal among other groups in the organization.	5
Technical communication is more connected than other groups.	5

In Figure 5.8, the first four illustrations (a–d) exemplify the tendency to see organizations as networked and technical communication as a communication hub within the network. Out of all eight sketches, illustration e is the only example that clearly focuses on the creation of a product: technical communication creates relationships across disciplines that are involved in a process culminating with the production of a set of deliverables. Illustration f focuses more exclusively on the technical communicator, who is pictured as a many-headed being (an image that occurred in the verbal data as well) who must negotiate complex sets of relationships.

The final two sketches—g and h—both picture technical communicators as facilitating conversations within organizations. The technical communicator coordinates and participates in some of the conversations that take place within organizations, and a possible goal for technical communicators is to ensure that the right people are involved in this exchange of information. We note, then, that communication here is not envisaged as a one-way exchange of information flowing from technical experts and managers, but is rather a more open-ended exchange of expertise and viewpoints.

We also captured extensive lists of verbal descriptions and metaphors that participants mentioned during the focus groups, and then analyzed these data, looking for commonalities. The metaphors provided by participants were rich and various, and we grouped them into broad categories to respect their variety. The biggest category, "emerging forms," includes metaphors and descriptions of technical communication as an emergent discipline that is changing and moving in new directions. Some talked about how those aspects of technical communication that can be commoditized are moving to offshore outsourcers, but the more

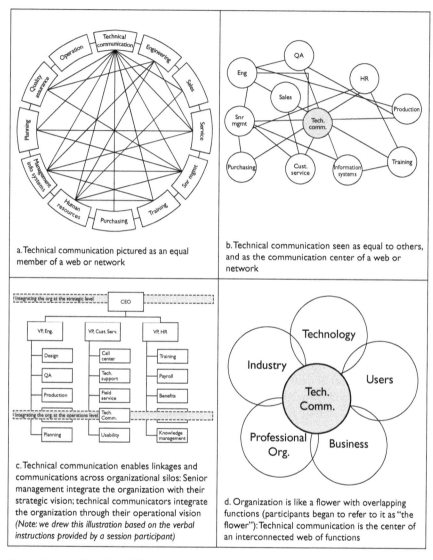

a. Technical communication pictured as an equal member of a web or network

b. Technical communication seen as equal to others, and as the communication center of a web or network

c. Technical communication enables linkages and communications across organizational silos: Senior management integrate the organization with their strategic vision; technical communicators integrate the organization through their operational vision (Note: we drew this illustration based on the verbal instructions provided by a session participant)

d. Organization is like a flower with overlapping functions (participants began to refer to it as "the flower"): Technical communication is the center of an interconnected web of functions

Figure 5.8, a–d Sketches of Four of the Eight Visual Models Suggested by Participants

complex communication processes are remaining within organizations—and it is here that technical communicators need to concentrate their energies.

Some participants talked about notions of empowerment and community, and suggested that we are moving toward a focus on complex, two-way communication exchanges. One person said: "We are the glue. We are a catalyst. We offer connections between areas, and we are the conduit. We provide the means for

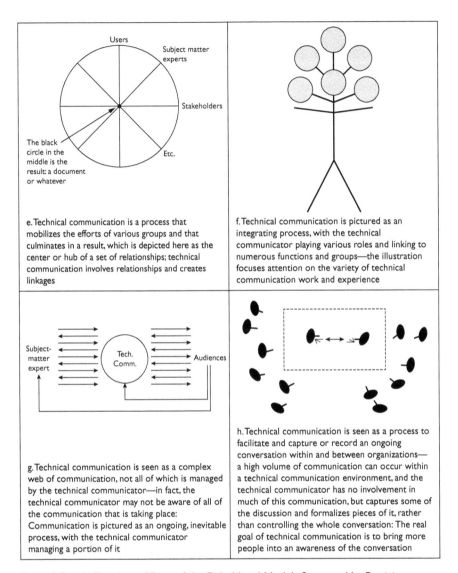

e. Technical communication is a process that mobilizes the efforts of various groups and that culminates in a result, which is depicted here as the center or hub of a set of relationships; technical communication involves relationships and creates linkages

f. Technical communication is pictured as an integrating process, with the technical communicator playing various roles and linking to numerous functions and groups—the illustration focuses attention on the variety of technical communication work and experience

g. Technical communication is seen as a complex web of communication, not all of which is managed by the technical communicator—in fact, the technical communicator may not be aware of all of the communication that is taking place: Communication is pictured as an ongoing, inevitable process, with the technical communicator managing a portion of it

h. Technical communication is seen as a process to facilitate and capture or record an ongoing conversation within and between organizations—a high volume of communication can occur within a technical communication environment, and the technical communicator has no involvement in much of this communication, but captures some of the discussion and formalizes pieces of it, rather than controlling the whole conversation: The real goal of technical communication is to bring more people into an awareness of the conversation

Figure 5.8, e–h Sketches of Four of the Eight Visual Models Suggested by Participants

others to communicate." Another said that technical communication now involves "facilitating a process of knowledge transfer and communication."

These metaphors suggest that technical communicators must focus less on texts and more on impacting the behavior of clients; technical communication is taking on an element of organizational development and facilitation; technical communicators may find that they are creating more experiences (such as simulations) and fewer texts. Technical communicators are like a cross-pollinator; they listen and facilitate; they act as catalysts to launch a process of discovery and conversation. Table 5.6 summarizes the categories into which we grouped the metaphors (the numbers in the right-hand column indicate the number of metaphors in the category).

Some of the specific words and phrases in the "emerging forms" category include the following: *evangelist, architect, visionary, image consultant, spin doctor, truth teller, explorer, detective, imposing order, coach and advocate, chess player, dynamic, diversity,* and *collaboration.* We called these "emerging forms" because when participants used these words, they were searching for metaphors and images to communicate a new or emerging conception of technical communication. This view sees technical communication as moving from user advocacy to user empowerment. The profession has moved from a focus on one-way communication through texts, to a focus on facilitating rich and evolving communication exchanges among diverse groups.

We labeled the second-largest category of metaphors "the performer and the director." We were struck by the tendency of all three focus groups to offer images of technical communication practice that relate to the performing arts. Some of the words and phrases in this category are: *juggler, circus ringmaster, producer, orchestra conductor, magician, quick-change artist, fortune-teller, vaudeville song-and-dance, clowns on a bicycle, lion tamer,* and *three-ring circus.* The word choices convey a sense of the technical communicator as a person who manages numerous details and activities, and who makes this complexity seem simple to others. Like a circus ringmaster, the technical communicator is responsible for the smooth flow of events—sequencing, integrating, creating variety, or facilitating flow. One participant suggested that "we are more ringmaster than project manager." Another person said that like a producer, the technical communicator "pulls pieces together to make something happen."

The third-largest category, "gods, monsters, and mysteries," included a variety of whimsical and obscure images. More than one participant compared the technical communicator to a Hydra—a many-headed god who, of course, must wear many hats to perform effectively. Another person said that to other people in the organization, technical communicators are a mystery: "The elephant in the room. They won't talk about us, they don't know what to do with us." Another person said: "I am a blob. . . . Or an amoeba, that can split itself. Technical

Table 5.6 Categories of Metaphor and Number of Metaphors within Each Category

Category of metaphor	Number of metaphors in the category
Emerging forms	58
The performer and the director	21
Gods, monsters, and mysteries	19
The negotiator and mediator	16
Disappearing forms of life	11
The powerful helper	8
The self-effacing helper	2

communication is a blob that is fluid and that morphs (like a lavalamp), and that separates and joins together. We bring communication value where it is needed."

These are some of the individual words and phrases in this category: Hydra, god, a goddess with many hands, the phoenix, building on quicksand, octopus, changeling, shape-shifter, spinning out of control, fireworks, the zipper ride at a carnival, clairvoyant, fungus, and iceberg. Some of these words emphasize the need for technical communicators to be highly flexible and adaptive; others emphasize that successful technical communicators must be powerful and insightful.

We termed the fourth-largest category "the negotiator and mediator," because many of the verbal images emphasized the intermediary role that some believe is central to technical communication work. One participant said that a technical communicator is a diplomat:

> A technical communicator is a diplomat: conducts communication among groups. Mediates and facilitates. Translates. Writes treaties, brokers compromises. Understands the agendas and politics and values. Also grasps what the client really needs. We must be able to say "no" when needed. Deal with different cultures and languages.

Another said that a technical communicator is like a translator: "The groups we interact with do not speak the same language. We must translate why the requirements are worth doing. And we must 'recommunicate' with each group." A third participant suggested the image of two warring castles:

> The metaphor of two castles, catapulting things to each other. Each side has guards and defenses. Mediator, translator, facilitator. Technical communicators are like diplomats making peace between the two castles.

Some of the individual words and phrases included in this category are: persuasion, SME [subject-matter expert] translator, shaking hands, de-babble-izer, facilitator, UN ambassador, crisis negotiator, translator, and interpreter. The intermediary role suggested by the fourth category perhaps harkens back to some of the past metaphors (such as the bridge) that have been used to explain technical communication.

The fifth-largest category, "disappearing forms of life," includes depictions of technical communication that appear to be waning. For example, one participant offered the following image: "The vanishing font monkey." This participant felt that technical communicators are seeking more respect and that some individuals are succeeding in this quest while others are failing.

Other participants said quite explicitly that perhaps technical communication is going to become obsolete. One person commented:

> Can they fire me because they no longer need me? Our goal is to give the end user direct access to tacit knowledge, or Horton's notion of "information

implosion," embedding usability in the product itself. Will we become obsolete? Certain aspects will: end-users will have more direct access to helpful information, and young people do not need as much help, as many instructions.

Another offered this metaphor: "Technical communication as the epicyclic gear. The cog in the middle that enables the other wheels to turn. We can disappear when we succeed."

Some individual words and phrases in this category are: *the Titanic, reporter, meat grinder, pasta machine, blueprint* ("allowing us to put things through repeatedly"), and *cleaning a dirty beach*. In this category, technical communication is depicted as a routine, operational task, or as a low-value activity that may not always be needed.

The final two categories, which we have labeled "the powerful helper" and "the self-effacing helper," relate to the conception that some technical communicators have of their discipline as being a "helping profession." The self-effacing helper sees technical communication as a support function: at one session, participants talked about nurturers, mothers, and secretaries. The powerful helper continues to focus on the helpful attitude and activity of technical communicators, but puts this in a different light. One participant suggested that technical communicators can be like psychotherapists: "Technical communication as psychotherapist dealing with multiple personality disorder. Separating and integrating each personality (where each personality represents a specific client group)."

Another said that technical communicators are empowerers:

> We are part of the process of empowering users more. We develop the wizards. We use technology to enable users to be self-helping. We advocate for the user with new technologies and in new ways. Business will wise up to the fact that the user forums are more valuable than the documents. So writers get laid off, and the new needs may not be filled by technical communicators.

Individual words and phrases to denote the powerful helper include: *teaching, shepherd, doing interventions* (heart-to-heart conversations), *midwife, flexibility,* and *adaptability*.

We wonder whether the three largest categories of metaphor are related to an emergent conception of technical communication practice as powerful, creative, and facilitative, while the four smaller categories relate more to older, perhaps receding conceptions of technical communication as playing a nurturing and intermediary function.

Alignment of the Findings

Overall, the findings gathered from our questionnaire, the word-association exercise, the facilitated conversations, and the creation of models and metaphors suggest that our research participants see technical communication as an activity that facilitates a flow of information through texts, conversations, and relationships. Technical communicators bring value first of all through the processes they manage or impact, through the relationships they create and manage, and through their ability to bring a diverse skillset to the aid of numerous organizational endeavors.

The questionnaire data indicated that almost 4 out of 10 technical communicators participate on teams 80 to 100 percent of their work time. It seems that at least some technical communicators spend about the same amount of time planning and facilitating communication processes as they do creating end-user documents and products. Of the technical communicators in our survey who had ill-defined work roles, many of them were nonetheless satisfied, a fact that suggests that technical communication work is becoming less routine and standardized and more flexible and adaptive.

The word-association exercise produced data emphasizing organizational or business goals and the people with whom technical communicators interact and to whom they provide services and products. The data may indicate that technical communicators place more emphasis on processes and interactions than on products and technologies.

The structured conversations yielded data suggesting that traditional technical communication deliverables remain the major focus of practitioners, but a wide variety of other deliverables are also now within the technical communication domain. Most of the technical communicators who participated in our study told us that they are responsible for producing a wide variety of deliverables and that the value of the discipline is related more to processes than to deliverables. Personal objectives include empowering users, creating communities, achieving a deeper understanding of user needs, thinking strategically, and becoming more influential.

Finally, the sketches and metaphors produced by our study participants show a vision of the profession that is non-hierarchical and highly networked. Most sketches focus on relationships and communication. Technical communication is depicted as enabling linkages within a network or webbed organization. Technical communication is often seen as a "center" that enables communication among teams and functions.

The metaphors often show technical communication as emergent or evolving: the profession is increasingly focusing on networking; on facilitative, two-way communication; and on impacting human behavior rather than texts. Some metaphors show technical communication as a type of performance or production whereby disparate activities and groups are woven by the technical communication producer or performer into a seamless whole. Others show technical communicators as flexible, adaptive, powerful, and insightful. Relatively few metaphors reveal technical communication as an activity that is essentially a matter of mediating between groups, or as providing help and support to others.

Conclusions

While technical communicators continue to work on traditional deliverables (manuals, training courses, and so forth), technical communication work involves increasing variety. Technical communicators are responsible for producing many types of documents, not just a few types (such as user manuals and online help). And they are increasingly involved in additional communication processes, such as the facilitation of communication flows on cross-functional teams, participating in quality improvement processes, and helping to improve the effectiveness of temporary teams. This finding is congruent with the finding of Rainey, Turner, and Dayton in 2005 that technical communicators continue to produce the deliverables that are traditionally associated with the discipline, while at the same time collaboration is increasing in importance (2005, 3). Our survey data show that at least 83 percent of communicators spend at least 20 percent of their work time on teams.

We also found qualitative evidence that technical communicators are coming to see participation in organizational processes as an important way to contribute to organizational goals. Their increasing involvement in planning and facilitating communication processes, not just products, increases the importance of interpersonal skills and the need to collaborate effectively with representatives of other disciplines. All of the many and varied models that participants created to represent their roles within the organization had one element in common: a representation of technical communicators as boundary-spanners who create and manage linkages within networked organizations.

Technical communication is undergoing a process of change, and experienced technical communicators appear to be energized by and committed to the opportunities and challenges that they face. The dominant metaphors used by these focus groups were of "emerging forms" in which technical communication is seen as a changing, expanding discipline in the process of redefining its purpose. We were also struck by the use of words like "empowerment" and "community," and we noted the many times technical communicators talked about the importance of gaining and exercising influence within their organizations. It is possible that the technical communicators' advocacy role is becoming enriched. To advocate effectively for users—to empower users as well as provide them with information—technical communicators see a need to extend and deepen their communication networks, and to become more adept at influencing business and technical colleagues on behalf of users.

These findings seem consistent with the views recently expressed by Moore and Kreth, who suggest that technical communicators could benefit by becoming more strategic and politically adroit:

> The days of being grammar cops, wordsmiths, and software applications experts are not over for technical communicators, but those skills are diminishing in value as the global information economy becomes more cost-

conscious, profit-driven, and focused on designing and delivering better experiences to individuals, groups, organizations, and entire cultures. Today, technical communicators who add value to their organizations do not merely write and edit documents.

(Moore and Keith 2005, 303)

A great majority of the workplace situations and stories we heard from our focus groups, and the visual models they produced, support Moore and Kreth's call for technical communicators to "make the transition to communication strategist" (2005, 303). Technical communicators are becoming strategic negotiators who bring disparate groups into conversations that are ultimately intended to benefit our user audiences.

All of these findings bring us back to the questions that prompted this two-year research project: What is the appropriate model and metaphor for the contemporary technical communicator? If the "bridge" no longer works as a representation of the communicator's role within the organization and with users, what image does work? We did not "discover" one image that depicts the complexity of these new roles. At the very least, though, a new model for technical communication would have to depict a non-hierarchical experience of work that combines sophisticated communication practices with the management of a complex web of relationships. This model would have to make clear that the primary objectives of technical communication practice include developing communication linkages within an organization to create an ever-expanding suite of communication products and events.

Next Steps

These findings need to be tested in new focus groups and with a new written survey of a larger population of technical communicators. We suggest that the STC membership is an appropriate survey audience. This survey should ask questions specifically about non-traditional technical communication tasks and processes, such as teamwork and business process development. Are many communicators— not just the experienced members of our sample— involved significantly in this kind of work?

These findings would also benefit from an extensive ethnography of a geographically diverse group of technical communicators, to assess the extent to which the shifts that we have noted are actually occurring in the day-to-day work experiences of practitioners. We would be interested in observing the variety that is apparently becoming characteristic of technical communication work, along with the more collaborative nature of the work. Are technical communicators fully integrated with cross-functional teams, and are they asserting themselves and their views on behalf of users and the discipline? Moreover, as a boundary-spanning discipline, are technical communicators encountering issues of identity as their roles evolve and expand, and are they encountering new types of resistance from the other disciplines with which they interact?

And finally, perhaps we should extend our collaboration in the next phase of this project and team with a graphic designer who would create out of these findings a memorable, accurate visual representation of contemporary technical communication.

Appendix 5.1: Questionnaire Administered To Focus Group Participants

1. Please rate your satisfaction with your current position as a technical communicator by responding to this statement: "I am extremely satisfied with my current position or situation."

5	4	3	2	1
Strongly agree	Agree	Neither nor disagree	Disagree	Strongly disagree

2. My satisfaction is higher with this position than with my last position or situation. True or false?

5	4	3	2	1
Strongly agree	Agree	Neither nor disagree	Disagree	Strongly disagree

3. In my current position or situation, my professional roles are well defined.

5	4	3	2	1
Strongly agree	Agree	Neither nor disagree	Disagree	Strongly disagree

4. I feel that I have the skills, training, and preparation for the roles required of me in my current position.

5	4	3	2	1
Strongly agree	Agree	Neither nor disagree	Disagree	Strongly disagree

5. Which of the following words describe your role(s) in your current position?
   - Writer
   - Information manager
   - Publications manager
   - Information architect
   - Editor
   - Other (please specify)

6. On a typical work day, these are the tools that I use:

7. On a typical day, these are the people I interact with (please list titles and responsibilities):

8. On average, I participate on teams this percentage of the time at work:
   - n 80%
   - n 20%
   - n 60%
   - n under 20%
   - n under 40%

9. During a typical week, what percentage of your time do you spend on these activities:
   - n % Creating documents for my user clients (for example, user manuals, instructional materials)
   - n % Creating documents for use by my team (for example, requirements documents, business cases)
   - n % Planning and facilitating communication processes (for example, information-gathering sessions, consensus-building meetings)
   - n % Other:

Acknowledgment

*This article, was originally published in *Technical Communication* (2006), 53: 395–415.

References

Albers, M. 2005. "The Future of Technical Communication: Introduction to This Special Issue." *Technical Communication* 52: 267–72.

Boreham, N., and C. Morgan. 2004. "A Sociocultural Analysis of Organisational Learning." *Oxford Review of Education* 30: 307–25.

Bushnell, J. 1999. "A Contrary View of the Technical Writing Classroom: Notes Toward Future Discussion." *Technical Communication Quarterly* 8: 175–88.

Chemers, M. M. 1997. *An Integrative Theory of Leadership*. Mahwah, NJ: Lawrence Erlbaum Associates Inc.

Creswell, J. W. 2002. *Educational Research: Planning, Conducting, and Evaluating Quantitative and Qualitative Research*. Upper Saddle River, NJ: Prentice Hall.

Davis, M. 2004. "Shaping the Future of our Profession." In *Power and Legitimacy in Technical Communication*, vol. 2, ed. Teresa Kynell-Hunt and Gerald J. Savage. Amityville, NY: Baywood.

Faber, B., and J. Johnson-Eilola. 2003. "Universities, Corporate Universities, and the New Professionals: Professionalism and the Knowledge Economy." In *Power and Legitimacy in Technical Communication*, vol. 1, ed. Teresa Kynell-Hunt and Gerald J. Savage. Amityville, NY: Baywood.

Hayhoe, G. 2003. "Inside Out/Outside In: Transcending the Boundaries that Divide the Academy and Industry." In *Power and Legitimacy in Technical Communication*, vol. 1, ed. Teresa Kynell-Hunt and Gerald J. Savage. Amityville, NY: Baywood.

Hughes, M. 2004. "Mapping Technical Communication to a Human Performance Technology Framework." *Technical Communication* 51: 367–75.

Kain, D., and E. Wardle, 2005. "Building Context: Using Activity Theory to Teach About Genre in Multi-Major Professional Communication Courses." *Technical Communication Quarterly* 14: 113–40.

Kouzes, J. M., and B. Z. Posner. 2005. *The Leadership Challenge*. 3rd edn. San Francisco, CA: Jossey-Bass.

Krueger, R. A. 1994. *Focus Groups: A Practical Guide for Applied Research*. 2nd edn. Thousand Oaks, CA: Sage Publications.

Lawler III, E. E. 2001. "The Era of Human Capital has Finally Arrived." In *The Future of Leadership: Today's Top Leadership Thinkers Speak to Tomorrow's Leaders*, ed. W. Bennis, G. M. Spreitzer, and T. G. Cummings. San Francisco, CA: Jossey-Bass.

Lipman-Blumen, J. 2001. "Why Do We Tolerate Bad Leaders?" In *The Future of Leadership: Today's Top Leadership Thinkers Speak to Tomorrow's Leaders*, ed. W. Bennis, G. M. Spreitzer, and T. G. Cummings. San Francisco, CA: Jossey-Bass.

Moore, P., and M. Kreth. 2005. "From Wordsmith to Communication Strategist: Heresthetic and Political Maneuvering in Technical Communication." *Technical Communication* 52: 302–22.

Morgan, M. G., B. Fischoff, A. Bostrom, and C. J. Atman. 2002. *Risk Communication: A Mental Models Approach*. Cambridge, UK: Cambridge University Press.

North, A. B., and W. E. Worth. 2000. "Trends in Entry-Level Technology, Interpersonal, and Basic Communication Job Skills: 1992–1998." *Journal of Technical Writing and Communication* 30: 143–54.

Orr, J. E. 1996. *Talking About Machines: An Ethnography of a Modern Job*. Ithaca, NY: Cornell University Press.

Rainey, K. T., R. K. Turner, and D. Dayton. 2005. "Do Curricula Correspond to Managerial Expectations? Core Competencies for Technical Communicators." *Technical Communication* 52: 323–52.

Rehling, L. 2004. "Reconfiguring the Professor–Practitioner Relationship." In *Power and Legitimacy in Technical Communication*, vol. 2, ed. Teresa Kynell-Hunt and Gerald J. Savage. Amityville, NY: Baywood.

Stanfield, B., ed. 2000. *The Art of Focused Conversation: 100 Ways to Access Group Wisdom in the Workplace*. Gabriola Island, BC: New Society Publishers.

Sullivan, D. L., S. M. Martin, and E. R. Anderson. 2003. "Moving from the Periphery: Conceptions of Ethos, Reputation, and Identity for the Technical Communicator." In *Power and Legitimacy in Technical Communication*, vol. 1, ed. Teresa Kynell-Hunt and Gerald J. Savage. Amityville, NY: Baywood.

Tebeaux, E. 2004. "Returning to our Roots: Gaining Power through the Culture of Engagement." In *Power and Legitimacy in Technical Communication*, vol. 2, ed. Teresa Kynell-Hunt and Gerald J. Savage. Amityville, NY: Baywood.

Waddell, C. 1995. "Defining Sustainable Development: A Case Study in Environmental Communication." *Technical Communication Quarterly* 4: 201–16.

Wilson, G., and D. F. Ford. 2003. "The Big Chill: Ten Years After the Master's Program." *Technical Communication* 50: 145–59.

Wojahn, P., J. Dyke, L. Riley, E. Hensel, and S. Brown. 2001. "Blurring Boundaries Between Technical Communication and Engineering: Challenges of a Multi-Disciplinary, Client-Based Pedagogy." *Technical Communication Quarterly* 10: 129–49.

Chapter 6

The Role of Rhetorical Invention for Visuals

A Qualitative Study of Technical Communicators in the Workplace*

Tiffany Craft Portewig

Editors' Introduction

Portewig's study of *visual rhetorical invention*, her term for the decision-making process by which visual elements are incorporated into technical documents, is an excellent illustration of contemporary social science research methods applied to technical communication practice. Though we believe that her article illustrates a study that is commendable for both its process and its results, we were especially keen to include it in this volume because of its thorough description and explanation of the qualitative methods she employs.

Like Carliner's article (reprinted as Chapter 4 in this collection), Portewig's research design is based on grounded theory. Although like Carliner she cites Strauss and Corbin (1998) as a source for grounded theory, she also makes use of the early work of Glaser and Strauss (1967) and the more recent work of Charmaz (2003). We mention these sources because grounded theory is often viewed as one of the more rigorous qualitative designs, and because a researcher interested in this approach is advised to understand the differing perspectives of Glaser, Strauss, and Charmaz, who are generally seen as its leading proponents.

Portewig provides an admirably clear and comprehensive account of her research design and methods. She explains the gaps in the current research literature that she hopes to address, she provides her research questions, and she comments upon how her research methods could be adapted for other technical communication studies. She states quite properly that she received ethics approval from an Institutional Review Board for the Protection of Human Subjects for her research project. This approval is required for all university-based researchers in North America whose research could potentially affect the psychological or material well-being of participants. She explains the data-gathering methods in detail, and she shares her reasoning for adopting this particular research design. The citations for her research methods provide interested readers with a starting point for further reading. Her discussion of the need to refine her interview questions during the research process is an excellent illustration of the emergent quality of many qualitative inquiries, and demonstrates that the adaptation of instruments to the emerging situation can be accomplished in a way that improves the study results.

Portewig also takes the time to describe in some detail the three workplaces where she situated her research. Such description is an important part of qualitative inquiries, because we can only assess the applicability of research findings to our own practice if we understand the context in which the research was carried out.

Her findings are both curious and compelling. She emphasizes that the visual rhetorical invention process is neither stable nor linear in the workplaces she studied, and often occurs through informal interactions. She also found that visual invention usually involves leveraging existing visuals from previously published documents. These findings demonstrate that qualitative research has the capacity to surprise us as it reveals the day-to-day realities of technical communication practice.

Portewig's article sheds light on the following themes raised by Giammano:

- What is a technical communicator today?
- How do we contribute to innovation?

References

Charmaz, K. 2003. "Grounded Theory." In *Qualitative Psychology: A Practical Guide to Research Methods*, ed. J. A. Smith. Thousand Oaks, CA: Sage.

Glaser, B. G., and A. L. Strauss. 1967. *The Discovery of Grounded Theory: Strategies for Qualitative Research*. Chicago, IL: Aldine.

Strauss, A., and J. Corbin. 1998. *Basics of Qualitative Research: Techniques and Procedures for Developing Grounded Theory*, 2nd edn. Thousand Oaks, CA: Sage.

Introduction

Since the 1980s, scholars have argued about the importance of technical communicators expanding their role from proficient writers to effective visual communicators as well. Notable technical communication scholars such as Barton and Barton (1985, 1990) and Kostelnick (1989, 1994) have contributed theories to explain visual concepts and elements, and research into ways that users perceive visuals in technical communication. In her seminal book on document design, Schriver (1997) contributed an understanding of research into analyzing the needs of users of documents and integrating text and words for a specific purpose. What is missing from scholarship, however, is an understanding of the workplace dynamics of those involved with visual communication—the writers, designers, engineers, managers, and others who plan, collaborate, and make decisions about what should be visually represented among the vast amounts of information in technical documents. Although clear guidelines for effectively designed visuals are readily available in textbooks and scholarship, we do not have actual inside accounts of exactly how visuals such as charts, graphs, diagrams, photographs, and screenshots get incorporated into technical documents in the workplace.

Uncovering these dynamics can help to assess our current approaches to teaching and practice.

This decision-making process, which I refer to as visual rhetorical invention, is an area of research that is essential to our development as visual communicators in the field. Invention, the first canon of rhetoric, lays the foundation for visuals that meet the needs of users. The rhetorical canons of arrangement and style have been our primary focus thus far, but this article encourages practitioners, educators, and students of technical communication to understand the importance of being rhetorically connected not only to the design of visuals, but to what information they choose to communicate in visuals (the subject matter). The three workplaces I studied show what this process can look like, prompting significant questions about current practices and our value as visual communicators. This study also serves to continue the conversation about our role as visual communicators and ways we can contribute in this area most effectively in the workplace.

Several research questions guided this study to understand the context, choices, tools, and negotiations involved in producing visual communication, including the following:

- What is the role of invention in the use of visuals in technical communication?
- What role do technical communicators play in the invention process?
- What is the context of the invention processes?

This qualitative study used grounded theory methodology, which focuses on generating theory from data rather than verifying theory, to investigate and theorize about these dynamics. A combination of methods was used to elicit details from practicing technical communicators about their role in that process. Furthermore, this article provides insight into the value of this methodological approach in studying visual rhetorical invention and the usefulness of grounded theory in conducting research in other areas of technical communication.

We have a variety of theoretical models, frameworks, and terminology for teaching and practicing visual communication (Barton and Barton 1990; Brasseur 1997, 2003; Cochran and colleagues 1989; Dragga 1992; Kostelnick 1989, 1996, 1998; Kostelnick and Roberts 1998; Kumpf 2000; Moore and Fitz 1993). Although these theories and frameworks deal with how to plan, design, and evaluate visuals, we need a clearer picture of how they translate and reflect practices in the workplace.

Previous research on rhetorical invention for technical communication has been conducted in areas such as writing, technical discourse, and engineering design. Research and scholarship has focused on technical writers (Allen 1978; Hall 1976; Miller 1985; Roundy and Mair 1983; Todd 2000), computer science students (Haller 2000), engineers (Hutto 2007; Selzer 1983), engineering students (Winsor 1994), and scientists (Campbell 1990; Hutto 2003; Latour and Woolgar 1986). This study fills a gap in our existing scholarship in that the focus is on the role of invention for visual communication rather than writing. Broadening the study of rhetorical invention to visuals reflects the continually changing landscape for technical communicators in the workplace.

In "Whose Ideas? The Technical Writer's Expertise in Invention," Regli argues that "if we do not work to articulate rich techniques for invention in the education of technical writers, we inadvertently reinforce the myth of the technical writer as a born scribe—a fortuitously gifted communicator who by instinct knows how to 'clean up' the products of the real 'inventors' of technical information" (2003, 71). Realizing the importance of invention to our development, this study seeks to understand the role of invention in visual communication and how technical communicators can contribute most effectively.

Methodological Approach

This research study was approved by the Texas Tech University Institutional Review Board for the Protection of Human Subjects. To answer my research questions, I adopted a mixed qualitative approach involving two methods, interviews and reflective analysis, that worked together to discover the dynamics of those who make decisions about what information should be visually communicated. I conducted interviews to gain an overall understanding of how visuals became part of documents and how technical communicators were involved in this process. I used reflective analysis, the second method, to collect data on numerous scenarios by having participants recount how visuals were incorporated into a published document that they had previously worked on. This qualitative approach was conducive to my research because its aim is to "discover variation, portray shades of meaning, and examine complexity. The goals of the analysis are to reflect the complexity of human interactions by portraying it in the words of the interviewees and through actual events and to make that complexity understandable to others" (Rubin and Rubin 2005, 202).

Other methods, such as a case study approach relying heavily on observation, might seem the obvious choice of method for studying visual rhetorical invention in the workplace. However, a combination of interviews and reflective analysis proved to be the most suitable method for two reasons.

First, in recruiting participants, for reasons including the proprietary nature of their work, they requested not to be observed or tape recorded but agreed to answering questions and reflecting on their decision-making process. Interviews were more effective than using a survey method to understand invention, because "interviewees are partners in the research enterprise rather than subjects to be tested or examined" (Rubin and Rubin 2005, 12). As Rubin and Rubin argue, "Unlike survey research, in which exactly the same questions are asked to each individual, in qualitative interviews, each conversation is unique, as researchers match their questions to what each interviewee knows and is willing to share" (2005, 4).

Second, invention is inherently difficult to study because it is not always verbalized and clearly identified as such. I found that work on visuals was scattered throughout a project cycle, which could last for months or close to a year, so it would have been difficult to observe it in action. Interviews allowed me to understand the dynamics of the process through targeted questions, whereas reflective analysis provided a way for me to analyze and reflect with participants on their decision-making process within the context of a product–documentation cycle. Furthermore, these methods enabled exposure to many more scenarios by studying three organizations and triangulating the results, rather than observing one company for an extended period of time and limiting my understanding of invention to a single perspective.

Methods for Collecting Data

Together, interviews and reflective analysis showed many examples of the decision-making process and were conducive to asking questions specific to the context in which the visual was used in the document. This section further describes the methodological approach, a justification of my methodology, and the methods and procedure used to collect and analyze data.

I began each interview by summarizing my research focus and asking my overarching research question: How do you decide what information should be communicated visually in a technical document (either print or electronic)? I then used my prepared interview questions to fill in any gaps in the participant's response. I took notes, using either a pen and notebook or a laptop computer, as participants responded to my questions.

The interview questions I used evolved as I conducted interviews because I used grounded theory (which I will discuss in depth in the next section) to guide my data collection and analysis. I formulated questions that helped me understand who the primary decision makers were and how technical communicators fit into this process; when in the process visuals are planned and designed; and how decisions are made about what information was communicated visually. Developing these interview questions forced me to focus my research and refine my way of explaining it, which resulted in the collection of valuable data regarding the decisions and design of visuals. The following is a list of my interview questions.

1. How do technical writers decide what information should be communicated visually in a technical document (print or electronic)?
2. Who makes the choices about what visuals to include in a document? Who creates visuals?
3. What types of decisions do you make when designing a visual?
4. Is there any brainstorming that takes place when creating visuals? Is it formal or informal?
5. How do you analyze your purpose and audience for a documentation project?
6. Do you have a planning process for creating a document? If so, in what part of the process do you plan the visuals in the document, and what is the medium (meetings, e-mail, face-to-face conversations)?

7.   What is the relationship between planning for text and visuals? For example, do you plan them together or separately?
8.   What factors affect the design of visuals for a document (for example, time, space, and so on)?

Because I wanted a more complete picture than interviews alone could provide, I needed a second method that could replace observation and cover details about the life cycle of a documentation project. The second method, reflective analysis, proved a useful tool for understanding how visuals fit into the entire documentation process. Reflective analysis is a method used in sociology and education to facilitate "in-depth reflections on practice" (Osmond and Darlington 2005, 3).

A variety of techniques can be used to aid in reflective analysis, including developing targeted questions and prompts that aid in reflection, think-aloud protocol, and reflective recall (Osmond and Darlington 2005). After I asked the interview questions, I adapted this method by asking each participant if they could provide an example of a document they had worked on and reflect on how the visuals for these documents were conceived, designed, and implemented. In some cases, the participant had something on hand that they could use, but if not, we scheduled a subsequent meeting time to discuss their document example.

Depending on the length of the document, participants would go through an entire document with me, describing and analyzing in detail how each visual was incorporated into the document. Alternatively, some participants chose to focus on one visual and took me through the evolution of that visual and how the idea had originated. I took notes throughout their reflections and was able to collect data on a variety of examples, ranging from visuals that were simply revised from previous versions to unique cases where visuals were given much more time and effort, such as several marketing posters for a new product. This method proved helpful in providing both description and analysis of how visuals fit into the documentation process and how they are conceptualized for technical documents.

Grounded Theory Approach

Grounded theory, a methodological approach adopted from sociology, guided my data collection, data analysis, and theorizing about the role of invention for visuals at the three workplaces studied. According to Strauss and Corbin, grounded theory methodology and methods are becoming "among the most influential and widely used modes of carrying out qualitative research when generating theory is the researcher's principal aim" (1998, vii). Developed by Glaser and Strauss (1967), grounded theory differs from many other methodologies in that it focuses on generating a theory from data rather than verifying theory. Charmaz (2003, 83), a qualitative researcher, identifies several key methodological procedures for carrying out a grounded theory inquiry:

- Simultaneous involvement in data collection and analysis phases of research.
- Developing analytic codes and categories from the data, not from pre-conceived hypotheses.
- Constructing middle-range theories to explain behavior and processes.
- Memo-writing; that is, analytic notes to explicate and fill out categories.
- Making comparisons between data and data, data and concept, concept and concept.
- Theoretical sampling; that is, sampling for theory construction to check and refine conceptual categories, not for representativeness of a given population.
- Delaying the literature review until forming the analysis.

These characteristics of grounded theory were well suited to my study, because I was able to inductively collect, code, and analyze data and theorize using a flexible approach. One example of how I was able to refine my methods is the case where I added an interview question in response to data I had collected in early interviews. After conducting several interviews, I realized that there were outside factors that greatly affected the decisions and design of visuals that were often beyond the control of the writer, designer, and engineer, so I needed a question that asked participants to identify these factors. Another example is that when I asked questions in interviews, I found that participants' responses were not really addressing the question. For example, one question that asked, "What is the decision-making process for designing visuals?" was narrowed to "Who makes the decisions about the design of visuals?" The latter version got more direct responses. Many questions I originally created assumed that there were formal processes in place regarding visuals, such as, "In what part of the planning process for a document do visuals get discussed?" I had several participants make comments, such as, "You seem to think this is a formal process," or, "You assume that we have an organized way of doing things." This question (and many others) assumed that there was formal planning associated with visuals, so I made the question more open-ended: "Do you have a planning process for creating a document? If so, in what part of the process do you plan the visuals in the document?" The use of grounded theory improved my interview questions and resulted in more accurate and detailed data.

I also used grounded theory to analyze the large amount of data collected and theorize from this data, using comparison and constant reflection. Instead of waiting to code my notes until I had collected data from all three companies, I coded them throughout my study at each company. I used the constant comparison technique of Charmaz (2003) to constantly compare the data I collected to previous data, so as to assure the accuracy of my coding and to refine my understanding of emerging concepts. To analyze my data, I used a technique called open coding, which involved developing categories and properties as I analyzed the data; categories being the larger groups and properties constituting the details in each category. Strauss and Corbin describe it this way: "Broadly speaking, during open coding, data are broken down into discrete parts, closely examined, and compared for similarities that are found to be conceptually similar in nature or related in meaning and are grouped under more abstract concepts termed 'categories'" (1998, 102). Using memo writing, I developed four categories, which I will discuss in the "Results" section, from coding data collected from interview questions and reflective analysis. These categories emerged by identifying recurring themes and trends in the data. The memo writing helped me to reflect on the data and further develop these categories.

Identifying categories from the data helped me to theorize about visual rhetorical invention by providing a conceptual way to understand my data and explain how those involved decide what information to communicate visually in a document. In the next sections, I provide background about the three companies studied and then present the results of this study on visual rhetorical invention.

About the Three Workplaces Studied

The study focused on three companies that ranged in size, geographic location, and number of technical communicators. This section provides a brief overview of each company, including details about the organization of technical communicators within each company and demographics about the employees who participated in this study.

Company A, headquartered in Texas, has more than 3,500 employees and direct operations in almost 40 countries. Company A specializes in virtual instrumentation for test, automation, and measurement. The Technical Communication Department is composed of more than 50 technical writers and a documentation production group that includes the art, desktop publishing, and publishing teams. The Technical Communication Department is part of Research and Development (R&D), and the writers are split up by product. They sit with the R&D staff for a product—which can include software programmers and/or hardware developers, technicians, product support engineers, a project manager, and a group manager—so they are in close proximity to the people they work with on a daily basis.

The 12 technical writers I interviewed came from a variety of educational backgrounds, including technical communication, literature, management information systems, radio/television/film, psychology, and journalism. The length of time the technical writers I interviewed had been working at Company A ranged from 1.5 to 18 years. Four technical writers specified that they had formal training in technical communication—either a formal degree in the field or outside professional training; the remaining seven had solely on-the-job training at Company A in technical communication. Of the 12 technical writers I interviewed, three were technical writing group managers, three were senior technical writers, two were staff technical writers, and four were entry-level technical writers.

Company B is a semiconductor company that creates system on a programmable chip (SOPC) solutions. It is headquartered in California and has 2,000 employees in 14 countries. Two main technical writing groups work at Company B: hardware and software, with about 10 writers in all. The projects technical writers work on are divided up by hardware and software rather than by product. Projects are assigned to writers as they become free. The technical writing group for hardware supports the applications engineers who write the initial draft of a document, and the software group works with the developers. The Technical Publications group, or "Tech Pubs," is responsible for formatting, making documentation clear, and ensuring good quality for both the text and visuals.

While at Company B, I interviewed a combination of engineers and technical writers. I chose to interview those who were most involved in the decision-making process; that is, senior applications engineers and senior and advanced technical writers. Advanced technical writer is an entry-level position, and senior technical writers require five years of experience at Company B. I interviewed three technical writers (two senior technical writers and one advanced technical writer) and four senior applications engineers. All of the applications engineers had bachelor's degrees in electrical engineering and at least one university course in technical writing. Their tenure at Company B ranged from six to 14 years. The technical writers I interviewed came from a variety of backgrounds, including a senior technical writer with a bachelor's degree in music and a master's in English. Another senior technical writer had a degree in rhetoric. The two senior technical writers had no formal training or coursework in technical writing or communication; they had only on-the-job training. The advanced technical writer had done a course in programming, and also had 20 years of previous experience as a graphic designer. The writers had been at Company B between two and seven years, with an average of four years.

Company C is located in Ohio and manufactures gas compression equipment for the oil and gas industry. The company of 500 employees had two technical writers who were part of the drafting department. Technical writers worked with other departments in the company, including design engineering, technical services, field services, and customer training. The lead technical writer held a two-year technical school degree in mechanical engineering. His training in technical writing came from years of on-the-job experience, plus training in the use of word processing and desktop publishing software. He had been with Company C for more than 11 years. The other technical writer I interviewed had a technical writing certificate. He had worked as a contract technical writer for 10 months at Company C and had five years of experience as a contract technical writer in several other industries, including avionics and aerospace. He had also worked as an avionics technician and operator.

The three companies studied produce a range of technical documents, products, and formats (electronic and print). Company A produces online help, manuals, and user guides. Company B publishes technical documents all together in one handbook for each product, and Company C creates one technical manual for each product and application manuals and customer technical bulletin updates. The most commonly used visuals among the companies include variations of illustration charts, block diagrams, screenshots, and tables describing technical information. Less common are photographs and graphs. Table 6.1 compares these types of documents and visuals among companies.

The variety of industries, demographics, levels of expertise, types of documents, and visuals represented in this study show important findings about practices related to visuals produced in the workplace. In the next section, I will present what I uncovered through interviews and reflective analysis of the visuals these companies produce.

Results

The data collected show the complexities and realities that characterize visual communication at these three workplaces. It is difficult to provide a general picture of how visuals are incorporated into documents because it was not a set or linear process, which is in itself an important insight. Visual rhetorical invention is not delegated to one phase of the documentation process but can vary depending on factors such as the writer's preference, the type and source of the document, and the type of product being documented. This quote from a technical writer at Company B reflects the situational nature of working on visuals for documents:

> When writing documentation, it depends on source material. If it is on the basis of existing internal documentation that I am making customer friendly, there will be a framework already. When I look at their documentation, I can see where a visual should be. I have a basic framework, do a diagram, and there would be an explanation of the diagram. Sometimes it happens at the end, once the text is written and I realize there needs to be a visual (in retrospect). It also depends on if it's totally from scratch or something already there.
>
> (Technical writer, Company B)

As this writer explained, the need for a visual might be identified at any phase of the documentation process, depending on the writing situation and the writer's own writing/design process.

Table 6.1 Comparison of Three Workplace Studies

Companies	Company A	Company B	Company C
Documents produced	Online help, hardware user manual, software user guide, installation guide, quick reference guide, and course manual	Handbook for each product, including data sheet, user guide, and a chapter for each feature	Technical manual, engineering reference sheet, customer technical bulletin, and application manual
Types of visuals	Tables, illustration charts, block diagrams, screenshots, and flow charts	Block diagrams, illustration charts, flow charts, screenshots, and tables	Illustration charts, photographs, tables, and graphs

The data show that visual rhetorical invention is also an informal activity that usually takes place between the technical communicator and engineer through face-to-face conversations or e-mail. Rather than taking place during meetings or planning sessions, discussions of visuals usually occur during the writing process; when, for example, the technical writer asks the developer to clarify things and may suggest a visual to address the problem. Sometimes new visuals are suggested through the document review process. If collaborators see a need to create an entirely new visual, they usually plan it together informally.

> Rarely is there brainstorming. Every once in a while we will be working on a document and have a short conversation with the engineer about a visual for a document. How detailed do we want to get? How much do we want to show? They are really informal discussions. I will go down and talk to the engineer or give them a phone call. It is usually because the engineer already knows what illustrations they want, so we just work with them to create illustrations.
>
> (Technical writer, Company B)

These informal discussions were prevalent among technical communicators, engineers, designers, and managers. Formal planning for documents, such as meetings and document plans, focus on higher level issues such as changes in the functionality of the product, new features, and the types of documents they would produce for a specific product or product line.

One of the more significant discoveries in this study that affects the process is that most visuals are leveraged from existing visuals—ones that were previously published in a technical document, visuals created by other groups within the company (including mechanical engineers, product engineers, application engineers, and drafting), and visuals created for other purposes (such as manufacturing, training, or marketing). These visuals are most often stored in electronic folders and are repurposed, "tweaked," or updated for a document. Technical communicators are most involved in revising or locating visuals already in existence: "Writers tend to be modifying something that already exists and in most cases they don't question 'do I need to add another picture here?'" (Group manager, Company A). According to the technical writers I interviewed, engineers come up with most of the concepts of visuals, because they know the product and the customers.

Leveraging from existing visuals is time and cost efficient: "We try to reuse things as much as possible, in documentation, a poster, and in other publications. It's the same information spun around. We try to get the most out of whatever we do" (Senior applications engineer, Company B).

Visuals are also reused for the sake of consistency:

> We try to do cut and paste from previous documents; it's familiar to the customer, and it's consistent. We want it to look similar to previous architectures

if it is similar. We don't need to change it if it's the same functionality or if it's the same family.

(Senior applications engineer, Company B)

Less often, new visuals are recommended and created when there is an entirely new product line or new features are added to a product, and for marketing purposes. A technical communicator's involvement with producing a new visual is dependent on factors such as whether the visual can be created using software they are familiar with; whether they are comfortable working with the subject matter; what the time frame for the project is; and where in the documentation process the project is. At Company A, for example, "Writers will work on some images, such as screenshots, simple shapes, and block flow diagrams" (Group manager).

Writers would decide to use a visual when they had difficulty explaining something in words. Technical writers use visuals "when a concept is difficult to convey, too wordy"; when "things . . . are too complex to put into words, it's easier to show a picture of hardware than describe in text." One technical writer's rule about whether to include a visual or not is "if it can be said easily with no room for misinterpretation, we leave it at that." Another writer uses the following guideline: "If it takes more than 10 steps to describe a concept, I will use a visual."

When I asked technical writers and engineers why they chose to use certain visuals, the overwhelming response was that "a picture is worth a thousand words." I repeatedly got that comment from technical writers who said they had not really thought about how they would go about deciding to use a visual. Engineers commented that they "just knew" from their education and experience. For them, it was a tacit type of knowledge, because they consider themselves visual thinkers and learners and had created standard types of visuals as part of their engineering training.

The above results reveal that visual invention can take place in any phase of the documentation cycle and that it is an informal activity influenced by the fact that most visuals are leveraged from existing ones. Operating within these parameters, a number of factors arbitrate the process of deciding what information to present visually in a document. In the next section, I will discuss four categories that emerged through data analysis from interviews and reflective analysis—these categories serve as important factors that shape the process and contribute to a theoretical understanding of visual rhetorical invention.

Factors that Shape Visual Rhetorical Invention

Table 6.2 presents four factors that reoccurred in the data from all three sites that participants said influenced the design, use, and choices about visuals. These factors include established conventions; technical knowledge of product and use of technology; social dynamics between groups such as technical communicators, engineers, and designers; and the resources allocated to a project, group, or product.

Table 6.2 Factors that Shape Visual Rhetorical Invention

Category	Properties
Conventions	Genre conventions (manual, quick-start guide, help)
	Product conventions (software, hardware)
	Environment of document use
	Documents that are localized (translated)
	Safety and international standards for visuals
Technical knowledge/ technology	Educational background
	Training in technical communication
	Technical knowledge of product
	Difficulty of learning graphic software programs
	Conversion of graphic files between software programs
Social dynamics	Degree of collaboration between engineer and technical communicator
	Role of technical communicator within organization
	Technical communicator's level of comfort with visual communication
Resources	Time frame of project
	Size of team dedicated to project (number of technical writers)
	Budget for project (affects the number and types of visuals)
	Time required for maintenance of visuals

Conventions

At these three companies, conventions for the use and type of visuals are in place according to product, genre of documentation, and the nature of the information being communicated. For example, there are common components to every compressor model that are illustrated: "Some of the illustrations in the front matter are the overview, numbering order, or cylinder. These are standard regardless of the manual. It's easier to show them than write three or four paragraphs to explain it" (Technical writer, Company C). If the document is going to be used in an industrial setting, for example, conventions are in place for the format of the information, which is usually a large technical poster.

Whether a document is going to be translated into other languages affects the use and types of visuals. Visuals are limited in these documents, because they must be translated and maintained, and those processes are time consuming and more expensive. In addition, safety and international standards influence many of the decisions about what information is illustrated. Because operating Company C's equipment has important safety considerations, certain information is always illustrated for the safety of the customer. They must also comply with international standards to sell their products overseas. Safety and international standards are set by outside governing bodies, such as the International Organization for Standardization (ISO). Many of these conventions are passed along through the company's style guide. Technical communicators are also acculturated in the use of their company's conventions for documentation through studying previously published documents and through repetition.

Technical Knowledge and Technology

Knowledge of technology (design and graphical software) and the technical nature of products affect the technical communicator's role in the invention process. Some technical communicators interviewed considered themselves "word" people and are not as comfortable with the visual aspect of documents, so they rely on the engineers, who are "visual thinkers." For example, at Company A, technical writers are involved with creating visuals for information that are more "straightforward" and/or can be created using tools they are more familiar with, such as FrameMaker. Visuals in many cases must also be converted between software programs, a step that requires a level of technical knowledge. For these reasons, the perception of visuals as time consuming and "extra work" came up in a number of interviews. Although many of these factors affect the actual creation of visuals, they also impact on the decision to use visuals in documents.

Social Dynamics

The technical communicator's attitude toward using and designing visuals corresponded to their training and involvement with visuals in many of the interviews I conducted. As already noted, most technical writers interviewed had no formal training in technical communication or in a technical field, but those that were trained in one or both of these areas were more assertive of their expertise and attempted to traverse beyond their perceived role of making text and visuals "look pretty." On the contrary, in the case of Company B, many of the technical writers were resistant to their increased responsibility for visual content; this company also had the fewest who were formally trained in technical communication.

The dynamics between engineers and technical writers also had a strong influence on how involved the technical writer was in decisions about visuals. Many of the engineers interviewed excluded technical writers from work on visuals because they did not see them as able to contribute technical knowledge to the process and/or were able to find appropriate visuals on their own from other sources. At Company B, there is a clear division between the expertise of the engineer and technical writer; application engineers write the initial draft of the documentation, and technical writers serve as editors of textual and visual content. At Company C, collaboration between technical writers and engineers was described as productive; writers have more involvement in creating content, partially because of their technical background and the fact that they are a small team. Several technical writers I interviewed have also gained the trust of engineers, which has allowed them to be more involved in the decision-making process, but these writers were especially interested in visual design and/or had backgrounds in graphic design.

Resources

In addition to conventions, technology, and the social dynamics within the organization, resources such as time, money, and people play a large role in the use of visuals in technical documents. Furthermore, these companies produce so many documents that resources are a significant factor. Products or product lines that are given more priority have greater resources in the form of more technical writers and a larger budget for documentation, which often corresponds to a greater number of and/or higher quality visuals. The reason for this phenomenon is that there are more resources for designing and maintaining these visuals. Technical writers are responsible for keeping visuals accurate and up to date, processes that are time consuming, so generally they try to limit the number of visuals in cases where resources for maintenance are an issue. Another reason for higher quality visuals is that the company wants to sell a product, so it creates documentation that is more visible to the user and other potential customers, such as a product poster that could be hung in an engineer's cube or office. Documentation will contain fewer visuals in some cases when it is designed for an established product that has a specific functionality, so there is not much concern that the user will "mess up."

These factors shape the invention process for visuals. The decisions about when to use a visual or what to illustrate are often already established through conventions and can be dictated by technical issues, workplace dynamics, and/or resources. In the final section, I discuss the implications of what this study has revealed.

Discussion and Conclusion

The purpose of this study was to understand, and construct a theory to account for, the role of visual rhetorical invention in the context of the workplace. I do not claim that these three workplaces are exemplary of all organizations or technical communicators in the field. However, my study is a starting point for broadening our scholarship and research from rhetorical invention for writing to visual communication. The results of this study point to a number of questions and reflections about our perceptions, teaching, and practices of visual communication.

First, perceptions about our expanded role, not only as writers but as visual communicators, are much more complex. Participants whom I studied faced challenges in terms of technology, training, collaboration, and their own sense of themselves as "word people." Only four of the 17 technical writers interviewed had formal training in technical communication, such as a bachelor's degree, master's degree, or technical writing certificate. This lack of technical communication training and familiarity with visual communication could explain the apprehension mentioned in interviews regarding the use and design of visuals. Technical communicators persistently think of themselves as word people, whereas engineers think of themselves as visual people. Technical communicators seem to default to thinking linguistically, as evidenced in using visuals "when words fail them."

If technical communicators are going to expand their role, they need to be trained as true communicators, of both text and visuals. Because seeking further education is not always possible, practitioners should pursue other forms of training, such as attending conferences, Web seminars, and workshops, which could provide professional development in the area of visual communication. The rhetorical expertise of technical communicators is crucial to effective visual communication, so training can ensure that they are part of the conversation.

Lack of technical knowledge is also a strong barrier to technical communicators contributing to visual communication, so increasing technical training could improve the situation. Engineers at many universities are required to take a technical writing course of some sort, so why do we not require our students to understand basic engineering concepts? Students need to be more exposed to technical concepts and technology, not just learn specific software tools. In a special issue of *Technical Communication* on the future of the field, Albers, the guest editor, differentiated between tools and technology: "More than dealing with issues of how to use one tool to perform a task, we need to teach and consider how using various tool features (such as styles) or technologies (such as single sourcing) affects the documentation process" (2005, 267). We need to recognize that technical knowledge and technology affect our role in decision making for visuals in the documentation.

Furthermore, we must train students and practitioners to respond to the practice of repurposing to ensure that visuals used fit the rhetorical situation. Technical communication textbooks teach readers to create entirely new visuals, but, as reflected in this study, they will often be called on to use existing visuals. This practice of leveraging from existing visuals changes the dynamic from determining what information should be explained, and designing a visual to fit the purpose, to finding visuals already created for that purpose.

We need to reexamine our teaching to reflect this shift in landscape from new to reused, and conduct research into the effects on users. Factors such as cost and efficiency are legitimate reasons for reusing visuals, but how can we ensure that visual invention remains a rhetorical practice and that these conventions are revisited and adapted over time? Kostelnick and Hassett argued that "conventions prompt rather than stifle invention," because they "invite adaptation and improvisation and allow designers to tailor them to specific situations" (2003, 5–6). However, they admit that "if conventions were etched in stone, they would undermine invention and seriously erode the designer's agency" (Kostelnick and Hassett 2003, 5–6). We must ensure that conventions are adaptable, responsive to user needs, and tailored for specific situations.

Invention is central to the rhetorical tradition that technical communicators adhere to, which is why we must understand how it functions for visual communication. The context and factors in which visual rhetorical invention operates at these three companies provide us with a foundation for further research and theory into the role of visual rhetorical invention in the workplace. If Schriver (1997) is right in suggesting that the integration of text and visuals is essential to effective document design, the results of this study may indicate that further research and theorizing on the real-world dynamics of visual invention deserves a place of prominence on our profession's research agenda.

Acknowledgment

*This article was originally published in *Technical Communication* (2008), 55: 333–420.

References

Albers, M. J. 2005. "The Future of Technical Communication: Introduction to This Special Issue." *Technical Communication* 52: 267–72.

Allen, J. W., Jr. 1978. "Introducing Invention to Technical Students." *Technical Writing Teacher* 5: 45–49.

Barton, B. F., and M. S. Barton. 1985. "Toward a Rhetoric of Visuals for the Computer Era." *Technical Writing Teacher* 12: 126–45.

———. 1990. "Postmodernism and the Relation of Word and Image in Professional Discourse." *Technical Writing Teacher* 17: 256–71.

Brasseur, L. 1997. "Visual Literacy in the Computer Age: A Complex Perceptual Landscape." In *Computers and Technical Communication: Pedagogical and Programmatic Perspectives*, ed. S. A. Selber. Greenwich, CT: Ablex.

———. 2003. *Visualizing Technical Information: A Cultural Critique*. Baywood, NY: Amityville.

Campbell, J. A. 1990. "Scientific Discovery and Rhetorical Invention: The Path to Darwin's Origin." In *The Rhetorical Turn: Invention and Persuasion in the Conduct of Inquiry*, ed. H. W. Simons. Chicago, IL: University of Chicago Press.

Charmaz, K. 2003. "Grounded Theory." In *Qualitative Psychology: A Practical Guide to Research Methods*, ed. J. A. Smith. Thousand Oaks, CA: Sage.

Cochran, J. K., S. A. Albrecht, and Y. A. Green. 1989. "Guidelines for Evaluating Graphical Designs: A Framework Based on Human Perception Skills." *Technical Communication* 36: 25–32.

Dragga, S. 1992. "Evaluating Pictorial Illustrations." *Technical Communication Quarterly* 1: 47–62.

Glaser, B. G., and A. L. Strauss. 1967. *The Discovery of Grounded Theory: Strategies for Qualitative Research*. Chicago, IL: Aldine.

Hall, D. R. 1976. "The Role of Invention in Technical Writing." *Technical Writing Teacher* 4: 13–24.

Haller, C. R. 2000. "Rhetorical Invention in Design: Constructing a System and Spec." *Written Communication* 17: 353–89.

Hutto, D. 2003. "When Professional Biologists Write: An Ethnographic Study with Pedagogical Implications." *Technical Communication Quarterly* 12: 207–23.

———. 2007. "Graphics and Invention in Engineering Writing." *Technical Communication* 54: 88–98.

Kostelnick, C. 1989. "Visual Rhetoric: A Reader-Oriented Approach to Graphics and Design." *Technical Writing Teacher* 16: 77–88.

———. 1994. "From Pen to Print: The New Visual Landscape of Professional Communication." *Journal of Business and Technical Communication* 8: 91–117.

———. 1996. "Supra-Textual Design: The Visual Rhetoric of Whole Documents." *Technical Communication Quarterly* 5: 9– 33.

———. 1998. "Conflicting Standards for Designing Data Displays: Following, Flouting, and Reconciling Them." *Technical Communication* 45: 473–82.

———, and M. Hassett. 2003. *Shaping Information: The Rhetoric of Visual Conventions.* Carbondale, IL: Southern Illinois University Press.

———, and D. D. Roberts. 1998. *Designing Visual Language: Strategies for Professional Communicators.* Boston, MA: Allyn & Bacon.

Kumpf, E. 2000. "Visual Metadiscourse: Designing the Considerate Text." *Technical Communication Quarterly* 9: 401–24.

Latour, B., and S. Woolgar. 1986. *Laboratory Life: The Construction of Scientific Facts.* Princeton, NJ: Princeton University Press.

Miller, C. R. 1985. "Invention in Technical and Scientific Discourse: A Prospective Review." In *Research in Technical Communication: A Bibliographical Sourcebook*, ed. M. G. Moran and D. Journet. Westport, CT: Greenwood.

Moore, P., and C. Fitz. 1993. "Using Gestalt Theory to Teach Document Design and Graphics." *Technical Communication Quarterly* 2: 389–410.

Osmond, J., and Y. Darlington. 2005. "Reflective Analysis: Techniques for Facilitating Reflection." *Australian Social Work* 58: 3–14.

Regli, S. H. 2003. "Whose Ideas? The Technical Writer's Expertise in Invention." In *Professional Writing and Rhetoric: Readings from the Field*, ed. T. Peeples. New York, NY: Longman.

Roundy, N., and D. Mair. 1983. "The Composing Process of Technical Writers: A Preliminary Study." http://www.jacweb.org/Archived_volumes/Text_articles/V3_Roundy_Mair.htm/.

Rubin, H. J., and I. S. Rubin. 2005. *Qualitative Interviewing: The Art of Hearing Data*, 2nd edn. Thousand Oaks, CA: Sage.

Schriver, K. 1997. *Dynamics in Document Design.* New York, NY: Wiley Computer Publishing.

Selzer, J. 1983. "The Composing Process of an Engineer." *College Composition And Communication* 34: 178–87.

Strauss, A., and J. Corbin. 1998. *Basics of Qualitative Research: Techniques and Procedures for Developing Grounded Theory*, 2nd edn. Thousand Oaks, CA: Sage.

Todd, J. 2000. "Burkean Invention in Technical Communication." *Journal of Technical Writing and Communication* 30: 81–96.

Winsor, D. 1994. "Invention and Writing in Technical Work: Representing the Object." *Written Communication* 11: 227– 50.

Chapter 7

A Work in Process

A Study of Single-Source Documentation and Document Review Processes of Cardiac Devices*

Lee-Ann Kastman Breuch

Editors' Introduction

Lee-Ann Kastman Breuch's "A Work in Process: A Study of Single-Source Documentation and Document Review Processes of Cardiac Devices" studies the processes of writing and reviewing product documentation at a company that has transitioned from a traditional documentation development process to single sourcing. Because qualitative methods are an ideal means of studying not only business processes but also the social dimensions of the individuals' work within the organization, her qualitative case study is a particularly apt basis for this article.

The subject of single sourcing has been on the periphery of most technical communicators' radars for the past decade or so because it has received a fair bit of scholarly attention. At the same time, little has been written on the subject to address the effects of single sourcing on the technical communicators themselves and the workflows of their organizations. For her exploration of the relationship between single sourcing and document review, Breuch has conducted a preliminary survey of a group of 12 technical communicators at a medical device manufacturer, interviewed them in some depth, and then attended three team meetings at which the new work flow and its impact on document review were discussed.

The article itself opens with a literature review of single sourcing and its effects on workgroups, and then summarizes how the organization Breuch studied made the transition. The "Materials and Methods" section notes that the author had no access to written artifacts because of corporate confidentiality policies, but we get the distinct sense that the opportunities to seek input from the participants meant that this did not hinder Breuch in any significant way. Her contacts with the writers in question seem to have yielded significant insights into the practice of single sourcing at the company in question. Breuch also describes the member checking technique that she used, where she would review her findings with participants and invite their comments and interpretations. The article concludes with the findings and a summary of significant themes that surfaced in the surveys, interviews, and observations, as well as a discussion and conclusion. Appendixes 7.1 and 7.2 include full transcripts of survey and interview questions.

It is interesting that Breuch's findings have a paradoxical quality. Her participants experience the process she studied as both structured and chaotic; efficient and inefficient; straightforward and complex. Qualitative methods are particularly adept at uncovering the paradoxes that people experience in complex social systems, and at offering researchers and participants alike the opportunity to make sense of these tensions.

Among Giammona's themes that are explored in Breuch's article, the most significant are the following:

- What is a technical communicator today?
- What is our future role in organizations?
- What should managers of technical communicators be concerned with?
- How do we contribute to innovation?
- What technologies are impacting on us?

This article is particularly helpful because it explores an important emerging business practice in the field, the technologies that support that practice, and its impact on individuals as well as on their organization.

Introduction

Single-source documentation is not a new phenomenon among technical communicators. As Carter has suggested, "the topic of single sourcing—producing documents designed to be recombined and reused across projects and various media—has been covered in great detail in the past several years in books, trade journals, conferences, and academic journals" (2003, 317). In a literature review about single sourcing, Williams has explained that single sourcing has been defined in books and articles in two ways: as a "single information source" that is reused without changes to content or as a single information source that can be modified in terms of content to fit multiple deliverables (Williams 2003, 321; see also Ament 2003). Bottitta and colleagues agreed, suggesting that there are "two types of single sourcing," which include a "single topic" that remains static in all deliverables or a "single set of source files" that can be changed for each deliverable (2003, 356). Williams (2003) stated that practitioners and scholars have begun to explore the implications of single sourcing, including an examination of the technologies used to support it (usually XML and content management systems), as well as the organizational changes that must be made (see also Ford and Mott 2007).

However, in a special issue of *Technical Communication* on single sourcing, Carter suggested that arguments about single sourcing thus far have articulated why and how single sourcing occurs, but not necessarily what happens to technical writers once single sourcing is implemented (2003, 317). In fact, Carter explored the possibility that technical writers may be outright resistant to single sourcing rather than supportive of the administrative arguments (cost saving and efficiency). Carter explains that "When production techniques are changed radically, workers often perceive this change as a threat to their jobs and their way of doing things" (Carter 2003, 318). Carter suggested that "radical change" is a key element of single sourcing (Carter 2003, 318).

Other studies have described the impact of single sourcing in similar dramatic terms. According to Rockley, "the two most significant changes in a single-sourcing initiative that often result in discomfort and resistance are structured writing and collaborative authoring" (2003, 350). Structured writing, he suggests, involves imposing more strict organizational structure and language consistency so that information modules can be interchangeable. This structure may be a departure for writers who are used to exercising individual freedom in their writing (Rockley 2003, 350). Rockley also explained that single sourcing may result in a "changing role of ownership" to collaborative, rather than individual, authorship:

> Although in the past, an author may have been responsible for a product document or a set of product documentation, an author may now be responsible for creating content for a set of common features that appear across multiple products. The author may no longer own the content for a whole document; the author may be responsible for a piece or a cross-section of a series of documents.
>
> (Rockley 2003, 352)

In a qualitative study, Bottitta and colleagues (2003) further explored these organizational changes, suggesting that when traditional writing teams engage in single sourcing, they experience a shift in roles that requires technical writers to go beyond text. That is, in a single-source environment, technical writers must negotiate social tensions and conflict as they work with others to create single-source documents. They assert: "Regardless of your team's size or structure, a successful transition to single sourcing demands that you understand and manage any associated organizational changes that might occur" (Bottitta and colleagues 2003, 355). Interestingly, similar claims have been made about the changing roles of technical writers in the context of cross-functional teams (CFTs). Conklin (2007, 210–11) suggested that the work of technical communicators has changed from focusing on product to focusing on social (especially interpersonal) interaction. He encourages technical communicators to recognize the social dimensions of their work that go beyond products. Similarly, Moore and Kreth have asserted that technical communicators must increasingly "manage situations, events, and the interactions and experiences of individuals, groups, and cultures" (2005, 303).

Like these authors, I am interested in the impact that single sourcing might have on the shifting roles of a technical writer, from individual author to collaborative author and project manager. However, to push the issue further, I am particularly curious about how single sourcing might intersect with or impact document review processes—a fundamental work activity among technical writers.

As Kleimann suggested, "Document review is an integral part of organizational writing that transforms individual products into institutional products" (1991, 520).

Document review is a collaborative activity in which individual authors solicit feedback from outside readers who provide multiple perspectives and suggestions for revision. Document review is one of the processes that technical writers must manage, both in terms of product and process. Kleimann asserted that "this review process is often a complex one because of the number of readers who must review a document, the number of times that document must be reviewed, and the time constraints under which writers and reviewers work" (Kleimann 1991, 520).

Review processes can be straightforward and linear, involving a supervisor and writer (Paradis and colleagues 1985), or they can be complex and messy when they involve multiple reviewers with no clear hierarchy (Kleimann 1991). As Conklin has discovered in his qualitative study of cross-functional teams, review processes can emerge as a source of tension for technical communicators who must manage ongoing feedback and "an obsession for perfection" among multiple reviewers who also have ownership in the creation of information (2007, 220).

Document review may also be an activity that can shape the roles of technical communicators—for better or for worse. For example, Conklin (2007, 221) discussed how there may be a tension between technical communicators as "typists" rather than "communication professionals" in cross-functional teams. This tension also may exist within document review processes, because outside reviewers may believe they have more ownership of content than the writers themselves.

Document review processes may indeed be an opportunity for technical writers to assert and shape their roles, especially within cross-functional teams. For example, in *The Dynamics of Writing Review*, Katz (1998) claimed that document review is an opportunity for both socialization of writers within a company (especially for new writers), and an opportunity for writers to gain agency—a concept that she describes as "individuation." Katz described "individuation" as "how newcomers become recognized for their distinctive talents and skills or how that recognition relates to their ability to gain authority, challenge the status quo, and become agents for change in organizations" (1998, 73).

In an examination of new writers, Katz described how one writer, "Darlene," gained authority by demonstrating writing competence and by being actively and thoroughly involved in review processes that occurred on multiple levels of authority in her company (1998, 81). Katz ultimately suggested that review processes are important opportunities for writers to negotiate positive roles through their writing and management abilities.

From this brief review of literature, we can see that one thing document review processes and single sourcing have in common is a tension between individual and collaborative authorship. Document review processes require that writers create content, yet receive and address any and all feedback from external reviewers. Writers are placed in the role of project managers as well as individual authors in these processes. In single sourcing, as Rockley (2003) explained, writers must comply with organizational decisions about writing structure; in addition, content may be stored in shared spaces where multiple authors may make necessary adaptations. Both of these practices influence the roles and identities of technical writers as individual authors. What happens when we examine the impact of both practices—document review processes and single sourcing—together? How do technical writers cope when such "radical change" happens simultaneously in both areas?

The case study I share in this article describes one such story of "radical change" that occurred both in terms of single sourcing and document review processes. Specifically, in this article, I share results of a case study about the document review processes of a technical writing team at a biomedical company. Twelve technical writers were observed, interviewed, and surveyed as they changed work practices from updating legacy documentation to generating single-source content modules about cardiac devices.

Results from the case study show both the benefits and drawbacks of changing to a single-source system. Every technical writer who participated in the study agreed that the manuals and documentation that emerged from the process were better, more thorough, and more accessible to readers. Although not all improvements to documentation were the direct result of single sourcing, certain aspects of the single-sourcing process did contribute to improvements, such as topic-by-topic review and consistent style standards. However, technical writers also expressed anxiety over the difficulty of the review process, specifically in balancing requests and schedules from multiple reviewers on the documents they were writing.

In addition, several writers commented that this approach required more collaborative writing than did their previous approach to documentation. The story of this technical writing team illustrates a kind of radical change in their work practices, both in terms of single sourcing and in terms of their day-to-day work practices involving document review.

Background of Study

Like many other companies, the company in this case study was facing pressures to increase efficiency and flexibility of documentation, which in this case included documentation of pacemakers and defibrillators. In response, over a two-year period, this team created a unique approach to single-source documentation of these devices that involved an internal content management system, a complete reorganization of content into reusable information modules, and an elaborate (but necessary) review system.

Each technical writer on the team was responsible for writing between six and ten "topics" related to the operation of defibrillators or pacemakers (each topic addressed a specific function or feature of the device in detail). Writers were responsible for writing a "baseline" or basic description of each topic and any additional versions of the topic that were needed for varying device models. Each topic was reviewed by an interdisciplinary team consisting of a subject-matter expert (usually an engineer), a representative from "regulatory affairs," a representative from "marketing," and an independent reviewer (often an experienced writer on the team).

Each topic required intense research and an iterative review process and was compiled into a large manual. Topics were written first for a clinician audience, and then they were written again for an audience of technical service representatives and field clinical engineers. Individual topics could be reused as necessary in manuals for appropriate related devices.

The idea of writing documentation in modular topics, according to one writer on the team, was to move away from a single, 500-page manual that relied on legacy content, and toward independent modules written about key functions that could be customized to multiple versions of the devices. The approach on the team became known as "the topics process." Approximately 230 "topics" were generated by the team that, when finished, contributed to more than 100 manuals.

Multiple versions of these manuals were necessary to address different device models and audiences (two distinct audiences included a clinician audience and technical service representative and field clinical engineer audience). "Topics" within these books were essentially "chunks" of information that could include great detail yet could be reused and customized for each version of the device and its respective documentation.

This way of writing documentation essentially required the team to rewrite all legacy content—a significant change in practice for experienced technical writers at this company. Why this change? According to the writing manager, there were three main forces driving the change:

1.   Several product lines were merging into a common technology platform, but there was no existing baseline documentation set.
2.   Feedback from clinicians suggested that previous manuals were "too technical."
3.   Engineers involved in document review complained that reviewing 500-page manuals was an inefficient use of their time.

In addition, throughout this transition, the company had to ensure compliance with the United States Food and Drug Administration (FDA) guidelines for design review, as specified by the FDA-mandated Quality Systems Regulation (QSR) for medical devices (FDA 1997). Compliance with QSR meant that all documentation of medical devices had to include documented feedback from a variety of reviewers, which in this case meant a minimum of four reviewers in the areas mentioned above.

This was not a new process for the writers in this company. However, the difference in this case was that reviewers in many cases examined new text instead of legacy text. They also reviewed smaller chunks of text (one topic at a time) rather than an entire manual. The smaller chunks of text led to greater scrutiny and increased feedback from reviewers. This was a change in the documentation process, and technical writers felt that the process was more collaborative than ever before. One writer described the change this way:

> The review process has evolved. Manuals were 500 pages long. Approved information [in the legacy manuals] could not be changed. People only reviewed the changed information. [In the] current project, [we are] rewriting everything and delivering information in smaller chunks. Reviewers have leapt at the opportunity.

The single-source, "topics-based" approach was a new approach for the technical writing team, and thus this study was a rich opportunity to investigate change in technical writing practices. Specifically, this study investigates the complexities of the transition from individually authored legacy manuals to collaboratively authored manuals with a complex, federally regulated document review process. As an outside researcher, I was primarily interested in technical writer roles as they intersected with document review processes and single sourcing. Consequently, my primary research questions addressed roles of technical writers:

- How do roles of technical writers and communicators change in a new single-source environment?
- How do technical writers manage the demands of a complex document review system within a single-source environment?

In addition to these primary research questions, I had questions about the details, benefits, and challenges of these technical writers' work practices. Thus, my secondary research questions included the following:

- What approach(es) did this technical writing team take toward single-source documentation?
- What benefits of a single-source documentation system does this team report?
- What challenges of a single-source documentation system does this team report?

Materials and Methods

To address these questions, I conducted a case study over a six-month period in which I (1) administered a survey about the company's document review processes, (2) conducted individual interviews with 12 technical writers on the technical writing team, and (3) attended three team meetings. A case study approach was most appropriate for this research because this study addressed only a group of 12 people. As MacNealy suggests, a case study "refer[s] to a carefully designed project to systematically collect information about an event, situation, or small group of persons or objects for the purpose of exploring, describing, and/or explaining aspects not previously known or considered" (1999, 197).

Because case studies involve small sample sizes, researchers do not intend to make generalizations from case studies, but rather to gather information from specific situations and individuals that may provide insight or generate new knowledge (MacNealy 1999, 197). Similarly, I approached this study not as a way to make generalizations about document review systems or single sourcing, but rather as a way to describe the experiences of this group of technical writers as they changed work practices to single sourcing. My specific intention was to address Carter's (2003) call for further research about how technical writers experience the transition to single-sourcing work practices.

The case study began with a qualitative, open-ended, online survey that included questions about the job/position of each writer and details of managing documentation review processes that were a regular part of their jobs. The survey addressed details such as how writers distributed "topics" drafts to reviewers, received feedback, and managed varying feedback from multiple reviewers. The survey also included questions about the technological tools that writers used to manage the documentation process (see Appendix 7.1 for a copy of the survey).

After the survey, I conducted an hour-long interview of each team member in which questions focused more deeply on the advantages and disadvantages of changing to a single-source documentation process and how the process affected writers individually. I used an interview script and asked all participants the same questions; in some cases, I asked additional questions to probe more deeply when clarification was needed. Participants were asked to describe their jobs and document review processes, as well as to give their impressions of how well the process of generating single-source documentation was working for their team (see Appendix 7.2 for questions).

Institutional Review Board approval was sought and granted for surveys, interviews, and observation of writers. In addition, the legal department of the biomedical company agreed to allow me to conduct surveys and interviews, and to observe the technical writing team.

I attended three team meetings over the course of several months. In the first meeting (which occurred at the beginning of the six-month period), I introduced myself to the team, proposed this research project, and discussed issues of informed consent. I explained that I was interested in learning more about document review processes that writers were using, as well as the technology they used to support it.

At this time, my initial interest was motivated by learning how professional writers engaged in document review processes and how technologies might facilitate review processes (I had previously conducted classroom research on how student writers engaged in peer review practices). However, during this first meeting and subsequent discussions with the writing manager, I learned that this particular team was in the middle of transitioning to a single-source documentation system.

I attended a second team meeting at the mid-point of my involvement in the case study. This meeting helped me to get better acquainted with team members and their rapport as a team. Many of the themes I had encountered during individual interviews were supported by team rapport discourse; I was able to listen to anecdotes about receiving feedback from designated reviewers. I also heard comments about technical writer roles and the difficulty of balancing writer authority with multiple reviewer comments.

The third team meeting I attended occurred at the end of my six-month period working with this team. At this meeting, I shared preliminary results of my study, which consisted of themes that had emerged from interviews and survey results. This meeting provided an opportunity to get feedback from the team about the accuracy of my documentation of interview and survey notes. I received some clarifications at this time and was grateful for the discussion.

Observations were also included in this case study. Because I was a frequent visitor during this six-month period, I had the opportunity to observe the workspaces of the participants, as well as some drafts of "topics" documentation and software used to manage the single-source process. For example, some participants allowed me to look at artifacts such as drafts of "topics," reviewer feedback on drafts, editorial comments, and formal compilation of reviewer feedback. However, I had no access to artifacts beyond what writers briefly shared with me; the company protected all in-progress documentation and did not allow me to copy or access any artifacts, including access to internal content management systems used to organize the documentation. Because I had no access to documentation or the internal content management system, I was unable to conduct any kind of content analysis on these artifacts, nor was I able to examine the internal content management system that was pivotal to their work. My case study was limited to the surveys, interviews, and observations (and no analysis of artifacts) that I conducted during my time visiting this company.

Survey and interview data were the most tangible data I collected. Surveys were conducted online, and I compiled data from spreadsheets generated by online survey software. I quantified results and logged all open-ended answers; however, survey data did not yield as much insight on my research questions as did interviews. Thus, I used survey data sparingly and only when relevant in the analysis.

Once interviews were completed, I transcribed interview data by referring both to my handwritten notes and tape recordings of interviews. I compiled a summary of all interview data first by creating a coding system for each participant so that identities were kept anonymous. I organized data according to questions from the survey and interviews, summarizing results from all participants for each particular question.

When all survey and interview data were organized, I began a qualitative analysis by examining data for themes related to my primary questions about technical writer roles in single-sourcing and document review processes. Specifically, I reviewed all survey and interview data related to (1) job descriptions and work practices; (2) benefits of single-sourcing and document review processes, and (3) challenges of single-sourcing and document review processes. In each of these categories, I identified a theme when similar comments were made by at least seven of 12 participants. Four themes emerged from the data:

1. *Streamlined work practices related to document review.* I asked participants to describe their work practices and found that each technical writer described a similar process. This process shaped the role technical writers played in the new single-sourcing system.

2. *Insights about the benefits of single sourcing.* Survey and interview data among all participants showed comments about the high quality of documentation produced from single sourcing and the thorough document review process. Many writers commented on the areas of personal growth they have gained from single sourcing. These insights addressed the benefits that resulted from the new single-sourcing system.

3. *Challenges related to managing document review processes.* When asked about challenges of the new document review system, all participating technical writers commented on the complexities of managing their "topics'" production schedules and, in particular, the schedules and feedback of multiple reviewers. These challenges, for many participants, created frustration and conflict about their roles as technical writers.

4. *Individual and collaborative authorship.* Many writers commented on the collaborative nature of the single-source, "topics-based" approach, starting with first drafts of topics and running through the final documentation. Although all writers recognized the need for this collective approach, a few voiced concerns about a loss of individual authority as a writer. These issues of autonomy and collaboration characterized attitudes about technical writer roles.

In the remainder of this article, I comment further on these themes, sharing data from interviews, surveys, and observations. In the next section, I provide details about the participants in the study.

Participants

Twelve technical writers participated in the study, spanning a range of job titles and experience at this particular company. Results from the survey indicated that participants included one technical writing manager, one editor, and ten writers of varying degrees of experience in the company. There were three distinct levels of technical writing positions among this team, and for purposes of confidentiality I refer to them as level 1, level 2, and level 3.

• A level-1 technical writer is the entry-level position in which writing and documentation are primary tasks. At the time I conducted this study, there was one level-1 writer.

• Level-2 technical writers had responsibilities that included writing documentation and special projects as they arose. Five level-2 writers participated in this study.

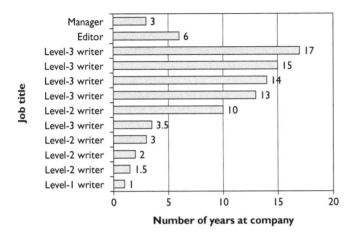

Figure 7.1 Technical Writing Team Member Experience at Company

- Level-3 technical writers have seniority on the team. As Figure 7.1 shows, level-3 writers often had 10 or more years invested in the company. The experience this team had with the company ranged from one to 17 years. Four level-3 writers participated in this study. Six of the team members were men and six were women.

Additional roles on the team include an editor whose primary responsibilities include developing and maintaining a style guide for all documentation, and a technical writing manager whose job it is to manage all team members, maintain professional and career development, and oversee the quality of the documentation processes and work. Table 7.1 shows a more detailed description of the technical writing roles on the team.

Findings

As mentioned in the "Materials and Methods" section, I reviewed all survey and interview data for information about how technical writers' roles were affected when the team moved to a single-source documentation system. I particularly looked for themes among data from surveys and interviews about work practices, benefits, and challenges. Four themes emerged: (1) streamlined work practices; (2) insights about the benefits of single sourcing; (3) challenges related to managing document review processes; and (4) individual and collaborative autonomy.

Streamlined Work Practices

When asked about their process for writing a topic, all participants described a similar approach, which generally included two broad phases: (1) an informal

Table 7.1 Technical Writing Team Roles and Responsibilities

Role	Responsibilities
Technical writing manager:	Manages all technical writing personnel and process development for the "topics" project. Develops procedures for technical writers and all management aspects of the project. Guides career and professional development of team members.
Level 1 technical writer:	Responsible for writing topics for single-source documentation.
Level 2 technical writer:	Responsible for writing topics for single-source documentation, but may also be asked to work on additional projects as they arise.
Level 3 technical writer:	Responsible for playing a lead role on teams and projects, defining and adjusting processes, training and helping less senior writers, and writing topics for single-source documentation.
Editor:	Responsible for generating and maintaining a style guide for use by all technical writers on the team; developing standards on a symbols development team for visual-only documentation; and working with other company divisions on matters of style and consistency.

phase in which the writer prepared the topic and received informal feedback of different kinds from two or more stakeholders; and (2) a formal review phase in which the writer prepared a polished draft of the topic and delivered it to a minimum of four reviewers for feedback. These stages are graphically depicted in Figure 7.2.

Informal Phase

As shown in Figure 7.2, the "informal phase" consisted of two main activities: preparation/research and informal feedback from at least two stakeholders. Writers began, always, by researching their topic. Many said that they reviewed research journal articles made available by the company, looked at legacy content manuals, or talked with subject-matter experts (SMEs) to learn about their topic. This stage was critical to every writer's process. Then, the writer generated an informal draft of a basic version of the topic. This provided a baseline for all variations of the topic. At this stage, typically the SME and a "lead writer" (often a senior writer who serves as developmental editor) reviewed the baseline topic for technical accuracy, conformance to writing guidelines, clarity, and ambiguity.

On receiving the feedback, the writer would make necessary adjustments, and the style editor would look for style and usage problems. After receiving feedback

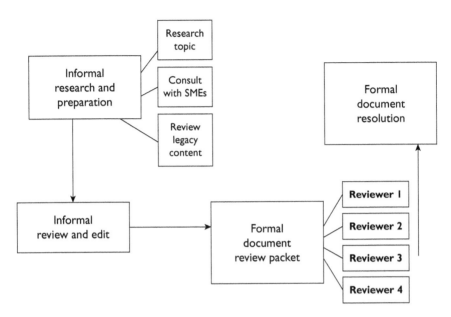

Figure 7.2 Document Review Process

on the informal review, the writer would continue to develop the topic, creating additional versions of the topic (called "instances") for various models of the device. Depending on the topic, a writer may need to create several versions or instances. (According to the writing manager, topics had five instances on average. Some had as few as two instances; others as many as 26 instances.) The writer could at this stage continue to receive feedback, particularly with subject-matter experts about the accuracy of the information. Several writers described the consultative nature of this informal phase of the process:

> [My steps included:] work up a draft with SME. I outline and annotate the outline; things go back and forth; I flesh out the content.

> One or two drafts to SMEs to get content right has worked for some folks.

> Informal phase [includes] developmental, informal, work with SME, [and regulatory affairs].

> The SME and writer figure out the context and also clinical situation [for the topic at hand]. The technical writer and SME collaborate from the beginning. There is a cycle of sending PDF files back and forth until they reach agreement.

Formal Phase

In the formal phase of the process, writers would compile a formal review packet that included, according to one writer, "[a] topic spec, cover letter, baseline topic, related instances (labeled), and a form for [reviewers] to check off." For record-keeping purposes, signatures were needed from all reviewers that they had indeed reviewed the topic. A formal review packet was created for each topic and would go out to a minimum of four reviewers including: (1) an SME/systems engineer; (2) a representative from regulatory affairs; (3) a representative from marketing; (4) an "independent reviewer" who was often another technical writer. Depending on the topic, a fifth reviewer might be involved. Reviewers were invited to provide any and all feedback on the topics. Feedback typically addressed content, style, and illustrations of each topic.

Once feedback was returned, writers were obligated to record and respond to every comment offered by reviewers. Many writers referred to this part of the formal process as "comment resolution." Most writers documented feedback either by creating a list or a table in which each comment was referenced and described, and in which decisions about each comment were articulated. Writers were obligated to send the comment resolution report and a revision of the topic back to reviewers, who were asked to review the topic again and sign approval on the final topic.

During interviews, almost all members of this technical writing team commented that the team followed a similar writing process. In fact, to help writers manage the single-source approach, the writing manager of this group wrote a detailed manual on the "topics" writing process. It was no surprise that writers described their processes so similarly; the two-part process was featured in this manual, along with suggestions for creating basic topics and their possible versions. The manager also shared that the "topics" approach was discussed in many team meetings, as were suggestions for creating it. And, as one writer explained, when the process was first introduced there were "regular weekly topic writer meetings—this was really done right. During these meetings we would all talk about issues of topic writing and how to resolve them. Writing topics is a very intense process." As I learned through the team meetings I attended, the team understood and accepted the topics process. Generally, team members' attitudes were very positive about the overall approach to single sourcing that they adopted.

Despite the commonalities, there were some individual differences among writers, especially in the ways each writer negotiated their relationships with reviewers. For example, some writers commented on the importance of establishing a good relationship with their SME, not only for the purposes of reviewing but also for creating content. One writer mentioned "[the] SME prefers to talk over the phone or face-to-face," whereas another writer said that communicating with the SME exclusively via e-mail was successful. Writers also varied in the ways they communicated with reviewers, although nearly all writers mentioned that they were somewhat dependent on reviewer schedules and preferences. Although some reviewers communicated comments via e-mail, some communicated via phone, and some, unfortunately, wrote comments using pencil or pen on paper that were difficult to read. Said one writer: "A lot of 'social engineering' is required in this process to 'develop relationships' and negotiate changes and convince reviewers when some of their changes are not needed."

Insights About the Benefits of Single Sourcing

Interview and survey data yielded many comments about the potential benefits of single sourcing for this technical writing team. A common benefit articulated by every writer was the acknowledgment that single sourcing yielded a better quality product than the 500-page legacy manuals the company had written in the past. Comments about quality included the following:

> In the end, it turned out to be effective. Good level of quality.
>
> It is a very thorough review. Doing books in the old way was inconsistent. Smaller things [topics] get through. People are taking the time and that's great. It is good quality.
>
> The end product would be better. . . . [this process] added so much value and quality to the process.

When prompted to explain, writers commented that the thorough review process required a closer examination of everything they were writing. As one writer said, "I like that [the review process] is thorough, structured, deliberate. Each word that goes into the topic is discussed and debated. Larger manuals do not have as much scrutiny." In fact, it was the combination of shorter textual modules with a more thorough review process that worked well in this case. That is, reviewing one topic at a time was easier than reviewing an entire 500-page manual, and suddenly reviewers paid closer attention. Said one writer, "Reviewers have been more responsive. Some reviewers like it and have more investment in the topics. We have been able to change things about documentation." Said another writer about the modular approach, "It's more efficient for reviewers."

Other insights about benefits of the single-source approach had to do with the personal growth of writers, both in terms of writing skill and their job satisfaction. Many writers suggested that the work was interesting and enriching to them personally. Said one writer, "It's a blessing to do new writing. I've been working with boilerplate and legacy content for most of my career."

Another writer commented that the shorter topics help writers get more invested: "I like working with the size of the document and length of the document—it lets writers get into the topic. I get a sense of ownership of content." Another writer mentioned that, for each topic she writes, she has a deeper appreciation for the content: "For me I think I have learned a lot; taking one topic has helped me learn a lot." Others mentioned investment in the emerging process created by this team: " I like being a part of a team for one end product; I like to learn about technical aspects; I like the challenge of learning new things; I like being part of a group that is exploring new ways of writing."

The writing manager of this team made similar comments about growth and professional development: "It is enjoyable to watch people grow and see them learning how something works. Seeing them go through the writing cycle is satisfying. Seeing process improvement is satisfying. Personal growth is satisfying." Another writer commented that as the process progressed, the team approach progressed, too: "We are collectively getting a better understanding of how to write these. We have a better sense of content and style."

Challenges Related to Managing Document Review Processes

Although writers articulated many benefits, they also articulated a number of frustrations. According to survey responses, two challenges rose to the top for this technical writing team. The first was coordinating feedback from multiple reviewers. All participants in the study agreed this was a most difficult challenge. One writer provided comments on the survey:

> Interpreting reviewer comments [is a challenge]. Some comments are very cryptic and it is difficult to guess their meaning or intent without contacting the reviewer again—extra time and effort I'd rather not have to expend. If a reviewer handwrites his or her comments, sometimes the comments are illegible—once again, requires more follow-up with the reviewer. Also, when a reviewer decides not to follow the process, it results in extra time and effort for me.

Interview data supported this finding from the survey. When asked about the ways in which reviewers returned comments, writers reported that reviewers provided feedback in multiple formats. Some reviewers used PDF commenting software to provide electronic feedback, which was very convenient for writers. Some reviewers listed comments within the text of an e-mail message and sent them to writers that way. Some reviewers handwrote comments directly onto a print copy of the draft; others handwrote and faxed copies of their handwritten copies.

Unfortunately, in those situations writers often reported that the comments were illegible and of no use. No matter how comments were delivered, writers were responsible for logging each and every comment that reviewers made and documenting their decision about which comments were incorporated into a revision of the topic. This responsibility was a difficult challenge for writers. As one writer put it: "It is difficult to know what I'm free to change and not change. I feel like I have to ask permission. I can't draw on my experiences as a writer. They [reviewers] look for expertise but they have the final say."

Beyond the very time-consuming activity of finding and logging all comments, writers also had to address situations when reviewers did not agree with what had been written. When such disagreements occurred, writers had ongoing consultations with reviewers. One writer said, "Working with people to reach an agreement is a challenge. A lot of people take stands on things and disagree." Although writers acknowledged that in some areas, reviewers had more technical expertise, some writers were adamant that they had greater expertise—such as in matters of editing and style—and they resented when reviewers provided feedback on such issues. Said one writer: "I don't accept stylistic comments!"

When such comments arose from reviewers, this writer documented them and simply overruled them. Another writer explained the difficulty of stylistic comments: "Stylistic and editorial comments are outside of [reviewer] purview. Edits can be incorrect." Other writers were not quite as bold. One writer suggested that he incorporated stylistic suggestions from reviewers unless they conflicted with the style guide that the team had created. In those cases, this writer would debate issues with reviewers.

In fact, debates with reviewers were described in detail by many writers, which supported the idea that balancing feedback from multiple reviewers was the most difficult challenge of the topic-based authoring environment. In one case, a writer described a debate that occurred over an illustration within a topic. The debate occurred not just with one reviewer but with many reviewers: was the illustration accurate? The writer and reviewers could not reach an agreement, so a meeting was called to discuss the matter in person. Another team writer described an additional story: "One time, an SME tried to rewrite everything." Perplexed by the situation, the writer reported: "I set up bi-weekly meetings with him and he would share comments. It worked so much better."

Debates about reviewer comments became such a challenge that the team designated a person on their technical writing team to address conflicts when they arose. If necessary, this person would call a meeting, in person, for the writer and all reviewers of a particular topic. Writers commented that such meetings were very helpful for everyone, and in fact, these meetings were a regular occurrence in the beginning of the transition (they became known as "Friday forums"). As the topics process became more familiar, both writers and reviewers began to feel more comfortable with the process and their responsibilities. When that happened, conflicts were more easily resolved via e-mail. Said one writer: "As time went on, everyone got better at resolving comments. Goals became clearer." A designated third party still mediated the disagreements, but most interactions occurred in writing via e-mail instead of in person.

However, coordinating feedback from multiple reviewers was only one of the challenges identified by writers. The other was the difficulty of scheduling the review process among multiple reviewers. Writers on the team overwhelmingly complained about the complexities of scheduling reviews. Said one writer, "There's always a delay." Many writers described the difficulties of trying to retrieve reviewer feedback, explaining that they had to work around vacations, illnesses, work-related travel, and overworked reviewer schedules that might interfere with retrieving feedback. They used words such as "inefficient" and "chaotic" to describe the scheduling hassles.

These scheduling difficulties, combined with the challenges of addressing disagreements about content, led the writing manager to declare that, for writers, the topics process "involves people managing" to a large extent. These challenges led to deeper reflections among technical writers about their roles. It seemed, for many, that their roles were changing from individual authors to collaborative project managers. This theme is explored in the next section.

Individual and Collaborative Authorship

A final theme that emerged from the data was a tension between individual and collaborative authorship. All writers on the team acknowledged their individual autonomy as writers, as well as how that autonomy was challenged in this topic-based authoring environment. For example, results from the survey indicated that writers saw themselves as both individual and collaborative authors. When asked to describe the ways they were involved in document review processes, 90 percent of all participants responded, "I individually write documents that will be reviewed by others," and 80 percent of all participants responded that "I collaboratively write documents that will be reviewed by others." Interviews helped to unpack these findings further.

According to one writer who helped design the topics process, there were hopes that the topics process would be "empowering to writers." He explained that as writers were assigned a topic, he hoped they would have ownership of that topic's content. "They need to feel like experts and that can be a motivator." He believed that "as writers feel more experienced, they will feel more empowered." Some writers agreed with the idea of individual investment and authorship. Said one writer, "I have to have a sense of ownership over [the topic I am assigned]." However, simultaneously, writers acknowledged the compromise implied by the single-sourcing system. The same writer observed: "But at the same time, you need to let go. You must conform to guidelines. It is more of a team effort and structured process. In the past, you owned the manual. Now it is a collective effort."

This acknowledgment of compromise and tension between individual and collaborative authorship was a dominant theme in interviews. Descriptions of collaboration were particularly apparent when writers discussed their relationships with reviewers. Said one writer of his relationship with an SME: "It is a 'co-author' partnership between me and the systems engineer." Said another writer: "This is a completely collaborative process. Things are done by building consensus, for better or worse here." The "worse" was articulated by some participants in terms of a loss of writer authority, which was intricately connected to the review process. Explained one writer, "Writers don't have much control over the final changes. Writers don't have authority."

Another writer agreed when she asked, "Where is my authority? Especially concerning stylistic matters? Reviewer comments trump technical writing authority." This frustration was voiced by only a few writers who participated in the study. Other writers acknowledged the struggle but handled the tension differently, such as by exerting authority over reviewer comments by simply "overruling" or "ignoring" stylistic comments, stating that reviewer comments about style "do not add value" to the documentation.

Nevertheless, it was apparent that all participants in this study felt the shift in their roles in this single-sourcing system to include more collaborative writing, especially because of the document review process. One writer described the shift as "painful but unavoidable." Despite the pain, writers also acknowledged the rewards of collaboration fostered by the document review process. Said one writer: "It is nice to have several reviewers dedicated to reviewing each topic." Another writer mentioned how the topics process required productive collaboration, not only among reviewers but among the technical writing team and style edits: "We have done an amazing job of writing with one voice throughout the topics documents."

Summary of Themes

Findings from the four themes indicated that writers felt both the joy and the pain of transitioning to a single-source documentation system. They acknowledged the joy in terms of improved documentation, deeper investment in their writing, and a sense of commitment and pride about being involved in such a process of change. They acknowledged the pain in terms of managing the document review process. Many writers expressed frustration regarding the "inefficiency" of the process, such as delays in retrieving reviewer feedback and the tedious nature of logging each and every reviewer comment, and they also expressed frustration over a perceived loss of authority regarding decisions over style and content.

As I reviewed all themes, it was clear that this group experienced both extremes of "radical change." To further illustrate these extremes, Table 7.2 shows one-word descriptions offered by technical writers to describe their team's single-source documentation and review process. These descriptions are in response to the interview question: "What adjectives come to mind when you think of 'topics' document review?"

Note the extremes that emerge here: "efficient" and "inefficient"; "structured" and "chaotic"; "straightforward" and "complex." A defining characteristic of this

Table 7.2 One-Word Descriptions of the Single-Sourcing Process

Challenging	Time-consuming	Inefficient
Interesting	Complex	Compliant
Painful	Excessive	Structured
Intense	Efficient	Labyrinthine
Frustrating	Lengthy	Straightforward
Chaotic	Logical flow	Hard work

team's experience was to see both the benefits and challenges associated with the transition to a single-source documentation system.

This shared understanding of benefits and challenges helped to shape the emerging roles of these technical writers. No more the lone writer working on a 500-page manual, technical writers in this team were now characterized by their abilities not only to write but also to work collaboratively with others and manage production and review schedules. With increased participation from reviewers, writers noticed an increase in their roles as negotiators of process, knowledge, and writing expertise. Their roles also changed by becoming more visible in the company, because they needed to interact with representatives from other departments on a regular basis. In a proactive fashion, the team took an active role in creating, managing, and shaping the single-source documentation system. In addition to experiencing radical change, they were creating radical change in their company.

Discussion and Conclusion

The themes that emerged from interview and survey data in many ways support the literature cited earlier in this study. Specifically, findings of this case study suggest that writers experienced changes such as a move toward a streamlined, structured writing process and collaboration with reviewers on the development of content. Rockley suggested that, indeed, technical writers moving to single sourcing may experience "significant changes" in terms of increased structure and collaborative authoring, and that those changes might cause discomfort (2003). In this case study, writers indicated that some changes were welcome, such as a structured writing process, whereas others were clearly more challenging, such as balancing more feedback from multiple reviewers.

An additional challenge for the writers in this study was the management of all aspects of document review, including coordinating schedules for review and discussing the topics with reviewers. Previous literature about single sourcing does not specifically address the dynamics of document review in the context of single sourcing. Certainly, many scholars have observed that single sourcing involves social tensions and conflict (see Bottitta and colleagues 2003). However, in this case study, the tensions and conflict were not produced by the transition to single sourcing but rather by the demands and complexities of document review.

The writers in this study found themselves, at times, in the nexus of conflict during the review and collaborative authoring process. In addition, the challenges of managing reviewer schedules and their idiosyncratic forms of feedback created tension that greatly impacted on the jobs of these technical writers. Thus, this case study illustrates that document review can be integral to single-source documentation, as well as a potential source of conflict and tension. Indeed, document review may serve as the catalyst for examining the collaborative nature of single-source documentation. Future studies might explore the relationship between single sourcing and document review in greater detail.

Although challenges of single sourcing certainly emerged from this case study, it is important to note that the writers simultaneously acknowledged the benefits of their single-sourcing approach and perhaps even recognized its power. Many writers reported that, despite the difficulties of managing multiple schedules and feedback, the process of writing in this way was stimulating and engaging and allowed them to become more fully immersed in the details of their writing. Many writers also reported satisfaction in being a part of a process of change. In these ways, we might consider the possibility that Carter's idea of "radical change" does not only provoke resistance.

The team in this case study took thoughtful—even methodical—steps toward creating and nurturing the transition to single sourcing. They experienced fully the growing pains this process required, but they emerged with a sense of pride about the outcome. In the words of one writer, "This is a success story, we hope. It is a huge accomplishment." Such attention and care given to single sourcing in this case illustrate a very positive way that technical writing teams can manage the organizational changes involved in transitioning to single-source documentation.

It is perhaps this point about management that contributes important implications about transitioning to single sourcing. We already know from this study and others that individual writers experience shifts in their work practices; we know that benefits and challenges emerge from the transition to single sourcing. However, the experience of this group of writers shows how a technical writing team can effectively manage these shifts and changes in a proactive fashion.

For example, the team held regular meetings throughout the transition to support writers and hear their concerns. They wrote a manual for the technical writing team that described the topics-based authoring environment that they had created, including their internal software. They created roles for team members to handle different parts of the process, including a "lead writer" who would examine first drafts of topics, a style editor, and a team member who would address conflicts between reviewers and writers. These efforts helped to manage the difficult transition to single sourcing, and they also strengthened team unity.

One effort that could be added to this transition is proactive training of those outside of the technical writing team, specifically document reviewers. Although many of the reviewers had conducted reviews before (such as on the 500-page manuals), the reviews of smaller "topics" invited more substantial feedback and greater involvement. It would be helpful to introduce reviewers to the "topics-based" process before they participate, and the training could also address issues such as deadlines and methods of feedback. Many participants on the team noted that such training would be helpful and might be the next step for the team.

Ultimately, this case study answers Carter's (2003) call to study how technical writers experience the transition to single-source documentation. Although it confirms some findings from previous literature, it shows that document review may be an important factor in understanding the collaborative role of technical writers in single-sourcing environments. As practitioners consider transitioning to single-source environments, they might consider how document review factors into their process and how to provide training for both writers and external reviewers. Training might provide an introduction to document review processes, preferred methods of feedback, technology used to support the writing process, and coordination of deadlines and schedules. By taking these steps, technical writers can take a proactive role in shaping their emerging roles as collaborators and project managers.

Appendix 7.1: Survey Questions

1. What is your job title?
2. In what ways are you involved in document review processes as a topics writer? (check all that apply)
   a. I individually write documents that will be reviewed by others
   b. I collaboratively write documents that will be reviewed by others
   c. I review documents for subject-matter accuracy
   d. I review documents for legal issues
   e. I review documents for foreign-language accuracy
   f. I review documents and provide "reader response"
   g. I review documents for style, grammar, or mechanics
   h. I manage and coordinate document review processes
   i. Other
3. What aspects of document review process(es) do you find most challenging? (check all that apply)
   a. Completing reviews on a schedule
   b. Archiving versions of documents in review
   c. Coordinating feedback from multiple reviewers
   d. Delivering and retrieving documents from reviewers
   e. Learning online tools for the review process
   f. Other
4. As a topics author, how do you like to deliver documents to your reviewers? (check all that apply)
   a. Distribute print copies to reviewers
   b. Distribute document to reviewers via e-mail attachment
   c. Distribute document in the text of an e-mail message
   d. Read document aloud to reviewers in a face-to-face meeting
   e. Post document on an internal groupware program
   f. Post document on an open-source Web site
   g. Post document on a blog
   h. Other
5. As a topics reviewer, how do you like to receive documents for review? (check all that apply)
   a. Receive print copies
   b. Receive document via e-mail attachment

    c.   Receive document in the text of an e-mail message

    d.   Hear document read in a face-to-face meeting

    e.   Retrieve document from an internal groupware program

    f.   View document on an open-source Web site

    g.   View document on a blog

    h.   I do not review "topics" or provide feedback as a reviewer

    i.   Other

6. As a topics author, what forms of written feedback from reviewers do you find most useful? (check all that apply)

    a.   Handwritten comments inserted directly in the text or margins of a print document

    b.   Handwritten comments in a separate note to the author

    c.   Electronic comments inserted directly in the text or margins of an electronic document

    d.   Electronic comments in a separate e-mail message to author

    e.   Electronic comments inserted as footnotes to an electronic document

    f.   Other

7. As a topics reviewer, how do you like to provide written feedback? (check all that apply)

    a.   Handwritten comments inserted directly in the text or margins of a print document

    b.   Handwritten comments in a separate note to the author

    c.   Electronic comments inserted directly in the text or margins of an electronic document

    d.   Electronic comments in a separate e-mail message to author

    e.   Electronic comments inserted as footnotes to an electronic document

    f.   I do not review other topics or provide feedback in a review process

    g.   Other

8. As a topics author, what kinds of meetings do you like to have with reviewers to discuss their feedback? (check all that apply)

    a.   Face-to-face meeting to discuss comments

    b.   Telephone meeting to discuss comments

    c.   Internet text chat to discuss comments

    d.   Multimedia conference with video, audio, and text capabilities to discuss comments and view text

    e.   E-mail exchange to discuss comments

    f.   Other

9. As a topics reviewer, what kinds of meetings do you like to have with authors to explain your feedback? (check all that apply)

    a.   Face-to-face meeting to discuss comments

    b.   Telephone meeting to discuss comments

    c.   Internet text chat to discuss comments

    d.   Multimedia conference with video, audio, and text capabilities to discuss comments and view text

e. E-mail exchange to discuss comments
f. I do not review other topics or provide feedback in a review process
g. Other

10. Which of the following technologies would you deem essential to your document review process? (check all that apply)
    a. Blog
    b. E-mail
    c. Fax
    d. Groupware
    e. Instant messenger
    f. Internet chat room
    g. Multimedia conference software
    h. Pen and paper
    i. Telephone
    j. Voice over the Internet Protocol (VoIP)
    k. Wiki
    l. Word-processing program
    m. Other

11. What specific online tools or programs would you recommend for document review? (open-ended)

12. What is one thing you enjoy about document review as a topics author? (open-ended)

13. What is one thing you find particularly challenging about document review as a topics author? (open-ended)

14. What is your gender?
    a. Female
    b. Male

15. How many years have you been using a computer?
    a. 1–5 years
    b. 6–10 years
    c. 11–15 years
    d. 16–20 years
    e. 21–25 years
    f. 26–30 years
    g. 30 years or more

Appendix 7.2: Interview Questions

1. What is your name?
2. What is your job title?
3. What does your job involve?
4. How many years have you been at this company?
5. When did you start topics writing?
6. Describe your topics review process in terms of delivering the topic to reviewers.

7. What technologies do you use for delivery?
8. Are you responsible for delivering topics?
9. Do reviewers have different delivery preferences?
10. Describe your topics review process in terms of receiving feedback from reviewers.
11. Do reviewers send feedback to you directly?
12. What technologies do they use to give you feedback?
13. What kind of feedback do they give you?
14. How do you document feedback from multiple reviewers?
15. Describe your topics review process in terms of making decisions about which feedback is integrated.
16. Do you follow up with reviewers about their feedback?
17. What do you do when feedback from reviewers conflicts?
18. How do you decide which feedback gets integrated?
19. What technologies do you use to help make decisions about which feedback is integrated?
20. How many reviewers do you work with for each topic?
21. Do reviewers change from topic to topic or are they static?
22. How long does the document review process for a single topic take?
23. What variables influence the length of a review process for topics?
24. How many review processes do you manage at the same time?
25. What adjectives come to mind when you think of "topics" document review?

Acknowledgment

*This article was originally published in *Technical Communication* (2008), 55: 343–56.

References

Ament, K. 2003. *Single Sourcing: Building Modular Documentation.* Norwich, NY: William Andrew Publishing.

Bottitta, J., A. P. Idoura, and L. Pappas. 2003. "Moving to Single Sourcing: Managing the Effects of Organizational Changes." *Technical Communication* 50: 355–70.

Carter, L. 2003. "The Implications of Single Sourcing for Writers and Writing." *Technical Communication* 50: 317–20.

Conklin, J. 2007. "From the Structure of Text to the Dynamic of Teams: The Changing Nature of Technical Communication Practice." *Technical Communication* 54: 210–31.

FDA: Center for Devices and Radiological Health. 1997. "Design Control Guidance for Medical Device Manufacturers." U.S. Food and Drug Administration Web site, U.S. Department of Health and Human Services. www.fda.gov/cdrh/comp/designgd.pdf.

Ford, J. D., and R. K. Mott. 2007. "The Convergence of Technical Communication and Information Architecture: Creating Single-Source Objects for Contemporary Media." *Technical Communication* 54: 333–42.

Katz, S. 1998. *The Dynamics of Writing Review: Opportunities for Growth and Change in the Workplace.* Stamford, CT: Ablex.

Kleimann, S. D. 1991. "The Complexity of Workplace Review." *Technical Communication* 38: 520–26.

MacNealy, M. S. 1999. *Strategies for Empirical Research in Writing.* Boston, MA: Allyn and Bacon.

Moore, P., and M. Kreth. 2005. "From Wordsmith to Communication Strategist: Heresthetic and Political Maneuvering in Technical Communication." *Technical Communication* 52: 302–22.

Paradis, J., D. Dobrin, and R. Miller. 1985. "Writing at Exxon, ITD: Notes on the Writing Environment of an R&D Organization." In *Writing in Nonacademic Settings,* ed. L. Odell and D. Goswami. New York, NY: Guilford Press.

Rockley, A. 2003. "Single Sourcing: It's About People, Not Just Technology." *Technical Communication* 50: 350–54.

Williams, J. D. 2003. "The Implications of Single Sourcing for Technical Communication." *Technical Communication* 50: 317–20.

Chapter 8

A Call for New Courses to Train Scientists as Effective Communicators in Contemporary Government and Business Settings*

Scott A. Mogull

Editors' Introduction

Mogull uses qualitative methods to look at the pedagogical approaches of eight experienced scientific communication instructors. He places this study in the context of the urgent need in the United States to enhance the capacity of the scientific community to communicate with the public. Though his methods and findings are presented in admirable detail, what is especially notable about Mogull's article is his clear intent to formulate recommendations for improving the curricula used for teaching scientific communication in universities. Moreover, he advocates for a mentoring role for technical communicators in both academic and industrial settings, and suggests that technical communication programs could be leveraged to provide communication education for scientists.

Some have claimed that the generation of new knowledge can serve at least three functions (Amara, Ouimet, and Landry 2004; Beyer 1997; Ginsberg and colleagues 2007; Kramer, Cole, and Leithwood 2004). Instrumental use of knowledge is the application of knowledge to the day-to-day work of practitioners; conceptual use involves arriving at a better understanding of situations or issues; and symbolic use involves the justification or legitimizing of a point of view or course of action. In using his research findings to create a case for involving technical communicators in the education of scientists, Mogull is engaging in the symbolic use of knowledge.

Like several other authors represented in this collection, and consistent with the recommendations of many standard texts on qualitative research, Mogull goes to some lengths to justify the qualitative methods he selected. He recruited eight participants through purposive sampling, which involves selecting participants according to predetermined criteria. He made use of a grounded theory research design, basing his work in part on Charmaz's (2006) more recent interpretation of grounded theory, but unlike most grounded theory studies, he did not seek to construct a theory to describe the teaching approaches of his study sample. He gathered data through semi-structured interviews, coded the resulting data, and used a member-checking technique to assure the quality of his findings.

It is worth noting that Mogull's member-checking procedure gave participants the opportunity not merely to verify the data that they provided through their interviews, but also to make revisions. Although none made substantive changes (perhaps because he did not provide them with transcripts until ten to twelve months after the interviews were completed), this strategy has become a common practice in many qualitative studies, and reflects the belief that one purpose of qualitative methods is to bring to light different voices and interpretations. Readers may recall the recommendation by Blakeslee and colleagues in their article in this collection (Chapter 2) that researchers should include in their reports the different and even contradictory views and interpretations proffered by research participants.

Mogull also provides a detailed description of his research participants. His study was not dependent on the contextual characteristics of a specific research site, but rather focused on the teaching approaches of his eight participants. To allow readers to assess the potential for transferring his findings to other scientific communication courses, he quite properly focused his "thick description" on participants rather than on places.

Mogull's article contributes to our understanding of the following themes raised by Giammano:

- What is our future role in organizations?
- How do we contribute to innovation?
- How should we be educating future practitioners?
- Where do we go from here?
- Does technical communication matter?

References

Amara, N., M. Ouimet, and R. Landry. 2004. "New Evidence on Instrumental, Conceptual, and Symbolic Utilization of University Research in Government Agencies." *Science Communication* 26: 75–106.

Beyer, J. M. 1997. "Research Utilization: Bridging a Cultural Gap Between Communities." *Journal of Management Inquiry* 6: 17–22.

Charmaz, K. C. 2006. *Constructing Grounded Theory: A Practical Guide Through Qualitative Analysis.* New York, NY: Sage Publications.

Ginsberg, L. R., S. Lewis, L. Zackheim, and A. Casebeer. 2007. "Revisiting Interaction in Knowledge Translation." *Implementation Science* 2: 34.

Kramer, D. M., D. C. Cole, and K. Leithwood. 2004. "Doing Knowledge Transfer: Engaging Management and Labor with Research on Employee Health and Safety." *Bulletin of Science, Technology & Society* 24: 316–30.

Introduction

The need to train scientists to be effective communicators in government and business settings has reached the awareness of the highest levels of our political system. In the "Scientific Communications Act of 2007," Congress proposed funding $10M a year specifically for the communication training of scientists. Major considerations mentioned in this bill are as follows:

1. With the increasing presence of science and technology in public policy issues, a greater national effort needs to be made to train scientists to engage in the public dialog.
2. Graduate training programs in science and engineering often lack opportunities for students to develop communications skills that will enable them to effectively explain technical topics to nonscientific audiences.
3. Providing training in communications skills development will ensure that United States-trained scientists are better prepared to engage in dialog on technical topics with policymakers and business leaders.
4. Given the enormous annual investment that the Federal Government makes in the US research enterprise, training scientists to interact with policymakers will improve accessibility to information and ensure that technical expertise is included in the public policy dialog.
5. Providing early career preparation for interaction with nonscientists will improve the ability of research scientists to engage in public/private partnerships to facilitate product development on the basis of US research discoveries.

(US Congress 2007, 2–3)

These issues addressed by Congress are consistent with concerns in our own field to prepare students to communicate technical information in the contemporary workplace (Hart and Conklin 2006). The Scientific Communications Act of 2007 proposes specific funding to train scientists to communicate for real world contexts. Through this bill, policy makers advocate the necessity of communication skills for the growth and success of a scientifically advanced society. The objectives specifically mentioned in the bill are to train scientists to interact with policy makers and business leaders, which has an important impact for technical communicators in academic, government, and business settings.

This study explores scientific communication courses in university settings by detailing the various approaches of eight experienced instructors of scientific communication through qualitative research methods. This research provides technical communicators with a descriptive analysis of the range of scientific communication training in various educational settings, thus serving as a foundation for the discussion of scientific communication courses in the academy. This research shows that, even when scientists have scientific communication training, preparation for contemporary needs may be quite varied—if those needs are addressed at all.

The majority of scientific communication training discussed in this study does not address many communication models (or genres) that are important for the majority of technical communicators working with scientists in professional settings. Recommendations are offered for new courses that address the formal scientific communication training of scientists in the academy. Hopefully, this paper will initiate a dialog between technical communication academics and practitioners to address the scientific communication skills necessary in industry that academic programs should teach.

Scientific Communication Education and Technical Communication

The literature suggests that current instruction in scientific communication at many universities does not adequately prepare scientists to communicate beyond standard academic models of scientific communication. However, any statement from current research must be qualified because primary literature describing scientific communication instruction at the university level is limited. The majority of the literature describes scientific communication courses based in science departments. The vast majority of articles are case studies written by instructors of science courses, which tend to fall into three general categories of instructional approaches. In the first category, instructors design their courses to simulate a full research investigation. Students are provided with a research question, which they investigate, gather data, and summarize in a traditional IMRAD (introduction, methods, results, and discussion) report and scientific presentation (Chaplin and colleagues 1998; Stukus and Lennox 1995).

In some cases, instructors incorporate limited exposure to other communication models—such as proposal or news writing (Henderson and Buising 2001). In the second category, students are exposed to scientific communication through reading and presenting primary literature (Houde 2000; Janick-Buckner 1997). In the third category, students are taught some scientific communication skills as a component of a science lecture course, so any scientific communication is secondary to science instruction (DebBurman 2002; Takao and Kelly 2003).

Scientific communication in government and business settings needs to be addressed. Thus, as I argue in this article, to achieve effective communication of scientific innovations in a scientifically advanced society, technical communicators in the academy and industry should collaborate in the mentoring of scientists' communication skills. The vision of this shared endeavor is discussed in the following section.

From the academic perspective, this paper argues that technical communication programs are ideally situated to provide the formal education to scientists. These programs are in a prime position to lead the educational objectives identified by Congress. As noted by scholars discussing the broader issue of science education reform, ideal education is an interdisciplinary approach in which academic departments collaborate according to their specialties. As stated by Zoller, effective science reform is "a systemic effort within which all components of the educational arena are involved and work together toward a common goal" (2000, 409). Furthermore, not only can academic technical communication programs prepare scientists to communicate for policy and business needs, but these programs can also cross-train technical communicators for a deeper understanding of scientific experimental design, data presentation, and data analysis.

The formal communication education of scientists directly impacts on technical communication practitioners.

Scholars in our field report that technical communication practitioners in organizations frequently coach and educate scientists on effective writing practices (Ford and colleagues 2004, 4). The demand on technical communication practitioners to mentor authors can be substantial. For example, through their observations of pharmaceutical product development and documentation practices, the researchers noted that "although there are some trained medical writers within most pharmaceutical companies, many of the authors are scientists and technicians first, and authors second" (Bernhardt and McCulley 2000, 23).These scientists "frequently had no formal training in scientific writing and rhetoric" and relied on the traditional scientific report, or the IMRAD format, as their sole model for writing (Bernhardt and McCulley 2000, 23). In many science-based companies, scientists write a significant amount of the documentation (Bernhardt and McCulley 2000). Many of these documents do not fit the IMRAD format, and thus it becomes the responsibility of technical communication practitioners to revise the documents and to educate the authors. Because technical communication practitioners have first-hand experience with scientific communication genres and they mentor scientists in government and business settings, practitioners should help to develop scientific communication courses for the preparation of scientists in contemporary settings.

Research Design

The primary objective of this study is to explore the instruction of scientific communication courses at the university level. Review of the literature showed limited descriptive research on this topic. The majority of publications that detail scientific communication instructional methods are case studies written by the instructors of such courses. Furthermore, the majority of these studies are published in science education journals and are written for instructors in science departments.

These published accounts have significant limitations to generalization. First, the communication training in academic science settings generally emphasizes the scientific communication skills needed to share scientific findings among the scientific community and lacks attention to other audiences and purposes. Second, most of the existing articles are case studies, which provide a single, isolated account of the instructional approach. Finally, most of these papers are written by the instructor of a course and therefore lack independent analysis.

A reasonable next step was to build on the existing literature and report on the instructional methods of a larger and more diverse sample. Qualitative research was selected as a method to gather rich, in-depth data without the potential bias imposed by a structured quantitative survey. Because this is a qualitative study, the data are descriptive; they are not intended to be a statistically valid sample of the population. Eight instructors of scientific communication courses were interviewed as a method for in-depth exploration into the subject (Charmaz 2006; Goubil-Gambrell 1992; Hughes and Hayhoe 2008). Interviews were conducted during the spring of 2007.

Methods and Theoretical Foundation

A preliminary search of scientific communication courses at academic institutions showed that they were generally located in either humanities (including English, technical communication, or writing departments) or science (which includes medical) departments. From these initial observations, a secondary research question emerged—to explore the similarities and differences of pedagogical approaches between humanities-based and science-based courses.

Participants were selected through purposeful sampling (Koerber and McMichael 2008). Initially, six instructors of scientific communication at the university level were identified by an Internet search for scientific communication courses. Three instructors were based in the humanities and three in science fields. Of the six instructors contacted, two agreed to participate. Of the first two participants, Participant G is based in a humanities department and Participant E is based in a science department (see Table 8.1; note that the letters assigned to participants are organized by level taught and not by order of participation).

To increase the sample size, additional participants were solicited via an e-mail request posted to the Association of Teachers of Technical Writing (ATTW) discussion list. The ATTW formed in 1973 to connect teachers of technical communication. In an e-mail to the author on March 10, 2008, list owner Sam Dragga revealed that the ATTW e-mail discussion list has slightly more than 1,000 active subscribers, nearly all of whom are educators. From a single ATTW e-mail request, 10 instructors of scientific communication courses responded. From these responses, six instructors met the study requirements as instructors of scientific communication courses at the university level and were successfully scheduled for telephone interviews within a month of the initial request. Because the ATTW e-mail distribution list has humanities origins, the majority of respondents were based in humanities fields. Participants A, B, D, and F are based in humanities departments, whereas Participants C and H are based in science departments (Table 8.1).

The research methods used in this study are from those of grounded theory. Grounded theory has evolved in somewhat divergent directions from its original construct in the late 1960s, and this study drew on some of the techniques from Charmaz's updated version of grounded theory (Strauss and Corbin 1998; Charmaz

2006). The original grounded theory, founded by Glaser and Strauss (1967) for sociology research, was a systematic process for qualitative research studies in which an exploratory approach, such as interviews, was used as a method of discovery. This study makes use of the grounded theory premise that ideas emerge from the data and that the researcher excludes any preconceived beliefs about the topic. This study also made use of the interviewing methods used in Charmaz's grounded theory, which are consistent with traditional interviewing techniques as described by Hughes and Hayhoe (2008). This study differs from conventional grounded theory in that it sought to explore and describe the main features of the teaching of scientific communication among a specific group of instructors, but it did not seek to construct a theory to account for these teaching practices.

Participants' interviews ranged from 30 to 75 minutes using a semi-structured format with open-ended questions. Interview questions provided an initial framework to the interview, but grounded theory allowed the interviewer flexibility to adapt to the participants' responses and explore statements in more detail. Moreover, grounded theory acknowledged the interviewer's role in the interview process and permitted modification of interview questions as different concepts emerged throughout the course of investigation (Charmaz 2006). All interviews were conducted on the telephone at the participants' convenience and digitally recorded. The recordings were transcribed verbatim and coded using verbal constructs (gerunds) to define the action from each statement (Charmaz 2006). These codes were categorized for comparison (Charmaz 2006).

For establishing the credibility of the findings, the results of this study were triangulated by member checking (Hughes and Hayhoe 2008). For member checking, each participant was sent an initial draft of the manuscript and requested to verify the results and interpretations proposed in the paper. Because member checking was requested 10–12 months after the initial interview, participants were also provided with a copy of the transcript from their interview. The majority of changes resulting from the member checking were slight modifications to the participants' personal description (specifically to be more precise or more general with respect to the institutional setting, depending on the personal preference of the individual). No substantive changes were made to the results or interpretation, which suggested that participants generally agreed with the findings of this study.

Participant Description

In this section, participants are described in detail. Comparison of participant differences is summarized in Table 8.1. The descriptions below detail that the participants have a significant amount of expertise in scientific communication, although the nature of this experience differs greatly.

Participant A teaches in the Writing Across the Curriculum Program at MIT, which is housed in the School of Humanities, Arts, and Social Sciences, but ties instruction to specific science classes. Participant A has been teaching a scientific communication course for five years. This course, which is associated with a

Table 8.1 Summary of Participant Differences

Participant	Level of scientific communication course	Primary department offering course (humanities or science)	Personal education (humanities or science)
A	Sophomore	Humanities	Humanities
B	Junior and senior	Humanities	Science/humanities
C	Junior and senior	Science	Science/humanities
D	Junior and senior	Humanities	Science
E	Senior	Science	Science
F	Senior and masters	Humanities	Humanities
G	Graduate	Science	Humanities
H	Doctoral and postdoctoral	Science	Humanities

sophomore biology laboratory course, meets eight times over the course of a semester. Participant A has a background in English and education.

Participant B teaches multiple sections of "Communication for Science and Research," which is offered by the English Department at North Carolina State University. Participant B has degrees in wildlife biology and English and has been a scientific editor. This participant has been teaching multiple sections of scientific communication for six years. The course is required by nearly all science disciplines at the institution.

Participant C is a scientific writing instructor located in the Department of Ecology and Evolutionary Biology at the University of Colorado. The scientific writing course is for junior and senior students in the department. Participant C has degrees in technical communication and applied ecology and has been teaching the course for eight years.

Participant D has an undergraduate degree in forestry and graduate degrees in natural resource management. Participant D has been an environmental writer for 40 years and currently teaches environmental writing in the English Department at the University of Houston Downtown.

Participant E is the director of a prominent viral research center in the University of California system. The participant is a professor of molecular biology and has previously been the chair of a science department. Approximately 20 years ago, Participant E began teaching a scientific writing course that was based in a science department and was required for incoming graduate students. Although the "faculty uniformly felt that this [scientific communication] course was one of the most important classes, it got eliminated [from the curriculum]." As a result of program reorganization, this class was no longer required of graduate students but began to be offered in two versions for undergraduate students. One class is for the undergraduates conducting scientific research and the other course is for students not performing any primary research.

Participant F teaches scientific writing in a writing department located in the California State University system. This course is designed for senior under-graduate students and master's students from science disciplines. Participant F has taught this course approximately eight times, has more than a decade of experience leading scientific writing training courses at pharmaceutical companies, and teaches an online scientific communications course for healthcare professionals in a department of regulatory affairs. Participant F has degrees in English literature and education with emphases in rhetoric and composition and has worked extensively as a writing consultant in the pharmaceutical industry.

Participant G has an undergraduate degree in English and a master's degree in journalism. Participant G was a science journalist for 25 years, writing for publications such as *Discover* magazine and Time Life Books. Participant G began teaching eight years ago at the University of Miami. This participant teaches undergraduate "Writing about Science" courses, which has students "translate" complex technical materials for general audiences, and a graduate scientific writing course offered through the school of medicine, which has students develop scientific papers and grant proposals.

Participant H teaches small, workshop-style courses for doctoral and post-doctoral science students who are writing articles for publication in peer-reviewed scientific journals. Participant H teaches primarily in the veterinary department at Utrecht University (a top European veterinary school based in The Netherlands) and at Wageningen University and Research Centre, and has a one-quarter appointment at the Institute of Social Studies. Participant H has previously taught courses at Copenhagen University, CODESRIA in Africa, Tilburg University, and Cambridge University, as well as having taught in industry. Participant H has been teaching scientific writing, primarily for veterinary students, for 20 years. Before teaching, Participant H was the editor of an international journal for nearly 15 years.

Scientific Communication Courses Design

This section begins with a review of each participant's scientific communication course. Each participant was asked to describe the major assignments and course objectives of his or her course. In the section below, the assignments are described along with the instructor's rationale. For a summary of these assignments, see Table 8.2.

Participant A: Sophomore Level

Participant A teaches scientific communication in a university Writing Across the Disciplines program. Unlike other scientific communication courses, this course is tied to a science course, specifically a sophomore biology laboratory course, and meets eight times over the course of a semester. Participant A describes the collaborative structure at the university: "integrated [instruction between departments] is key there, so the staffing is kind of a consultancy model. We're really a writing fellows model."

The overall goal of the scientific communication course is to teach the elements of the scientific paper. Participant A explains, "The instruction is really broken

Table 8.2 Summary of Major Assignments in Scientific Communication Courses

Participant	Level of scientific communication course	Major assignment(s)	Audience
A	Sophomore	IMRAD paper	Scientific
B	Junior and senior	IMRAD paper	Scientific
		"Micro" literature review	Scientific
		Proposal from an RFP (request for proposal)	Funding agency
		News article or informational brochure	General
		Poster	Scientific or general (decision made by student)
C	Junior and senior	Literature review	Scientific
		Proposal	Funding agency
		Web page	General
D	Junior and senior	White paper	Advocacy group CEO
		News article	General
		Informational description	General
E	Senior	IMRAD paper or scientific observation	Scientific
F	Senior and master's	IMRAD paper	Scientific
		Proposal	Funding agency
		Scientific poster	Scientific
G	Graduate	IMRAD paper	Scientific
H	Doctoral and postdoctoral	IMRAD paper	Scientific

down in terms of the elements of a research article. So it's to understand those elements. Intro, methods, results, discussion, conclusion, tables and figures, titles and abstracts." Participant A emphasizes to students that the data do not "speak for themselves"; rather, writing in science is a deliberate act of persuasion. The students learn to evaluate rhetorical situations, focus on purpose and audience, and make decisions from these situations. Participant A states:

> We really try to impress on students how to think rhetorically about the kinds of choices writers—scientific writers—are making, and why. So we read examples. They write examples. The main element of it is they write a scientific article.

Although students are enrolled in a biology laboratory, Participant A stated that the writing faculty did not feel that writing up predetermined laboratory results was a successful way to teach scientific communication. Rather, the students select one of several defined semester-length projects. They can select to: (1) take data from key scientific studies (such as Gregor Mendel's pea plant data or Oswald Avery's data on DNA transformation) and write the data in a contemporary style; (2) conduct a textual/rhetorical analysis of four articles on a particular topic; or (3) write up their own data if they have done primary research. Participant A also attempts to convey the importance for a scientist of communication and writing and have the students learn to apply the skills to independent research—forming the foundational skills to fulfill future needs in school and as scientists.

Participant B: Junior and Senior Level

Among the participants, Participant B has possibly taught the most sections of scientific communication courses. Participant B teaches a junior- and senior-level, humanities-based scientific communication course that is required by nearly every science department at the institution. Participant B notes, "I do have quite an array of science majors in my class." The course is structured to analyze scientific communication genres initially and then transition to writing science communication genres.

For the first assignment, students do a rhetorical analysis of an IMRAD scientific paper. For this assignment, students review a published paper and evaluate it on the basis of the criteria for each section of the paper. If the paper does not strictly adhere to the IMRAD format, students must discuss how it differs and why. At a more detailed level, students also evaluate the style of the writing, such as verb tense and the use of active or passive voice.

The second assignment, dubbed a "micro-review," is a paper that is a short version of a mini-review. The paper requires students to research and cite eight to ten primary sources—synthesizing the information as is done in a scientific literature review. Participant B explains:

> They [students] may not use Web sites, they may not use books, they may not use anything else, and they must go to the primary literature and go to the peer-reviewed journals. They do struggle with this because many of them haven't had to do that kind of hardcore literature searching before. A lot of them have not been exposed to how to synthesize the information, how to identify trends in the literature, come to conclusions about what we know and don't know based on what exists.

Students select their topic from their field of study, which allows the students to familiarize themselves with the primary literature in their own field. Participant B states:

> I have them choose a journal that publishes review articles in their field of study to target as if they were submitting to that journal. So they have to print out the citation guidelines and the manuscript formatting guidelines, as well as a sample review article from that journal, and write their article as if they were submitting to that journal.

The diversity of topics and journal standards is a little overwhelming for the instructor, but Participant B believes that students get more value from researching and writing papers in their own discipline. A major theme in Participant B's course is audience analysis:

> I do have colleagues who . . . when they assign this kind of a paper, they'll choose a journal—something broad, like *Science* . . . or *Nature*, and have the students write as if they're writing to that journal. But I find that it actually benefits the students more if they can really see what people in their discipline are writing, [and] more specifically, to that more specific audience.

Furthermore, different science disciplines have different conventions. Participant B focuses on the following:

> How the material is presented in terms of the audience's expectations—in other words, the discipline. So, for example, in geology, research articles look very different from those articles in biochemistry or organic chemistry versus those in other disciplines. So, is it structured in a way that meets the needs of that particular discipline and the journals in that discipline?

For the third assignment, student teams find a Request for Proposal (RFP) from a specific funding agency and develop a proposal in teams. Halfway through the assignment, the teams provide a status report to the class in the form of a formal oral presentation. The students must address the topic and describe the funding agency (in terms of goals, values, and fit between the topic and the needs of the agency). Unlike the other participants, Participant B notes that it is crucial for students to learn to work together in groups. She states, "I happen to think that in the sciences it's crucial that students learn to work together because almost all science is collaborative. It's rare for a scientist to work in isolation."

Students can write their final two assignments for a more general audience. The fourth assignment asks students to write a magazine article that presents a scientific concept to a general audience or an informational brochure similar to what would be handed out at a pharmacy or doctor's office. This assignment not only addresses a general audience, but students must also address visuals and layout appropriately. The fifth and final assignment is a poster session, and students may select to design a poster for a scientific or general audience.

Participant C: Junior and Senior Level

Participant C teaches a course for scientific communication based in a science department. The department decided to administer its own writing courses, which were previously offered through the university writing program, to better meet the curricular requirements of the students. Although the focus of the interview was on the course this participant teaches, this course fits into a larger context of communication instruction offered by the department.

The course that Participant C teaches is for juniors and seniors in ecology and evolutionary biology. The primary objective of the course is to cover "the trends and the common practices of writing scientific material to other scientists and other intended audiences that scientists have to address." Participant C describes the approach as follows:

> Starting out actually with a review of published literature on research that the students show themselves to be interested in, moving from that into a research proposal. Building, of course, on what they have found from their literature review, and then after that moving out into communicating with a lay audience by designing and writing a rather basic Web page that's intended to be informative for a lay audience.

In this course, Participant C discusses the organization of a scientific paper, in conjunction with the literature review but does not have the students write an IMRAD paper. When asked to describe why the IMRAD paper is not a focus of this course, Participant C states:

> Because to a large degree it is dealt with in other classes within a department's curriculum. A lot of our . . . lab reports follow that classic introduction, methods, results, and discussion kind of format. In, of course, a very small scale way. On top of that, the actual development and production of publishable quality research reports is something that's dealt with in our graduate writing seminars. So it's—I guess that my choice of not dealing with it as a writing assignment in the undergraduate class is based primarily on the context in which this class is actually located.

Participant D: Junior and Senior Level

Participant D teaches an environmental writing course. Environmental writing is a topic that the other participants have not addressed yet seems to be an important component of the Scientific Communications Act of 2007. Participant D notes, "The whole environmental field is advocacy groups of various sorts—one side or another." In the environmental writing course, Participant D has each student select a controversial environmental topic and research the primary literature.

Each student has to select an organization with a vested interested in the topic, such as Greenpeace, and write a white paper for the executive officer of the organization that would be used as background information for speeches, press releases, and other communications. The participant requires that each claim presented in the white paper must be documented in peer-reviewed articles— selecting the most authoritative sources. For this assignment, Participant D tells the students the following:

> Everything you wrote, you had to be able to document in peer-reviewed articles or authoritative sources. . . . In other words, you can't just invent

things, which is the tendency in the environmental field. You must be able to back it up because you don't want your boss to be embarrassed by someone showing that it's complete fraud.

The second assignment is for students to write a second white paper on the same subject but for an organization with an opposite perspective on the issue. The third assignment in the course is to write an in-depth news article, such as an article that would appear in the *Wall Street Journal*, one that is fair and balanced. For the final assignment, students write the text for an exhibit that would be seen at a museum or similar attraction.

Participant E: Senior Level

Participant E, who is a scientist and is based in a science department, has a somewhat different approach to the scientific communication course. Participant E's course developed from a graduate-level scientific communication class. Although Participant E believes that having scientific communication at the graduate level is beneficial, the graduate scientific writing course was changed into two courses for advanced undergraduates—where it fit into the curriculum. One course is for students conducting primary research and the other is for students who are not.

In both courses, Participant E focuses on developing students' scientific thinking skills—clearly presenting arguments in a defined structure for a specific, scientific audience. Participant E uses writing as a tool to structure student thinking and a means to develop the critical thinking skills for enculturation of students into scientific disciplines. In the course for students conducting primary research, the main objective is to prepare students for the American Association for the Advancement of Science (AAAS) meeting, a large, general scientific conference.

In student writing, Participant E focuses extensively on experimental design, controls, and distinguishing the types of information (specifically distinguishing data from interpretations and conclusions). Participant E believes that clarity of thinking impacts on the writing. The students have freedom of the content, but are required to follow the IMRAD structure precisely. In fact, Participant E states:

> For example, if you just take a section of a results paragraph from a paper, the first, second, third, fourth, fifth sentence almost always follows a particular pattern. They [the students] just can't see what those patterns are when they first start reading them, but this is what I'm asking them to adhere to in terms of structure of writing.

Participant E takes a mentoring approach to the course—not expecting students to achieve the style correctly on their first attempt, and returns the papers to the students until they have mastered the style. In this scientific communication course, students practice both scientific writing and oral presentation of scientific arguments.

Although conducting research is a critical process for scientists, not all undergraduates have the opportunity to participate in primary research. For these students, Participant E teaches a separate scientific communication course. This course has two major assignments—the first is a personal experience of significance that leads to a scientific topic. This is modeled after the Sam Scudder essay, in which the observation of a fish is a pivotal scientific moment in Scudder's education and career. This essay models the same scientific thought process that distinguishes data from interpretations and conclusions.

For the second major assignment, students write a formal, structured science essay from primary literature. Consistent with the research-based scientific communication course, Participant E emphasizes the distinction between observations and interpretations, and facilitates students' development of scientific arguments.

Participant F: Senior and Master's Level

Participant F teaches scientific communication to senior and master's students in scientific fields. The course is offered through a writing department. Participant F both focuses on a rhetorical analysis of the scientific paper and presents a range of other communication models. Participant F describes the objectives of the course as follows:

> The course goals and objectives of that class are to practice writing as research activities . . . analyze effective and ineffective written and visual workplace communication as focusing specifically on academic scientific writing.

In this class, Participant F takes a rhetorical approach to evaluating effective and ineffective written and visual communication—with a focus on scientific papers, grant proposals, scientific posters, and the oral discussions that accompany poster presentations. Most students enrolled in this course are involved in primary research and focus on presenting their own data. Students not involved in research do alternative assignments, such as an analysis of a published scientific paper.

Participant F provides a thorough overview of scientific communication models. This participant emphasizes "written, oral, and visual scientific communication; structuring and designing information; obviously clear and concise writing; [and] making professional and ethical decisions." A major focus of the course is on how to develop a "good scientific story" rather than presenting all the data.

Participant G: Graduate Level

Participant G began teaching a scientific writing course based in an English department. For the undergraduate course, Participant G developed a science writing course in which students take complex scientific material and simplify the information. Participant G stated, "Having undergraduate students, who are not yet fully in the discipline, writing about complex materials in a kind of translation mode was cognitively pretty challenging for them."

After teaching the science writing course for several years, Participant G began offering scientific communication seminars for graduate students, postdoctoral students, and faculty. Participant G stated, "A guy at . . . [the] office of research . . . had this perception that their postdocs, their faculty, and various people were not up to speed with their writing and his chief concern was money." The seminars led to the development of a formal scientific communication course offered through the interdisciplinary biological sciences program. Participant G currently teaches scientific communication for graduate biomedical and medical students who have participated in several laboratory rotations.

The scientific communication course emphasizes scientific papers. Participant G states, "This course is strictly for scientific communication with an emphasis on papers in all their guises, short reports and letters, and various other things. And then we do a little bit of grant writing and we do a little bit of graphical stuff, but the focus is really on papers." Participant G summarizes the course objectives as follows:

> Our goal and aim is to get these students who are not terribly well versed in the history of their fields and the history of rhetoric in their field familiar with the modes and norms of scientific communication. And so the idea is not only to work with them on individual pieces of writing, but also to raise their understanding and awareness of the important role that communication plays in science.

Specifically, Participant G states, "the major product is not the scientific paper as a whole," but rather to learn rhetorical analysis and to practice writing each section of the paper. Participant G states:

> What I try to do is show them how each section has a format and what the expectations of each section are from the higher order and lower order. Everything from what verb tense you use and what voice you use in the materials, the method section, to what does the materials section do.

Participant G states that having students write an entire scientific paper is valuable only with advanced students who are doing research and need to report on and write about their own research. Participant G believes that having mock data for a complete IMRAD paper is not particularly useful and would be difficult to design for the entire class. In fact, Participant G states that graduate students do not even fully understand the details of what their classmates are doing because students have different areas of expertise at the graduate level.

Participant G notes the diverse capabilities of science students' communication skills. "They basically haven't done any writing—the bulk of them—because they specialize in science." Participant G works individually with each student to improve his or her own writing. "The particular aims are to take their writing from wherever they are, their scientific writing, and move it up a couple of notches, and for every student that's going to be somewhat different." These issues tend to be universal to science students, but such issues can be compounded with language barriers for international science students. At this institution, the scientific communication course includes many foreign students—including Chinese, Korean, Spanish, and various other European students. Although some students understand English only as a second language, other programs at the university support these non-native English speaker issues.

Participant H: Doctoral and Postdoctoral Level

Participant H teaches a scientific writing course to small groups of doctoral and postdoctoral science students at European universities. Unlike other participants, Participant H provides scientific communication instruction for scientists who are writing up their data for publication in scientific journals.

Conceptually, Participant H separates the scientific writing process by distinguishing between developing the scientific argument and writing the paper. "The first half [of the course] is completely about planning a paper and figuring out what you want to say and actually working on a first draft, although it probably won't be finished in that period of time." In this part of the course, Participant H includes a series of workshops focusing on the content of each section of the paper, helping the students determine the content to include in each section of their papers. The students study the underlying patterns of the scientific paper down to the paragraph and sentence level.

During the second part of the course, students focus on writing sections of their paper. Participant H works individually with students to structure their own writing, which is individualized, because for many students, English may not be their first language. However, Participant H clarifies that this "is not teaching an English course." The emphasis is on the structure—rather than on the mechanics or scientific content. Participant H states:

> I'm not looking for people to be able to do things that are absolutely grammatically perfect and that sort of thing because we can always find somebody who can help with that sort of thing . . . I want a strong structure in there. I want it to be reader friendly. For the papers that are written in my class, I can't assess the scientific content. That's not my job. They either have supervisors, or they have a promoter, or they have colleagues who do that. But I'm looking for something that has a reasonably strong structure and also that it is readable for an intelligent person.

The final product of this course is a paper for publication, with each student developing his or her own approach to the scientific writing process. Rarely do students complete a paper during the 10-week course, so Participant H continues working with the students until the paper is accepted for publication.

Analysis of Scientific Communication Courses

The data gathered in this study (Table 8.2) provide a descriptive characterization of scientific communication courses in the academy—from sophomore to postdoctoral levels, in both humanities and science departments. Although the results of this study were intended to be descriptive and not statistically significant, the data suggested that the majority of scientific communication courses included in this inquiry focused on academic communication models, specifically the IMRAD paper, in which scientists share information among the scientific community.

When scientific communication genres other than the IMRAD paper were introduced in the curriculum, the majority of these assignments emphasized the communication needs of scientists in academic settings (for example, proposals and scientific posters). A few instructors introduced news articles as a genre to practice writing to a non-scientific audience— although, retrospectively, the need for scientists to write news articles has been questioned. Only one assignment, the white paper, was a project that would be commonplace in government and business.

This study showed that, at least in the sample studied, most scientific communication courses did not prepare students to communicate scientifically in contexts other than academic settings. (This might be partly because of the search for "scientific communication" courses, which one participant defined as scientist-to-scientist communication. However, other participants had more of a general definition of "scientific communication," with an emphasis on the content being communicated.)

The implication of this research is that many universities may need to develop new courses specifically designed to meet the goals of the Scientific Communications Act of 2007, in which students pursuing a career in science are trained to communicate technical information effectively within government and business settings. This finding represents an important opportunity for technical communication practitioners and academics to collaboratively define the objectives of scientific communication courses. Practitioners are called on to identify the communication skills and genres needed by scientists in contemporary government and business settings. The academic programs can then provide courses around those needs.

An important point for consideration is that scientific communication courses should not merely be a repurposing of existing technical communication courses. Participant G emphasized this point by stating, "I think it is very different— scientific writing versus technical writing—they [aren't] the same thing." To be effective, technical communicators need to identify the specific needs of scientists in government and business settings. The next step for this research would be collaborative discussions between practitioners and academics in which both groups discuss the communication needs of scientists. For example, what documents or genres do scientists create to share information among their scientific team—within the organization (for example, documents to management or to transfer technology to other teams), with government regulatory agencies, or with potential customers? Through collaborative efforts, technical communication practitioners and academics can serve an important role in the initial and continuing enhancement of scientists' communication skills.

The assignments in Table 8.2 illustrate how this collaboration might impact on educational objectives. Take, for example, the news article mentioned as an assignment by Participants B and D. This assignment is one of the few assignments listed that addresses a general audience. However, a practitioner might be able to provide feedback by saying that scientists in their organizations do not usually write news articles, but the scientists may contribute to writing a press release. The documents produced in scientific communication courses could be directly reflective of the communication skills needed in contemporary workplace settings.

Technical communication academics holding a critical place within universities could provide this professional training across science departments and throughout the entire institution. To continue with the press release example, technical communication academics could leverage the vast array of resources and business activities that take place in higher education, to create a real-world assignment. Academics could have students interview science faculty, possibly overlapping with student laboratory rotations at the graduate level, and the students could write a press release for the academic institution.

A recurring point mentioned by several participants was that they had little guidance in the development of the scientific communication course. In one of the more candid statements, Participant G mentioned, "They [the department] brought me in to teach undergraduate scientific writing . . . the curriculum was poorly developed and they sort of said, 'Do what you want.'" On the basis of this research, the most influential variable in the development of a course seemed to be the individual experience of the instructor. Many of the instructors emphasized the scientific communication contexts that had been prominent in their own careers. Participant D, with 40 years of experience as an environmental writer, focused instruction on controversy in environmental writing and simulates professional assignments, such as the white paper. By contrast, participants with a strong scientific background in the academy, such as Participant E (who is the director of a viral research center), focused entirely on scientific communication contexts in academic settings.

Another factor that apparently influenced the design of scientific communication courses was the unique context of each class. Scientific communication courses have a broad range in the level of students—which, determined here, had more of an impact on course design than location in a humanities or science discipline. Participants in the interviews taught courses designed for students at sophomore, junior, senior, graduate, and postdoctoral levels. The level of scientific communication instruction was an important consideration for the instructors because scientific knowledge developed considerably during this time frame.

At the sophomore level, students transition from having essentially no knowledge of primary research data, having been most familiar with the communication style that they find in their academic textbooks. As students advance into the graduate levels, they develop proficiency for reading the primary literature and reach the point where they have their own research data. Therefore, scientific communication skills taught too early in a student's education tend to lack significant context for student comprehension. This may be, in part, the rationale that Congress used to specify that professional scientific communication is needed at the graduate level.

The interviews with the participants in this study help to provide further anecdotal support for location of such programs within the academy. Participant C noted that undergraduates at the junior and senior levels have difficulty understanding the primary scientific literature and still need to learn more about research methods to make this area clearer for them. Participant D, who also teaches juniors and seniors, agrees, stating that one must "make sure that they [the students] don't get in over their heads on the science part and the terminology." It is possible that students' own understanding of the information, and comprehension of the primary literature, may limit their ability to present information clearly.

From a more humanities-based perspective, the undergraduates may not be cognitively ready to grasp professional communication skills. For example, an undergraduate's mental model of textbook scientific communication may conflict with communication goals, such as audience analysis, because science undergraduates may mistake the textbook style for expert communication. In fact, textbook writing may be closer to writing science for a general audience. According to Participant C, science students have difficulty realizing that conventions in each model are actually products of the audience's needs and communication objectives. Students can acquire and apply the stylistic standards of a communication model in weeks, but understanding the rationale behind the genre is considerably challenging for the students. Participant C states, "It [the communication style] seems to be very recognizable and rather quickly understandable. But figuring out the actual purposes behind the genres—that's something that they have some considerable challenge with." Participant C notes that students may have difficulty adapting their styles to different contexts.

In contrast to Participants C and D, who teach juniors and seniors, Participant F, who teaches seniors and master's students, feels that the students leave the class with a good understanding of how to present scientific information to different audiences, stating, "They [the students] really understand it at the end."

The research data were evaluated for potential differences between humanities-based and science-based scientific communication courses. Although the sample size in this study was small, the results do not seem to indicate any clear influence on curriculum that results from this difference in location. Instructors in both kinds of departments mentioned similar course objectives. Participants B, C, D, and E provide a useful comparison because they teach nearly the same level of course, and two courses are humanities based (B and D), whereas two (C and E) are science based (Table 8.1). These instructors generally described similar approaches. As Participant C explained about the class:

It's designed to get people to not only look at how to write a particular genre in science, but why they are being called on to write those genres and to think about the strategies that they can use to make those genres successful—and actually getting scientists and science majors in general to think about the potential audiences for the information, for the knowledge that they're generating through their scientific work. That is something that my entire department feels is very important for science curricula in this day and age.

One of the reasons why departmental location had less influence than expected could be attributed to the confounding variable of the instructor's own educational background. Note that all four instructors compared for the junior/senior level courses (Participants B, C, D, and E) had science degrees, and that B and C have degrees in both science and humanities. However, instructors with a high degree of science experience, such as Participant E, emphasized the clarity of the scientific argument in the course. Participant E stated:

It's a way to structure their [student] thinking, to make it coherent, and to introduce formally what it is to do critical thinking the way scientists do it, and that has to do with the structure of scientific writing—the separation of observation from interpretation (the results and the discussion concept), because almost always, students come into the class in a . . . mode of thinking and communication that mixes those two, so when they see data, they tell you the interpretation. The exercises we have, and so forth, is to formally structure this mental separation, this rigorous separation, and that is the basis of their defense, because this is what I think it means. Here's the data, here are the controls for the data. That's basically how you defend your science, but it's not an inherent skill that they have, and it's not something that they've had developed, certainly not in writing classes, and certainly not in humanities classes.

By contrast, instructors who had more training in humanities emphasized the communication context—in particular more audience analysis—and, although the logic of the argument was important, several humanities-based instructors mentioned that they could not evaluate the technical details of the argument. The following is a representative comment from one of the humanities-based instructors:

One of my big concerns with my course is making sure that students leave my course understanding that they have to know their audience in every specific context, and they have to give the audience what the audience wants. It's not about how they necessarily want to say something or present it; it's more how the audience wants to have it presented.

An interesting point emphasized by all scientific communication instructors based in humanities departments, or those with a humanities education, is that they do not feel that good writing instructors need to be content experts. Rather, as Participant B states, "So what a good writing teacher has to have is not that great depth of understanding of the material, but a good teacher has to have the ability to read the material and not be intimidated by it."

However, as illustrated by Participant E's comment (above), scientists do not seem to separate the scientific validity of their argument from the communication style. This may be why Participant E had reservations about humanities-based scientific communication instruction. The ideal situation may be more of a collaborative model, such as the description provided by Participant A. Participant A teaches in the Writing Across the Disciplines Program, a humanities-based program that is tied to a science program. This might provide a skeleton framework for possible innovation at graduate-level science education.

Conclusions

As described in the Scientific Communications Act of 2007, Congress has indicated that it is important for scientists to communicate with policy makers and business leaders in a scientifically advanced society. Because the majority of scientific communication training described does not address many of the communication genres that are required of scientists in professional settings, this study suggests that new courses are needed to focus on the communication skills of scientists in contemporary government and business settings.

Technical communication programs are ideally situated within the academy to provide this training. However, course development should not be a mere repurposing of other technical communication courses. Rather, practitioners and academics should work together to identify the unique needs of scientists to develop these courses. Hopefully, this article will initiate discussion between practitioners and academics to address these needs.

Acknowledgment

*This article was originally published in *Technical Communication* (2008), 55: 357–69.

References

Bernhardt, S. A., and G. A. McCulley. 2000. "Knowledge Management and Pharmaceutical Development Teams." *Technical Communication—IEEE Transactions on Professional Communication* 43: 22–34.

Chaplin, S. B., J. M. Manske, and J. L. Cruise. 1998. "Introducing Freshmen to Investigative Research—A Course for Biology Majors at Minnesota's University of St. Thomas." *Journal of College Science Teaching* 27: 347–50.

Charmaz, K. C. 2006. *Constructing Grounded Theory: A Practical Guide Through Qualitative Analysis*. New York, NY: Sage Publications.

DebBurman, S. K. 2002. "Learning How Scientists Work: Experiential Research Projects to Promote Cell Biology Learning and Scientific Process Skills." *Cell Biology Education* 1: 154–72.

Dragga, S. 2008. Electronic mail conversation with author, March 10.

Ford, J. D., S. A. Bernhardt, and G. Cuppan. 2004. "From Medical Writer to Communication Specialist: Expanding Roles and Contributions in Pharmaceutical Organizations." *AMWA Journal* 19: 4–11.

Glaser, B., and A. Strauss. 1967. *The Discovery of Grounded Theory*. Chicago, IL: Aldine Publishing.

Goubil-Gambrell, P. 1992. "A Practitioner's Guide to Research Methods." *Technical communication* 34: 582–91.

Hart, H., and J. Conklin. 2006. "Toward a Meaningful Model for Technical Communication." *Technical Communication* 53: 395–415.

Henderson, L., and C. Buising. 2001. "A Research-Based Molecular Biology Laboratory: Turning Novice Researchers into Practicing Scientists." *Journal of College Science Teaching* 30: 322–27.

Houde, A. 2000. "Student Symposia on Primary Research Articles." *Journal of College Science Teaching* 30: 184–87.

Hughes, M. A., and G. F. Hayhoe. 2008. *A Research Primer for Technical Communication*. New York, NY: Lawrence Erlbaum Publishers.

Janick-Buckner, D. 1997. "Getting Undergraduates to Critically Read and Discuss Primary Literature." *Journal of College Science Teaching* 27: 29–32.

Koerber, A., and L. McMichael. 2008. "Qualitative Sampling Methods: A Primer for Technical Communicators." *Journal of Business and Technical Communication* 22: 454–73.

Strauss, A., and J. Corbin. 1998. *Basics of Qualitative Research: Techniques and Procedures for Developing Grounded Theory*, 2nd edn. Thousand Oaks, CA: Sage.

Stukus, P., and J. E. Lennox. 1995. "Use of an Investigative Semester-Length Laboratory Project in an Introductory Microbiology Course." *Journal of College Science Teaching* 25: 135–39.

Takao, A. Y., and G. J. Kelly. 2003. "Assessment of Evidence in University Students' Scientific Writing." *Science & Education* 12: 341–63.

U. S. Congress. 2007. H.R. 1453–110th Congress. 2007: Scientific Communications Act of 2007. GovTrack.us. http://www.govtrack.us.

Zoller, U. 2000. "Teaching Tomorrow's College Science Courses—Are We Getting It Right? Preparing Students to Become Informed and Responsible Participants in the Decision-Making Process." *Journal of College Science Teaching* 29: 409–14.

Chapter 9

Proceeding with Caution

A Case Study of Engineering
Professionals Reading White Papers*

Russell Willerton

Editors' Introduction

We selected Russell Willerton's "Proceeding with Caution: A Case Study of Engineering Professionals Reading White Papers" for inclusion in this collection because it provides an excellent example of a case study, but also because, of all the genres of technical communication, the white paper has probably received less attention than any other by researchers and theorists. Nevertheless, it is an important means by which high-tech companies publicize and sell their products and services, and thus seems to be a very worthwhile subject.

As Willerton notes, despite the fact that technical communication as a discipline constantly admonishes practitioners to consider their audiences, there have been relatively few studies of workplace reading activities. In his study, he examined five consultants at an engineering firm to discover how they read white papers. His research design consisted of two parts: approximately 30 hours of observation of the five consultants, followed by discourse-based interviews in which he asked each one to choose a white paper, read it, and employ the "think-aloud" protocol to provide their thoughts about what they were reading. Since four of the five consultants chose to hold their interviews in their offices, Willerton also got a sense of the interruptions and distractions that were common in their reading of such documents.

The results of Willerton's study can be summarized as follows:

* why the consultants read white papers;
* how they find white papers;
* how they regard white papers;
* how they evaluate the source of the paper;
* how they evaluate the content;
* how they read white papers.

Unlike most of the chapters within this volume, then, Willerton's is more concerned with the consumption of technical communication artifacts than with their production. Still, it is possible to see several of Giammona's major themes play out in Willerton's study, albeit indirectly. The following are the most significant:

- What forces are affecting the field?
- How do technical communicators contribute to innovation?
- How should we be educating future practitioners?
- What technologies are impacting on us?
- Does technical communication matter?

We'd like to note the strength of Willerton's case study design because this method has sometimes been criticized as producing little more than a collection of anecdotes. The lengthy observation period that Willerton employed provided significant information and insights about the context of the five consultants' work in their companies. The discourse-based interviews presented the author with numerous details and specific examples of insights, reactions, and thought processes of the consultants with whom he worked.

Introduction

The number of white papers in high-tech industries continues to grow. As they are currently used, white papers represent an intersection of technical documentation with marketing communication. Companies offer their white papers to readers through many avenues, including their websites, click-through advertising, and paid placements in search engine results. Despite their prevalence in the marketplace, white papers are not prominent in technical communication literature. Only a few recent professional communication texts mention white papers at all, and the descriptions in these books do not reflect current practices. Sammons (2003) and Johnson-Sheehan (2005) focus on white papers as documents that provide an overview of background information on a topic; this is a longstanding definition of white papers, but it does not align with the type of white paper most prevalent in today's high-tech marketplace. Andrews (2000) mentions only briefly the combined persuasive and promotional purposes for which many, if not most, white papers are used today. While Markel (2010) shows the links between a white paper's information and an organization's persuasive purposes, Stelzner's book on writing white papers provides a thorough definition: "A white paper is a persuasive document that usually describes problems and how to solve them. . . . It takes the objective and educational approach of an article and weaves in persuasive corporate messages typically found in brochures" (2007, 2). In fact, many companies offer white papers in exchange for the reader's contact information, to develop sales leads (Weil 2001; Stelzner 2007).

White papers are often used to aid decision making. Because technology markets are so competitive and because investments in technology are so expensive, businesses must carefully consider available options. White papers allow vendors to describe their products and services to potential customers, and they allow potential customers to get the type of technical information they need— information that would not be available or would not be provided in appropriate technical detail by "traditional news sources" (Lange 2003b). White paper consultant Jonathan Kantor notes how white papers have shifted over recent decades to aid decision makers:

> It has gone from a purely IT perspective to a combined IT [and] C-level [senior executive] decision-maker perspective. As a result, the content, which used to be very highly technical, today has to be much more marketing-oriented in nature. Not too much marketing so it becomes fluff, but it has to clearly explain technical terminology; offer executive summaries, graphics of charts, tables, and bulleted information. This way, high-level business decision makers can read and understand it, and act on it.
>
> (Quoted in Lange 2003a)

Bitpipe, a syndicator of web-based technical content, has sponsored several surveys of its readers to document how executives and IT decision makers use white papers. One 2004 study with more than 300 respondents, conducted with Forbes.com through a third party, found that 63 percent had used white papers or case studies (stories about implementations of a system) when evaluating high-tech products and services; 78 percent had passed a white paper to a colleague; 68 percent had contacted a vendor after reading a white paper; 50 percent had used them to compare vendors and products; and 93 percent said that a good white paper enhances their view of the vendor who produced it (Forbes.com and Bitpipe 2004).

It is increasingly likely that technical communicators will read white papers, write white papers, or both (Willerton 2007). Thus, this article provides a case study of a specific technical communication experience; it also contributes to the technical communication literature by providing insight on current practices. It sheds light on the complexities of a dual-purpose, informing-and-persuading document for an audience that has learned to read with skepticism. Specifically, this case study examines these research questions in the context of a particular workplace: Why and how are white papers read by professionals at work? How do readers select white papers? How do readers evaluate white papers?

Theoretical Approaches to Professionals' Workplace Reading Practices

The field of technical communication encourages writers to address readers' needs, but few in-depth studies of working professionals' reading activities have been completed. Redish, in an oft-cited article, distinguishes between three types of reading: reading to learn, reading to do, and reading to learn to do (1989). This three-part framework provides writers with some helpful distinctions among reading tasks, but Redish's article focuses on the writing of instructions and tutorials. Most white paper readers strive to learn new technical information while reading, but companies provide white papers for the ultimate purpose of marketing their goods and services; readers must separate information from what Killingsworth and Gilbertson call *brochuremanship* (1992). Insights on how readers accomplish this may be found in applications of schema theory and in Friestad and Wright's (1994) "Persuasion Knowledge Model."

Schema Theory and the Reading of Technical Documents

For several decades, researchers have used schema theory to describe how humans process what they read. Kant saw schemas as mediating "the external world and internal mental structures; a schema was a lens that both shaped and was shaped by experience" (McVee and colleagues 2005). Psychologist Frederic C. Bartlett wrote in 1932 that a schema is "an active organization" of past experiences to which new information is compared (1932/1964, 201). Readers categorize information and attach it to schemas instantaneously (Duin 1988, 185). Blyler (1991) shows how schemas help readers infer, reason analogically, and learn. According to Taylor and Crocker (1981), a schema may be regarded as a structure comprising an individual's knowledge in a given domain (quoted in Areni and Cox 1994, 337). Redish likens schemas to mental models that aid understanding (1993, 27). Faris and Smeltzer point out that schema theory is often discussed "in machine terms such as slots, default values, input, and output" (1997, 9); information that fits goes into appropriate slots, and inferences are made for non-fitting information. Some propose that schemas are fixed, while others propose more flexible views (1997, 9). Kent writes that schemas "facilitate the communication process by serving as common ground between writer and reader" (1987, 248).

Leinhardt and Young (1996) developed two heuristic schemas to describe how historians read texts and evaluate their significance. While these schemas were developed outside the realm of professional communication, they provide insights into how professionals in specific fields may read documents. The first of Leinhardt and Young's schemas is the *identify* schema, by which the reader classifies, corroborates (questions the document's veracity), sources (considers the authorship and any biases inherent to the document), and contextualizes a document. The second schema, *interpret*, involves two readings: a *textual* read of the words in the text and the surface structure, and a *historical* read that invokes the reader's interpretive stance and notions of the purpose of history (Leinhardt and Young 1996, 446–49). A case study with three published historians supports Leinhardt and Young's two-schema model.

In a book chapter originally published in 1988 and recently republished electronically, Bazerman (2000) reports on his study of seven physicists, which he conducted to gain insight on how reading journal articles relates to scientific activity. (Another account of this study appears in Bazerman 1985.) Bazerman observed that the physicists' purposes for reading—they read to aid their individual research agendas—and their schemas of background knowledge were intertwined: each shaped the other (2000). Bazerman's research, discussed later, provides a model that may be applied to understand how schemas affect the reading of white papers.

The Persuasion Knowledge Model

Decades ago, marketing professor Peter Wright discussed the need to investigate consumers' intuitive theory about the "tactics that are used in the game of marketplace selling-and-buying" (1986, 1), and he called this theory the "schemer schema." Consumers invoke their own schemer schemas as they cope with persuasive messages. Wright calls the schemer schema a "control schema" that guides the consumer's responses to specific promotional messages (1986, 2). Friestad and Wright (1994) incorporate the schemer schema into a larger category called "persuasion knowledge"; their Persuasion Knowledge Model, or PKM, visually represents how that knowledge is used in response to persuasion attempts.

The PKM depicts two entities meeting in the context of a "persuasion episode," which may occur interpersonally or through an advertising medium. These entities may be visualized opposite one another with the persuasion episode linking them in the space between them. One entity is the agent; the other is the target or consumer. Each entity possesses topic knowledge (beliefs about the topic of the message) and persuasion knowledge (beliefs about how persuasion works). The target possesses some knowledge about the agent, and the agent possesses some knowledge about the target. The target's knowledge is mediated through persuasion coping behaviors, while the agent's knowledge channels through the persuasion attempt (Friestad and Wright 1994, 1–3).

While both the agent and the consumer possess persuasion knowledge, Friestad and Wright focus on the consumer's knowledge:

> Persuasion knowledge performs schema-like functions such as guiding con-
> sumers' attention to aspects of an advertising campaign or sales presentation,
> providing inferences about possible background conditions that caused the
> agent to construct the attempt in that way, generating predictions about the
> attempt's likely effects on people, and evaluating its overall competence.
> Furthermore, persuasion coping knowledge directs one's attention to one's own
> response goals and response options.
>
> (Friestad and Wright 1994, 3)

Friestad and Wright acknowledge the role of culture in shaping an individual's "commonsense" persuasion knowledge. In the PKM, a commonsense belief is one that is shared widely. This persuasion knowledge "grounds consumers' and marketers' understanding of the social psychological processes at work and provides the basis for empathetic, respectful, and efficient interactions between them" (Friestad and Wright 1994, 7). Over time, consumers will add to and modify their commonsense knowledge and their repertoires of coping tactics (Friestad and Wright 1994, 7, 12).

To identify more of consumers' persuasion knowledge in the PKM, Kirmani and Campbell (2004) investigated the persuasion knowledge of individuals who had face-to-face interactions with store sales staff. Their findings may shed some light on readers' experiences with reading marketing materials from a vendor. They write, "In the buyer–seller context, persuasion is central to the relationship. . . . Therefore, consumers expect both assistance and persuasion when they encounter marketing agents" (2004, 574). Kirmani and Campbell conducted three qualitative studies (two with interviews, one with an experiment) about consumers' face-to-face experiences with marketing agents. The authors identified two broad categories of consumer tactics for their interactions with sales agents: seeker strategies, in which consumers engaged salespeople in order to achieve their goals for finding out desired information and/or completing purchases, and sentry strategies, for "guarding against an agent perceived as impeding goal attainment" (Kirmani and Campbell 2004, 577). Kirmani and Campbell also identified two broad categories for consumers' relationships with the salespeople: cooperative and competitive. "Informants described cooperative relationships as 'helpful' and competitive relationships as 'pushy.' Cooperative relationships were typically associated with seeker strategies, and competitive relationships with sentry strategies" (Kirmani and Campbell 2004, 578). The influence of schemas and of persuasion knowledge on the processes involved with reading white papers will be discussed later in this article.

Case Study Site and Research Design

Telecommunications service providers have grown tremendously in recent years, with developments in cellular telephony, messaging, and Internet telephony leading to demand for new services. This industry is flush with new technologies and new ideas, with significant competition among vendors at all levels; in short, this is an industry in which people are likely to read white papers. The consultants at a telecommunications consulting firm that I will call Tel-Work meet researcher Mary Sue MacNealy's definition of a purposeful sample (1997a); as professionals who read white papers in the course of doing their work, these readers have the characteristics necessary to answer my research questions.

Case Study Theory

A case study provides one method of answering the main research questions in this article from the perspective of a specific group of people in a particular industry. As social scientist Robert Yin writes, the case study method is an effective research strategy for use when "a 'how' or 'why' question is being asked about a contemporary set of events over which the investigator has little or no control" (1994, 9). MacNealy points out that case studies are "a valuable research tool, especially for areas which haven't received much research attention" (1997a, 200).

Yet the case study method sometimes suffers from a negative stereotype. Yin writes, "The case study has long been stereotyped as a weak sibling among social science methods. Investigators who do case studies are regarded as having insufficient precision (that is, quantification), objectivity, and rigor" (1994, xiii). This negative stigma has not escaped the technical communications field. MacNealy bemoans "case studies" in technical communication that are merely stories of someone's experience and not results of a formal research process (1997b, 182).

Yin (1994) describes the best case studies as empirical and systematic, and categorizes case studies as explanatory, exploratory, or descriptive in nature. In descriptive studies, the researcher "gives up the goal of trying to control variables" and attempts to preserve the natural setting when collecting data (MacNealy 1997a, 44). Sensitivity to context is important because a case and its context are "one and the same" (Wells, Hirshberg, Lipton, and Oakes 2002, 346).

The case study method is a framework (Yin 1994) that may be employed in a variety of situations. Kathleen M. Eisenhardt writes, "Case studies typically combine data collection methods such as archives, interviews, questionnaires, and observations. The evidence may be qualitative (for example, words), quantitative (for example, numbers), or both" (2002, 9). In addition to providing descriptions, case studies can be used to test theory or generate theory (Yin, 1994); a researcher generally needs multiple cases to effectively generate theory about particular phenomena (Eisenhardt 2002). Case studies should not be used to generalize about populations (Yin 1994).

Case Study Site

Tel-Work offers a full range of consulting expertise to assist clients with all aspects of designing, building, and generating revenue from a telecommunications company. Tel-Work's clients tend to be small- to medium-sized telephone service providers that serve rural areas, some of whom will enter a market to compete with an established provider such as AT&T or Verizon. The engineering office John led was a stand-alone entity before it merged with another firm in the late 1990s to create Tel-Work. Tel-Work has additional offices in other cities focusing on financial affairs and on regulatory issues, as well as several field offices that support ongoing projects. Tel-Work has occasionally produced its own white papers. Shortly before I started my research, Tel-Work had produced a white paper on VoIP, or voice over Internet protocol. VoIP involves sending voice calls over the Internet as data packets, as opposed to the traditional system of establishing a circuit between two telephones over the public switch telephone network, or PSTN. VoIP allows consumers to avoid paying some charges, such as long-distance fees, and has the potential to help telephone service providers reduce their costs (Alliance Datacom 2004).

The members of my sample were selected by Tel-Work's president of engineering; I will call him John. (All names are pseudonyms.) Tel-Work's engineering group includes John, whose registration as a professional engineer in many states allows Tel-Work to consult widely; the senior vice-president; the directors of network planning and CAD (computer-aided design), of network engineering, of wireless engineering, and of outside plant; those who report to the directors and support their respective areas of work; project managers; and contract administrators and other support staff. John recommended that I work with the senior vice-president and the four directors who report to him. These five are not necessarily the only people at Tel-Work who read white papers, but they all had insight from their positions as managers, and they did represent all technological areas in which Tel-Work consults.

Tel-Work's Consultants

In this section, I will provide a brief description of each consultant participating in the study and his responsibilities (all are male).

- Roger is director of network planning and the CAD group. Months before we met, he had moved from a senior project management position in the outside plant group to his current position; he had spent nine years with Tel-Work, comprising his post-collegiate career. Roger had a bachelor's degree in industrial engineering.
- Andy was director of wireless telephony engineering. He had spent nine years with Tel-Work. He joined the company as a network administrator while in college and moved into wireless technology after graduating with a bachelor's degree in electrical engineering.
- Tim was in charge of facilities and outside plant (involving manual labor like digging trenches and pulling cable). Tim had worked at Tel-Work for seven consecutive years and nine years in total. Tim had previously served in the armed forces. He studied building trades, began his undergraduate work while in the armed forces, and was completing a bachelor's degree in business.
- Steve is the senior vice-president for engineering. He frequently interacted with clients and developed proposals and business plans for them, while also overseeing the work of the engineering group. He had joined the company 22 years previously as a technician and worked his way up to his executive position. He is an active member of professional groups serving Tel-Work's typical clients, including the Organization for the Promotion and Advancement of Small Telecommunications Companies (OPASTCO). Steve earned an Associate's degree in electrical engineering technology.
- Peter is director of network equipment. Peter's career with the company spanned 23 years. He had entered telecommunications after earning an Associate's degree and teaching electrical engineering technology at a technical school.

Case Study Design

This study is descriptive, as it attempts to preserve the natural setting I observed and to provide a "rich understanding" (MacNealy 1997a, 44–45) of how white papers are read in a specific place of work. Like many qualitative studies, this study is exploratory; it begins to investigate an activity that has not previously received much attention.

This study has two phases. In the first phase, I spent about 30 hours over three weeks observing the consultants' regular work *in situ* to understand the context of their work and their reasons for reading white papers. Following the plan approved in my initial meeting with the consultants, I conducted several observations of them in their offices, taking a seat and observing as unobtrusively as possible. I also observed an RFI (request for information) meeting that Tel-Work hosted about VoIP (voice over Internet protocol). In the engineering profession, it is common for a potential customer to send an RFI to a vendor to get answers to specific technical questions. In this case, interested clients paid to attend this two-day RFI meeting in Tel-Work's training facility, in which they could hear presentations from several vendors about VoIP products. Several Tel-Work consultants attended this meeting and interacted with their clients.

In the second phase, I conducted discourse-based interviews in which each consultant read a white paper of his choosing and provided a think-aloud protocol. Because my research questions were well specified, I developed a framework that prestructured (Miles and Huberman 1994, 83) the discourse-based interviews (Odell, Goswami, and Herrington 1983). My interview protocol included questions about determining a white paper's credibility, about dealing with a white paper's mix of persuasive and informational material, and about the frequency with which each consultant reads white papers. My case study design was approved by the Texas Tech University Institutional Review Board (IRB).

Research Exchange

Some types of field research may be characterized as participant observation. If an observer enters a new environment, he or she certainly becomes part of it to some degree; the observer's presence affects that environment and will help to shape the data collected (Lincoln and Guba 1985). A researcher's level of participation within an environment will vary according to the situation and the goals of the research. Sociologists Patricia A. Adler and Peter Adler (1987) provide a spectrum of three levels of researcher participation that I found helpful. This spectrum includes peripheral membership researchers (PMRs), active membership researchers (AMRs), and complete membership researchers (CMRs). Of these three, the PMR is the most detached from the group being studied. I was neither an employee of Tel-Work nor an expert in any of the areas of telecommunications in which Tel-Work conducts business, and thus I was "blocked from more central membership" (Adler and Adler 1987, 37). I was an outsider, an academic, and an invited guest. I was an observer and not an active participant in the firm's activities. However, because I had negotiated a "research exchange" with the firm, I had limited, peripheral membership in the group.

The research exchange (Adler and Adler 1987, 40) provided a way to reciprocally benefit Tel-Work to acknowledge the privilege of conducting research there. John explained that he would like a report detailing how his consultants get, use, and store information. He explained that knowledge management is a great challenge for Tel-Work, because the company is geographically distributed and because what he called "silos of expertise" had developed over time. Being able to effectively use and share knowledge is part of Tel-Work's "value proposition" as a true full-service telecommunications consulting firm, he continued. Tel-Work had recently lost three key employees, including a vice-president of engineering and a director of network design engineering. I later learned that this loss of expertise and personnel led Tel-Work to decline some potential consulting engagements.

Collecting and Interpreting Case Study Data

Methods of Observation and Data Collection

When I observed the consultants working in their offices, I recorded my observations in a loose-leaf notebook. When making field notes, I used categories endorsed by Doheny-Farina and Odell (1985): observations of events; interpretations of events; and methodological notes about future research activities to pursue. I regularly looked over my field notes after leaving Tel-Work, clarifying details I had observed and identifying interpretations to investigate further.

To avoid disrupting the consultants' billable work with the recorded white paper reading sessions, I scheduled time with each consultant. Before these reading sessions, which are discourse-based interviews, each consultant signed a consent form. During each session, I used my IRB-approved interview protocol as a prestructured framework for obtaining comparable data from each consultant. I used a video camera to record three of the five sessions and used a microcassette recorder for the other two. I later dubbed the audio from the videotaped interviews onto microcassettes and had transcriptions made. I also took notes by hand during each reading session. Because each session involved the same core questions, I then grouped the consultants' responses to each question. I placed responses to extemporaneous questions and comments under the core questions that related most closely.

I allowed each consultant to choose where the recorded white paper reading session would occur, knowing that the place could impact on the activity of reading. One consultant chose a conference room, but the rest chose their offices. Through conducting interviews in their offices, I got a sense of the intrusions that these readers face: e-mail alerts; telephone calls from colleagues, customers, and family; and visits from colleagues all became part of the reading sessions in consultants' offices.

Data Analysis

Following Lincoln and Guba (1985), I sought to use member checking to ensure the accuracy of my interpretations and to give my informants the opportunity to participate further, if they chose to. One of the primary results of my research was my development of the "schema as landscape," which will be discussed later. I discussed it in person with Andy after watching him read a white paper, because I recorded his reading session last and was developing the idea at the time. To solicit feedback from the consultants, I e-mailed each one a copy of the summary report of my observations that I provided to John, the president of engineering. However, none of the consultants responded to my e-mails. I attribute this to the significant increase in business Tel-Work experienced during the summer of my observations, and to the heavy workload the consultants bear. Andy's experience is probably typical of the other consultants. I happened to meet him weeks after I had sent out my report. Andy thanked me for sending the report and said he enjoyed reading it. He said he didn't get to finish it, however, because the telephone rang when he was on the second page (of four), and he hadn't found time to pick up the report again.

Results

Why Consultants at Tel-Work Read White Papers

Engineering consultants at Tel-Work read white papers in order to learn about new technologies and about new applications of existing technologies. They usually seek out white papers upon hearing of a trend in the industry or after being prompted by a question from a client. As Roger, the network planning and design director, put it, "They are coming to us because they don't know and we should know." Peter, the network equipment director, said that clients can call at any time, and that they expect "clear, concise, and informative" responses to their questions. In the first phase of this study, I observed Peter working in his office for about an hour and a half, in which he fielded several phone calls from clients. The consultants said that white papers can help them find answers to technical questions.

Another reason the consultants read white papers is that their industry changes constantly. Several commented that the industry tends to run in a cycle of around six months. Both Roger and Steve, the vice-president, emphasized that keeping informed of trends is part of the value Tel-Work provides for its clients. Peter commented that white papers provide technical information in a timely fashion; they are better than books for providing lots of information in a short period of time. For example, he said that he used to keep a book of tele-communications terms on his desk, but now the pace of change had outstripped that book's usefulness.

Two consultants mentioned the possibility of using ideas from white papers to generate additional business with clients. Peter pointed to a list of benefits in the white paper he was reading, and said those benefits could be shown to a client. As Tim, the outside plant expert, pointed out, sometimes consultants have to tell clients what questions to ask in order to upgrade their equipment or to take their businesses in new directions; a white paper can be a springboard for additional consulting work.

Tim also mentioned the possibility of using a white paper for support in decision making, especially if the client disagrees with the consultant. However, he could not cite a specific example of this from his previous experience—and as several consultants pointed out, Tel-Work makes recommendations, but the clients make final decisions. Additionally, clients are not the only people to make decisions about telephone networks. Local zoning boards and other entities have tremendous influence over whether networks can be built or expanded, and many people on these boards lack specific expertise in telecommunications. Consultants could conceivably use white papers to inform these individuals about specific issues.

White papers are also valuable because they provide quick access to information and analysis that the consultants would otherwise have to find on their own time; they help consultants stay focused on billable activities. As Roger explained it, every hour spent in non-billable research costs the company two hours: the opportunity cost of one hour that could have been spent on billable activity, and the overhead cost of one hour spent on the non-billable research.

How Consultants at Tel-Work Find White Papers

Consultants at Tel-Work usually use the Web to find white papers. Four of the five consultants said they usually enter key words into a search engine, such as Google or Yahoo!, to find a white paper. Sometimes they go to industry-specific websites or to portals such as bitpipe.com or lightreading.com. They also receive white papers via e-mail. Peter said he doesn't often search for white papers because he receives so many through direct e-mail from vendors, industry research committees, and other sources. Other white papers are forwarded by colleagues; every consultant mentioned receiving information from the company's director of corporate communications, Gretchen.

Tel-Work's Corporate Communications Director

Because Gretchen's name came up whenever I asked the consultants about their sources of technical information, I interviewed her over the telephone. Gretchen lived in a distant state and worked by telecommuting. She reported to the Tel-Work headquarters, in still another city, but worked with people throughout the company. Her official duties included writing internal and external newsletters, managing advertising and public relations for the company, planning programs for clients, and working for clients directly on a billable basis. Unofficially, she provided what she called an "informal clipping service" for Tel-Work. This activity was not a part of her formal job description; she monitored many sources of information in order to keep herself informed about the industry, and she simply shared the information to be helpful. She said some Tel-Work employees had cancelled subscriptions to industry news sources because they got similar or better information from her for free. Most of her information came from industry press and her many personal contacts.

The consultants clearly valued the service Gretchen provided—to the point that at least one mistakenly believed that Gretchen's clipping service was an official part of her job at Tel-Work. Gretchen told me that when she searches for information, she focuses on industry trends more than technological trends (although the two are certainly closely related). She is mainly concerned with industry news and regulatory developments; she did not necessarily look for white papers on technical subjects, although sometimes she found them and passed them along. Nevertheless, Gretchen is an important source of information for the consultants at Tel-Work. White papers are often shared through informal networks (EMarketer 2004), and this sharing helps the company that wrote the paper to gain credibility with readers throughout the network (Willerton 2007).

The White Papers Read by Consultants in Recorded Sessions

I had asked consultants to select white papers that were relevant to their work. The white papers they chose came from various points in the network of information sources they commonly used. Roger got his paper from a Web site for another consulting company; he found this site through an e-mail from Gretchen. Tim did a key word search on Yahoo!; Andy, the wireless telephony engineer, used the portal www.bitpipe.com. Peter read a paper that was e-mailed to him by a prominent vendor. Steve asked Roger to suggest a paper, and Roger gave him a copy of the paper he had read. (Steve had intended to search for a white paper on his own, but due to extenuating circumstances before our appointment he consulted Roger instead. Because white papers are often passed around within an organization, I considered his choice of a white paper as authentic as the others.) Each consultant indicated that he selected his white paper because it was about a hot topic in his area of work and because he thought the source was trustworthy. Only one consultant, Andy, was unfamiliar with the vendor behind his white paper; however, his previous experience with the portal hosting the paper had led him to try something from a new vendor.

How Consultants Regard White Papers

Consultants approach white papers with some skepticism, aware that white papers are used to promote companies, products, and services. Two consultants even said that white papers are "tainted" when they are produced by, or at the request of, a vendor. This quote from Peter, the network equipment expert, illustrates the limitations that several consultants noted concerning the information found in white papers.

> When it comes from a vendor, that's what you have to understand, it's not neutral. . . . Sometimes the vendors are pushing like a politician as he is

pushing for his own policies. And that's what sometimes the white papers are addressing—it's coming from a vendor's perspective. And so when you see the very good information—generalized information, but you also have to understand who wrote that paper, where it's coming from and the perspective may not necessarily be a total package. But if you know where it's coming from, you understand it and you accept it and you take it from that perspective. And so you read the next paper which may not necessarily agree with everything that's been said from that perspective. You evaluate them and you come to your own conclusions.

Roger, Tim, and Peter all mentioned that when they read white papers on a certain subject, they read papers by multiple vendors and compare the information they find. In general, the consultants do not consider a stand-alone white paper a sufficient source of information.

While consultants are wary of vendors' white papers, they still value them as a source of technical information. As Tim, the outside plant expert, stated, "You want to keep an open mind as much as possible, with a slight leaning toward the understanding that the information provided is going to attempt to sway you to buy a product. It may make a very valid point that this application is the best way to go, said product is the best way to go." Tim pointed out that usually more than one product exists that could perform a specific function. Thus, an application described in a white paper by Acme Company, which will favor Acme's products, could be developed with a product from Beta Company that might be cheaper, more durable, more easily configured, or otherwise different.

The consultants prefer to read white papers that cite outside sources, because the citations lend credibility and support further research. They also resist white papers describing "vaporware" that has not passed field trials. One reason they resist is their lack of time; the consultants don't want to spend time reading about a product that might never reach the marketplace. Andy screens for vaporware by looking for data from the testing of the product, and by looking for a timeline showing the projected time to market. Another reason, which Steve cited, is to protect Tel-Work's clients from investing in unproven, potentially unreliable technology.

Evaluating the Source of a White Paper

When I videotaped each consultant reading a white paper, only Andy, director of wireless engineering, read a white paper from a source he did not already know. The other consultants reported that they usually know something about the source of a white paper they read. While few white paper readers might know the writers of the papers they read, consultants may be expected to have a larger range of industry contacts than other professionals who tend to work in-house for their employers. The consultants reported that if they didn't know much about a white paper's source, they would use online information to profile the source or ask around the office to see what others' experiences with that source have been. Thus, the concept of agent knowledge in Friestad and Wright's (1994) PKM relates to these consultants' white paper reading habits.

The consultants are aware that each vendor will use white papers to put its products and services in the best possible light, so they read white papers critically. Previous experiences with vendors have taught the consultants what to look out for, shaping their persuasion knowledge. When I observed him in his office, wireless telephony director Andy mentioned that sometimes he accompanies clients to vendor presentations to "count the lies" and ensure that clients don't end up buying a system that lacks important elements or requires extensive upgrades. Even white papers by third parties are subject to critique, as outside plant director Tim pointed out. The consultants stated that papers focusing more on selling than explaining are likely to get tossed aside, as will papers that contain questionable information.

Evaluating the Content of White Papers

The consultants evaluated the credibility of each paper's information by continuously comparing it to their existing knowledge of the field. When reading white papers, the consultants reported comparing the *new* information they read with the *given* information they already possess. This given knowledge might be called their schema of background knowledge. While this schema likely helps them to understand what they read, the consultants also use their schema of background knowledge to judge whether the information is credible. White papers do not receive anonymous peer reviews like academic articles in physics; furthermore, the white paper readers in this study acknowledge that vendors carefully craft their white papers to market their products and services. So, white paper readers are probably more concerned with ascertaining a paper's credibility than the readers of physics articles in Bazerman's 2000 study.

For the consultants, comparing the information from a white paper with their schemas of background knowledge was a sentry strategy (Kirmani and Campbell 2004). As Andy, the director of wireless technology, said, he continuously refers to his background knowledge while he reads to see if any "BS flags" go up in his mind to indicate that the information might not be true:

> Very rarely do I get a paper where I don't know anything that they're going to talk about in the paper. I'll have at least some rudimentary understanding in the beginning of where they're going to go with the paper. And so they're making statements about things that I at least have some knowledge about. If they don't contradict anything that I'm aware of, then they kind of gain credibility. . . . In this [white paper], when they started out, they made several statements in the first section and they were all statements that I had seen before and understood and agreed with; the paper didn't have anything that set off a flag. So that kind of gained some credibility with me there. But I find

that if I read through it and . . . if I find something that just completely contradicts something that I believe, that raises a big flag at that moment.

The consultants shared the view that companies tend to use white papers to promote their products and services; they are always looking out for what Roger called "sales fluff"—exaggerated claims lacking technical substance. To paraphrase another statement from Andy, "There's a lot of smoke out there; I look for actual fires."

As they are screening information to ensure its plausibility, the consultants evaluate its usefulness; they want information that expands their technical knowledge and helps add value to the services they provide for their clients. Roger (the network planner) and Steve (the senior vice-president) read the same white paper, and each cited reasons why that paper was not particularly useful to him. Roger pointed out that the paper, which discussed potential applications for VoIP in commercial telecommunication networks, included national market data; because Tel-Work serves smaller, regional clients, these data were not especially useful. Steve said that the information was not useful because it was outdated.

For consultants at Tel-Work, white papers tend to have a short "shelf life" because they are generally seen as leading indicators of where technology might go. The general sense I got from these telecommunications consultants was that after about six months, the ideas from a white paper tend to either become part of the mainstream of technology or else they disappear. Additionally, once a white paper is about six months old, its technical information might have become outdated or simply incorrect. (This amount of time will vary from one industry to another. See Willerton 2007, 196.)

Methods of Reading White Papers

For the videotaped sessions, each consultant printed out a hard copy of his white paper. Most said that they will skim white papers that are online or in electronic format, and that they print out a document when they want to read it more closely. Each one read the paper's introduction and table of contents closely, and most said they would stop reading if a paper couldn't catch their interest or couldn't sustain it.

I was surprised that four of the five consultants read their white papers sequentially, from beginning to end. I had expected them to skim and flip through their papers as they looked for specific information, much as the physicists in Bazerman's 2000 study did. Tim did not read sequentially and displayed some reading habits I would expect to be more typical in a workplace situation. Tim went to the references first, to evaluate the document's credibility. Next he read the table of contents, followed by the summary and the introduction. Then he skimmed through the remainder of the paper by sections, but he did not read through his paper's extensive appendices (comprising about half of its 40 pages).

My observations in all five consultants' offices during the first phase of this study clearly showed that Tel-Work consultants have a lot of paperwork to deal with—contract documents, binder-size government loan applications for clients, large format CAD drawings, and so on. Thus, electronic copies of white papers (usually in PDF format) are convenient for consultants to store on their computers and to forward to co-workers or clients. And yet, it can be more difficult to skim through an online document or to browse and find specific details. Each consultant had a separate way of handling white papers; most reported a mix of paper and electronic copies.

Peter (network equipment director) and Andy (wireless telephony director) each separately described a Catch-22 with storing white papers: If they keep only electronic copies, they are likely to lose track of them or forget where they are stored. If they keep hard copies, they might be more likely to remember them in the short term; however, that contributes to office clutter and eventually makes printed white papers more cumbersome.

"Schema as landscape"

Each reader in this study has a view of the technological landscape that affects his reading of white papers. This view is selective—the individual narrows it or broadens it as he sees fit. The individual controls the depth of field, to borrow a photographic term. Ideas that are new, pressing, or otherwise deserving of attention are in the foreground; ideas that are older or foundational are in the background. New ideas are on the periphery; ideas that don't concern the reader are outside of his horizon, outside of the landscape's frame.

Primary elements of the Persuasion Knowledge Model are also found in this landscape: the consultants compared what they read against their existing subject knowledge; their persuasion knowledge was constantly engaged; in some cases, they had direct knowledge about the agents behind each white paper. The placement of the frame may change over time or in response to clients' needs. See Figure 9.1.

So, if a white paper has potentially useful content and checks out after comparison with the schema of background knowledge, the reader adds those ideas to his conceptual landscape. The schema as landscape also serves a gatekeeping function. If the white paper does not appear potentially useful or relevant, does not pass the background knowledge-schema credibility test, or does not pass the persuasion knowledge test because the sales pitch dominates, it does not register on the landscape horizon.

This landscape is similar in many respects to the "maps" developed by the physicists in Bazerman's study:

> [In] deciding whether to look further into an article, the reader is actually placing the article within his or her personal map or schema of the field. As in Steinberg's famous drawing of a New Yorker's map of the world, the items

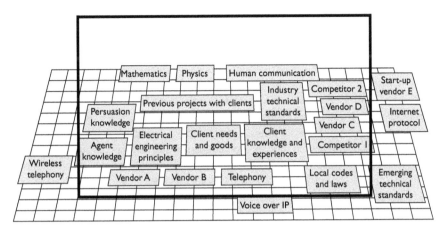

Figure 9.1 Sample Schema as Landscape Model

This sample "schema as landscape" model incorporates elements of the maps described by Bazerman (2000) and the PKM described by Friestad and Wright (1994), while reflecting the conditions in the marketplace where each consultant works.

are given various importance or size based on the observer's perspective— in this case the reader's own work. Some items look large and must be investigated in detail, whereas others seem to fall off the end of the known personal universe.

(Bazerman 2000, 242)

This schema as landscape is similar to the "purpose-driven schema" Bazerman describes in the physicists he observed. Bazerman says the schema holds "personally organized knowledge. This schema extends beyond textbook knowledge of accepted facts and theories to include dynamic knowledge about the discipline's current practices and projections of its future development" (2000, 239). However, the schema-as-landscape model is updated with the audience for high-tech white papers in mind. It more explicitly acknowledges that some items or concepts will remain outside the frame because they are either not useful or not relevant to the reader; items outside the frame are there intentionally and are not necessarily like the blank spaces outside a map representing *terra incognita*. (The schema as landscape focuses on the context for the activity of reading, but it does not comprehensively account for a reader's process of comprehending a particular text as does the Landscape model proposed by van den Broek and colleagues (2004).)

For each consultant, relevance is related to his assigned duties and his contacts with clients. For example, because Steve is an executive, he needs to maintain a landscape that is more expansive than those of his direct reports. His employees might construct landscapes that are not as expansive as Steve's, because their specific job duties require them to focus their attention on specific facets of the industry; however, their landscapes will perhaps show greater depth and detail. Bazerman even makes reference to the idea of a landscape with the physicists he studied: "[Scientific a]rticles, in their challenge to existing statements, foment new work. Plausible new methods, evidence, claims, and interpretations change the landscape against which the researcher plans and realizes research purposes" (2000, 249–50).

The consultants' purposes for reading white papers might sometimes be landscape-driven, as it was for the physicists in Bazerman's study; reading journal articles was a proactive, programmed, predictable activity driven by their need to produce their own research. For the consultants, however, the purpose in reading any document is often customer- or market-driven; customers make specific requests, or new technologies enter the market, and the consultants act on them. For these consultants, reading white papers can be a reactive, sporadic activity, and their needs for new knowledge are greatly influenced by specific questions from clients.

The landscape is subject to change over time, as is the schema of background knowledge. Both Tim and Andy specifically mentioned the need to keep an open mind and to be willing to change beliefs and opinions. With Bazerman's map metaphor (2000), changes in schemata require redrawing of maps. With the schema as landscape, the scene is recomposed. Thomas Kent comments on how schemata can change over time:

> Schemata mediate between the individual and the external world, and individuals continually modify them as they receive new information from the environment while simultaneously interpreting their environments through the schemata they already possess. Schemata, therefore, may be seen as a kind of dialectic, transactional process that facilitates and promotes meaning production.
>
> (Kent 1987, 245)

Tension Within the Schema as Landscape

The consultants face a tension as they read white papers: they want their landscapes to be accurate and to contain all the information that will be relevant and useful to their clients and to themselves as professionals, and yet they don't want to be swayed unduly by a vendor's message. Each wants his landscape to accurately reflect developments in the industry; however, the consultants can't believe everything they read or hear from a vendor, and they don't want to add specious information to their landscapes.

Nor is each consultant's landscape ever complete. Peter emphasized that the industry changes constantly, and that knowledge is infinite, not finite. Andy also commented that he is sometimes wrong about a new technical development, and that he will do further research to clear up any technical issues he does not understand clearly.

White Papers and the Consultants' Activities

The physicists in Bazerman's study are all members of the same discipline or "communal endeavor" (2000, 235), and reading physics articles is a disciplinary activity. The consultants and vendors interact through writing and reading white papers because they are participants in an economic marketplace. While communities do develop within marketplaces—and Tel-Work personnel cited many personal relationships that they have with vendors and other colleagues—the communities are based on economics, not on the advancement of an academic field. For physicists and other academics, advances in knowledge are shared so that the whole community may benefit. The consultants' work, however, occurs in a market economy in which specific knowledge may lead to competitive advantages; knowledge becomes a commodity that is sometimes distributed only to those who pay for it (or to those who at least give over their contact information). Additionally, the marketplace allows vendors to write white papers that favor their products; the fact that vendors do not write white papers that are fully objective seems to be part of the social contract between writers and readers. This social contract is reflected in each reader's persuasion knowledge.

Conclusion

For the consultants in this study, white papers are just one information source among many. White papers provide useful technical information, but the consultants recognize that it is packaged selectively to suit the source's purposes of self-promotion—thus making white papers incomplete sources of information. Before advising clients to invest in a technology explained in a white paper, consultants will want to see typical technical documentation (spec sheets, drawings, and so on), and they will want to confirm the technology's function by attending field trials or by contacting customers who have already purchased it.

The physicists in Bazerman's study used their background schemata to select their reading material and to assess its usefulness, and the schema as landscape aided these white paper readers similarly. An additional gatekeeping function of the consultants' schemata is to assess a white paper's credibility. Because the consultants do not have a peer review system in the marketplace similar to that in academia, they must perform their own screening functions. When the consultants read, they continually separate reliable claims from puffery. The schema as landscape, which incorporates insights from the Persuasion Knowledge Model (Friestad and Wright, 1994), may help explain how white papers in high-tech industries are read.

Suggestions for White Paper Writers

While I would not generalize about all white paper readers after observing this group, some things I learned may help technical communicators reach white paper readers more effectively.

- Focus on providing useful information. Use a soft-sell approach (King 1995); avoid blatant salesmanship (Graham 2004).
- Realize that timely information will be more valuable. Experienced white paper writers agree that white papers should be revised or removed when information is out of date (Willerton 2007).
- Do your outside research (Graham 2004), and then cite these outside sources; they lend credibility. References may be especially important for white papers that challenge accepted ideas or cause the reader to rethink foundational concepts.
- Employ principles of effective document design. Incorporate headers, footers, section headings, and other features of document design to help your readers scan it and navigate through it.
- Because workplace readers are busy, use visuals to illustrate important points (Dragga and Gong 1989). However, omit graphics that convey little or no meaning (for example, most clip art).
- Do not use too many acronyms, even when your audience is highly technical; provide definitions and, if needed, a glossary of terms.
- Make sure that a white paper in PDF format prints out cleanly in black and white and that data displays are not color-dependent (Appum Group n.d., 2; Gordon and Gordon 2004, 6).

Acknowledgment

*This article was originally published in *Technical Communication* (2008), 55: 370–82.

References

Adler, P. A., and P. Adler. 1987. *Membership Roles in Field Research*. Newbury Park, CA: Sage.

Alliance Datacom. 2004. Voice over IP technology. www.alliancedatacom.com/voip.asp.

Andrews, D. C. 2000. *Technical Communication in the Global Community*, 2nd edn. New York, NY: Prentice Hall.

Appum Group. n.d. "Are You Making Your White Papers Work Hard Enough?," from www.whitepapercompany.com.

Areni, C. S., and K. C. Cox. 1994. "The Persuasive Effects of Evaluation, Expectancy, and Relevancy Dimensions of Incongruent Visual and Verbal Information." *Advances in Consumer Research* 21: 337–42.

Bartlett, F. C. 1932/1964. *Remembering: A Study in Experimental and Social Psychology*. Cambridge, UK: Cambridge University Press.

Bazerman, C. 1985. "Physicists Reading Physics: Schema-Laden Purposes and Purpose-Laden Schema." *Written Communication* 2: 3–23.

———. 2000. *Shaping Written Knowledge: The Genre and Activity of the Experimental Article in Science*. WAC Clearinghouse Landmark Publications in Writing Studies. http://wac.colostate.edu/aw/books/bazerman_shaping/. Originally published 1998, Madison, WI: University of Wisconsin Press.

Blyler, N. R. 1991. "Reading Theory and Persuasive Business Communications: Guidelines for Writers." *Journal of Technical Writing and Communication* 21: 383–96.

Doheny-Farina, S., and L. Odell. 1985. "Ethnographic Research on Writing: Assumptions and Methodology." In *Writing in Nonacademic Settings*, ed. L. Odell and D. Goswami, 503–35. New York, NY: Guilford Press.

Dragga, S., and G. Gong. 1989. *Editing: The Design of Rhetoric*. Amityville, NY: Baywood.

Duin, A. H. 1988. "How People Read: Implications for Writers." *Technical Writing Teacher* 15: 185–93.

Eisenhardt, K. M. 2002. "Building Theories from Case Study Research." In *The Qualitative Researcher's Companion*, ed. A. M. Huberman and M. B. Miles, 5–35. Thousand Oaks, CA: Sage. Originally published 1989 in *Academy of Management Review* 14: 532–50.

EMarketer. 2004. "What Execs do with White Papers," March 22. http://www.emarketer.com/Article.aspx?1002714.

Faris, K. A., and L. R. Smeltzer. 1997. "Schema Theory Compared to Text-Centered Theory as an Explanation for the Readers' Understanding of a Business Message." *Journal of Business Communication* 34: 7–26.

Forbes.com and Bitpipe. 2004. "2004 Forbes.com & Bitpipe Study: Readership & Usage of White Papers & Case Studies by Corporate & IT Management." http://wp.bitpipe.com/resource/org_973204426_74/2004_Forbes_and_Bitpipe_Study__Bitpipe.pdf.

Friestad, M., and P. Wright. 1994. "The Persuasion Knowledge Model: How People Cope with Persuasion Attempts." *Journal of Consumer Research* 21: 1–31.

Gordon and Gordon. 2004. "The State of the White Paper, 2004." www.gordonandgordon.com/downloads/State_of_the_White_Paper_2004.pdf.

Graham, G. 2004. "White Papers Help Vircom Build Mindshare, Win Awards, and Sell Software (+19 Tips You Can Use to Follow Their Lead)." www.softwareceo.com.

Johnson-Sheehan, R. 2005. *Technical Communication Today*. Boston, MA: Longman.

Kent, T. 1987. "Schema Theory and Technical Communication." *Journal of Technical Writing and Communication* 17: 243–52.

Killingsworth, M. J. 1985. "The Essay and the Report: Expository Poles in Technical Writing." *Journal of Technical Writing and Communication* 15: 227–33.

——, and M. K. Gilbertson. 1992. *Signs, Genres, and Communities in Technical Communication*. Amityville, NY: Baywood.

King, J. M. 1995. *Writing High-Tech Copy That Sells*. New York, NY: Wiley.

Kirmani, A., and M. C. Campbell. 2004. "Goal Seeker and Persuasion Sentry: How Consumer Targets Respond to Interpersonal Marketing Persuasion." *Journal of Consumer Research* 31: 573–82.

Lange, L. 2003a. "Balance, Writing Skill are Keys in Producing White Papers." http://techrepublic.com.com/5102-6299-5062647.html.

——. 2003b. "Why White Paper Popularity is Rising." http://techrepublic.com.com/5100-6299-5086932.html.

Leinhardt, G., and K. McCarthy Young. 1996. "Two Texts, Three Readers: Distance and Expertise in Reading History." *Cognition and Instruction* 14: 441–86.

Lincoln, Y. S., and E. G. Guba. 1985. *Naturalistic Inquiry*. Beverly Hills, CA: Sage.

MacNealy, M. S. 1997a. *Strategies for Empirical Research in Writing*. Boston, MA: Allyn and Bacon.

——. 1997b. "Toward Better Case Study Research." *IEEE Transactions on Professional Communication* 40: 182–96.

Markel, M. 2010. *Technical Communication*, 9th edn. Boston, MA: Bedford/St. Martin's.

McVee, M. B., K. Dunsmore, and J. R. Gavelek. 2005. "Schema Theory Revisited." *Review of Educational Research* 75: 531–66.

Miles, M. B., and A. M. Huberman. 1994. *Qualitative Data Analysis: An Expanded Sourcebook*. Thousand Oaks, CA: Sage.

Odell, L., D. Goswami, and A. Herrington. 1983. "The Discourse-Based Interview: A Procedure for Exploring the Tacit Knowledge of Writers in Nonacademic Settings." In *Research on Writing: Principles and Methods*, ed. P. Mosenthal, L. Tamor, and S. A. Walmsley, 221–36. New York, NY: Longman.

Redish, J. C. 1989. "Reading to Learn to Do." *IEEE Transactions on Professional Communication* 32: 289–93.

——. 1993. "Understanding Readers." In *Techniques for Technical Communicators*, ed. C. M Barnum and S. Carliner. New York, NY: Macmillan.

Sammons, M. C. 2003. *The Internet Writer's Handbook*, 2nd edn. New York, NY: Longman.

Stelzner, M. A. 2007. *Writing White Papers: How to Capture Readers and Keep them Engaged*. Poway, CA: WhitePaperSource Press.

Taylor, S. E., and J. Crocker. 1981. "Schematic Bases of Social Information Processing." In *Social Cognition: The Ontario Symposium*, ed. E. Tory Higgins and colleagues, 89–134. Hillsdale, NJ: Erlbaum.

van den Broek, P., M. Young, Y. Tzeng, and T. Linderholm. 2004. "The Landscape Model of Reading: Inferences and the Online Construction of a Memory Representation." In *Theoretical Models and Processes of Reading*, 5th edn, ed. R. B. Ruddell and N. J. Unrau, 1244–69. Newark, DE: International Reading Association. Originally published 1999 in *The Construction of Mental Representations during Reading*, ed. H. van Oostendorp and S. R. Goldman. Mahwah, NJ: Lawrence Erlbaum.

Weil, D. 2001. "White Papers: B2B Email Marketing's Best Friend." www.clickz.com/experts/em_mkt/b2b_em_mkt/article.php/844601.

Wells, A. S., D. Hirshberg, M. Lipton, and J. Oakes. 2002. "Bounding the Case within its Context: A Constructivist Approach to Studying Detracking Reform." In *The Qualitative Researcher's Companion*, ed. A. M. Huberman and M. B. Miles. Thousand Oaks, CA: Sage. Originally published 1995 in *Educational Researcher* 24 (5): 18–24.

Willerton, R. 2007. "Writing White Papers in High-Tech Industries: Perspectives from the Field." *Technical Communication* 54: 187–200.

Wright, P. 1986. "Schemer Schema: Consumers' Intuitive Theories about Marketers' Influence Tactics." *Advances in Consumer Research* 13: 1–3.

Yin, R. 1994. *Case Study Research: Design and Methods*, 2nd edn. Thousand Oaks, CA: Sage.

Chapter 10

Using Web 2.0 to Conduct Qualitative Research

A Conceptual Model*

Christopher Thacker and David Dayton

Editors' Introduction

Thacker and Dayton argue that Web 2.0, with its ability to enable a variety of ways in which users can interact within virtual worlds, can be used to enhance and strengthen the participative nature of interview-based qualitative research. Their article reports on the development of a conceptual model for an interactive Web site to capture the technical communicator experience of single sourcing and content management through a survey, interviews, and site visits. Their goal is to develop a framework for research conducted on a Web 2.0 platform that can be used to operationalize a qualitative inquiry through the formation of a research community. These communities would be made up of investigators and participants, where the investigators create a structured context for accumulating data, and where all members of the community (researchers and participants alike) have full access to the data.

This article is a clear instance of the idea put forward by Blakeslee and colleagues in Chapter 2 of this collection, calling for new ways to share authority between researchers and participants (and Thacker and Dayton explicitly acknowledge the debt their work owes to that of Blakeslee and colleagues). With Web 2.0, they argue, the people who access and use Web sites are gaining more and more influence over the content of the sites, and are able to interact in real time. For research projects, this approach introduces the possibility of creating a platform that can efficiently capture and organize data from a screened sample of participants, and that allows these participants to take on the identity of active research collaborators rather than passive research subjects.

Thacker and Dayton point out the central role played by qualitative methods in technical communication research, and review the ways that Web technology can assist the research endeavor. For example, a Web-based research community could allow for the timely interaction between researchers and participants to clarify and further develop data gathered through interviews or journaling. It might also promote collaboration among teams of researchers, and it could host electronic tools that would facilitate the organization and extraction of data.

The authors suggest that the main barrier to participative research is that researchers are busy people with limited resources, as are research participants.

Though many acknowledge the value of participatory research, it is often simply not feasible for researchers and participants to engage with each other after the data have been gathered. Web 2.0, they argue, offers a way of overcoming this barrier.

The article also describes their effort to implement and test their Web 2.0 conceptual model by constructing and using a first-hand reports (FHR) Web site. They describe the choices that they made as they looked for a suitable software platform for their site, and report their progress so far in constructing the site. We can look forward with anticipation to their next publication, which we hope will report on the full results of their test of the conceptual model.

This article is a clear demonstration that research methodology is not a fixed and finished endeavor; research methods and the technologies that support them are constantly evolving. Thacker and Dayton show us how technical communicators are likely to take advantage of and contribute to innovative uses of emerging technology, which is one of the themes highlighted by Giammona. They suggest that emerging, interactive Web technologies could have a significant impact on the way that academics conduct research into technical communicator experience, and the way that practitioners investigate the needs and constraints of their audiences and users. The article, like its subject matter, has an intriguing unfinished quality, as it is a progress report on the ongoing development and testing of a conceptual model.

Introduction

Web 2.0 refers to innovations that have enabled entrepreneurs to reinvent the Web by making it more interactive and participatory. Web 2.0 sites such as Facebook, Linkedin, YouTube, and MySpace have experienced phenomenal growth, energized by the desire of people with shared interests to socialize and regularly exchange information, opinions, and other content. By combining instant Web publishing, social networking tools, user-generated content, and communal tagging, rating, and commenting—all within an easy-to-use content management system—Web 2.0 has the potential to increase the richness, dynamism, and ultimate impact of interview-based qualitative research.

To explore this potential, we have developed a conceptual model for a research Web site designed to collect structured accounts of technical communicators writing about their experiences and opinions related to single sourcing and/or content management methods and tools. This novel data-collection method is part of a research project supported by a grant from the Society for Technical Communication; to date, the project has gathered data through an online survey and through interviews and site visits. The first-hand reports (FHR) Web site, as we call it, will complement traditional data-collection methods by using Web 2.0 technologies such as those in use at the well-known social networking sites MySpace.com and Facebook.com.

Of course, those Web sites constitute a new form of grass-roots mass communication; the research Web site we envision will operate on a much smaller scale. Indeed, keeping the scale small and the focus limited is an important constraint—and a big advantage—in our conceptual model. In the FHR Web site, informants will be members of a virtual community that forms to share information on the specific focus of the Web site, which in our test case is first-hand information and opinions about single sourcing and content management in technical communication. Members of the virtual community we envision will be those who have applied and received approval from the principal investigator running the site. Each informant will have his or her own first-hand report space that will include a detailed professional profile and an in-depth account of that person's experience with and knowledge about the topic. Each first-hand report will be composed in response to prompts presented by the project's principal investigator. Each informant will have the option of creating a blog, which will be accessible only to other informants. The site will also have a public community message board for site members and visitors to exchange information and opinions about single sourcing and content management.

We believe that the kind of Web site we envision has the potential to alter radically how researchers collect and make sense of first-hand accounts from research informants. The site's principal investigator or research team will structure and moderate information sharing, but any member of the community will be able to search and analyze the information collected on the Web site. Thus, the distillations and interpretations of information published by the site's research team may be supplemented, or even contested, by participant-investigators with different perspectives.

The primary purpose of this article is to present our conceptual model for an FHR Web site and to discuss issues that need to be resolved in order to make such a site feasible. We also feel that it is important to discuss the potential of such Web sites to facilitate the truly participatory, multi-vocal qualitative research that scholars in our field have envisioned and advocated for some time (Blakeslee and colleagues 1996). We begin by glossing the role and value of technical communication research, emphasizing the centrality of qualitative methods to constructing our discipline's body of knowledge. Next, we briefly review the benefits and limitations of conducting qualitative research using Internet-mediated communication and explain why we think Web 2.0 technologies have the potential to enhance the benefits and minimize the limitations. We then present a high-level description of the conceptual model for the first-hand reports Web site as originally envisioned (Thacker and Dayton 2008). After that, we sum up and discuss the compelling advantages we see in this method of qualitative data collection compared to traditional interview-based methods. Finally, we discuss the most obvious barriers to implementing an FHR Web site, some of which are technical and some of which are social and institutional. In the final section, we describe why we decided to use Ning, a free Web service for building social networking sites, and how this decision will impact on our ability to implement the most important features in our FHR model.

Role and Value of Technical Communication Research

Our understanding of technical communication has grown tremendously over the past three decades because of insights gained through research, which has increased the discipline's practical, scientific, and scholarly body of knowledge (Smith 1992; Rainey 1999; Hayhoe 2006). According to Allen and Southard, technical communication researchers desire primarily to understand the motivation, attitudes, and behaviors of users (readers), content developers (writers), and their intertwined communication practices (1995, 33). Technical communication research often tries to shed light on how communication designers and their audiences interact with the technologies and media used to create and deliver the communication products.

The discipline of technical communication creates knowledge about the work of practitioners and its impact on employers, audiences, and other stakeholders by examining aspects of technical communication primarily through an interpretivist lens—constructing and interpreting reality by collecting empirical data through qualitative methods: interviews, case studies, focus groups, and field work. According to Blakeslee and colleagues, "one important goal of such research is to improve our understanding of the settings and individuals we study through accounts that describe the rhetorical practices of our participants in ways that are meaningful and useful to them and to ourselves" (1996, 126). Blakeslee and colleagues "argue for judging how meaningful and worthwhile our accounts are from how well they inform practice and on what they teach us" (1996, 126).

Internet-mediated communication is now widely recognized as a productive site for generating research questions as well as a means of collecting information about activities and attitudes not otherwise related to online discourse (Gurak and Silker 2002; Kastman and Gurak 1999). Researchers have used first-generation Web technologies such as e-mail, online chat, listservs, message boards, and threaded discussions to study the rhetorical dynamics and communication patterns in cyberspace, as well as to query informants about activities and attitudes from "real life." Compared with face-to-face communication for gathering qualitative data, Internet-mediated communication is cheaper, faster, and more convenient for the researcher because the information does not have to be transcribed to produce a text for analysis. On the other hand, certain of these advantages can also be limitations: asynchronous, text-based communication is less immediate and, often, lacking in the depth and assured understanding that emerges from real-time dialog, with its confirmatory back-and-forth exchanges to clarify and probe for details and examples.

Managing the data collected through Internet-mediated communication has also been something of an obstacle for qualitative researchers. Although online communication methods may enable lots of information to be collected rapidly from many people, the unstructured nature of the texts thus collected makes the data analysis process laborious and time consuming. One of us knows this first-hand, having moderated an online discussion group for a qualitative study (Dayton

2001). Although the information generated by the online discussion, which extended over several months, was rich in factual details, provocative opinions, and occasional brilliant observations, coding and sorting the data so that it could be reduced to a set of generalizations took many, many more hours than were required to collect the information.

To improve qualitative data gathering using Internet-mediated communication, researchers would benefit greatly if they had a Web site that enabled them to:

- impose some uniformity of structure on and embed metadata in the textual information as it is collected;
- facilitate timely interaction to clarify and elaborate the texts first presented by informants;
- provide data-exploration tools built into the primary data-collection platform;
- enable teams of researchers to work closely together to collect and analyze information presented over time by many informants.

The technology to build such Web sites already exists, and it is being implemented widely on the Internet today under the rubric of Web 2.0.

How Web 2.0 Works

Treese defined Web 2.0 as an incremental set of changes to existing Internet technology (2006, 16). By combining instant publishing, interactivity, social networking, Web services, communal tagging and rating, and content management, this new generation of technology has changed the Internet into a participatory medium (Treese 2006). These technologies have the potential to mitigate the limitations associated with first-generation Web technologies. Qualitative researchers will benefit especially, because they will be able to foster more in-depth communication with and among people from whom they are seeking information on a particular question, problem, or topic. Equally important, Web 2.0 has the potential to enable researchers to manage their communications and analyze the information they collect much more efficiently than current Internet and computer-based methods.

Web 2.0 changes the flow of communication to a bottom-up model. Web developers now create multiple input channels that allow users to communicate in real time to post feedback or comments or even edit a Web site's content instantaneously. These richer interactive channels on blogging and social networking sites have contributed to the growth of online communities, "social aggregations that emerge from the Net when enough people carry on those public discussions long enough, with sufficient human feeling, to form webs of personal relationships in cyberspace" (Rheingold 1993, 5).

Web 2.0 conveys one overarching concept: the Internet is becoming more user-centric. Rather than simply using the Internet as a repository of information, users are driving content. Hart-Davidson (2007, 9) lists important trends in user behavior that influence the concept of Web 2.0: that users produce, organize, and share content; that they interact with Web sites as content aggregators and even content creators; and that they pursue social goals as well as work goals.

Web 2.0 has revived the Internet as a participatory medium, where users "[create] network effects through an 'architecture of participation'" (O'Neil 2005, 1). Users are no longer passive, but instead are actively involved with creating content, setting agendas, and interacting in online communities. This level of participation is made possible by a host of technologies that have come to define Web 2.0. Among the most commonly used by social networking sites are content management systems (CMSs), wikis, blogs, and Really Simple Syndication (RSS).

CMSs are client-server Web applications that separate the content of a Web site from the design patterns used to present the content to users. Content is stored in a database, and pages on the Web site are dynamically assembled through Web forms that allow designers and content developers to manipulate all the usual elements of the site's architecture, screen layout and design, and navigation. CMSs significantly increase the efficiency of Web site maintenance by allowing multiple users to author, modify, record, and delete data without requiring specialized knowledge of hypertext and programming languages.

Wikis are basic Web sites run by CMSs; every wiki provides a simple, easy-to-use mark-up language, which allows users of all proficiency levels to contribute, edit, and delete content. Wikis allows users to control content while keeping the Web site's look and feel consistent (Fuchs-Kittowski and Köhler 2005).

Blogs (from Weblogs) have also gained an incredible level of popularity. Blogs are online journals that are frequently updated through a CMS. Blogs foster two-way communication. Readers can review blogs and leave comments and feedback to guide future discourse. Readers can subscribe to receive new blog content automatically, and bloggers can comment on one another's blogs using a special system for interlinking.

RSS feed is analogous to a subscription service. Users are able to have new online content such as news stories, blogs, and threaded discussions sent automatically to their computer or Web-enabled device (PDA, cell phone). A Web application collects this content and presents it to subscribers for instant access.

Social networking sites such as Linkedin, MySpace, and Facebook have become big business by enabling users to create a Web presence and personal profile using a variety of the tools just mentioned. These social networking sites feature simple CMSs that allow users to instantly update their personal sites. Users can present a variety of personal and professional information in their profiles including photos, video, music, groups, interests, resumés, and curricula vitae to other people within the network. Additionally, these sites allow users to create personal blogs with RSS feeds.

In sum, Web 2.0 technologies offer an innovative and accessible toolkit for researchers who wish to collaborate with practitioners in studying workplace technologies, organizational contexts, and any important issue or concern related to work practices. First-generation Web technologies have seen limited use for collecting qualitative data because of constraints on immediacy and interactivity and because of the data management problems that we have discussed previously. We are confident that Web 2.0 technologies can be assembled to create dynamic community Web sites that transform the qualitative research process, making it richly participatory and thus more relevant, trustworthy, and useful. In the next section, we describe the main technologies and aspects of interaction design that make up our conceptual model of an FHR Web site.

Conceptual Model for a First-Hand Reports Web Site

The primary purpose of the Web site that we plan to build is to collect first-hand reports from technical communication practitioners writing about single sourcing and/or content management methods and tools. What we are calling the conceptual model of the FHR site is simply our starting-point design schematic. The original version is detailed in an unpublished master's thesis (Thacker 2007). In the next section, we provide a general description of the site's technology and architecture. We use the future tense, even when discussing features that we now think we will not be able to implement exactly as described, for reasons which we explain in the closing section of the chapter.

Social Networking Focused on a Narrowly Defined Topic

The FHR Web site has three primary functions: (1) to collect information from numerous people on a relatively narrow topic; (2) to enable easy search and retrieval of the knowledge base thus created; and (3) to build the social cohesion and communication that characterize healthy virtual communities. The FHR Web site will implement those functions by combining and reconfiguring basic tools for interactivity and communication used by popular social networking sites (Linkedin, MySpace, and Facebook, for example).

Users of the FHR Web site will interact with fellow community members and the information on the site through a relatively small set of key functions. These are described below.

- *Profile:* A personal profile space that features all of the user's relevant demographic and psychographic information. The profile is also the main gateway for users to interface with an informant's first-hand reports, blogs, and discussion.
- *First-hand report:* Structured narrative accounts generated from a uniform sequence of prompts created by the research team.
- *Search:* A search engine that queries the site's database to find relevant information from the inputs and preferences of the user.
- *Forum:* The site's threaded message board that is accessible to the public. The message board is designed for members of the community to discuss topics in

an open forum, interacting with guests—visitors to the site who have not joined the site or have not been approved for membership in the site by the principal investigator (PI) and/or the research team.

- *Help:* A wiki that addresses common issues pertaining to the site. The wiki will be initiated and maintained by the research team, but any member will have the ability to comment on any page, and some members who volunteer for the role will be able to edit pages.

The first-hand reports will be the primary means of collecting information. To generate the reports, the site will present newly registered users with a series of prompts—directive, content-defining questions. Each prompt will have a text-entry box for the informant's response, including a rich text editor like those found in popular Web-based e-mail applications. Responding to the prompts, informants will fill in factual details and compose experiential narratives that will have a common structural framework.

Each member's first-hand report will be stored in a personal profile and blog space on the Web site. The researchers and other site members will be able to communicate with informants privately through internal messaging or by posting comments appended to the first-hand reports. Only site members vetted and approved by the site's research team will be able to create first-hand reports and search the reports of others. However, non-site members will be able to interact with site members on public-facing message boards, where the research team will seed and moderate discussions by site members and visitors.

The FHR site's search function will be designed to allow members of the community to query the site's knowledge base for specific information. The general search capabilities will be useful and usable, but not as robust as those provided to the site's research team, who will use a more complex and feature-rich interface enabling more granularity in specifying terms and conditions for a search. Academic and practitioner researchers who are members of the site community may request access to the more robust search application, which would enable them to explore the knowledge base in depth on particular research questions. Although not as powerful, the search tool available to all community members will still allow them to explore specific questions, search for patterns in the first-hand reports, and formulate questions for discussion in their blogs and/or in the public message boards.

Ease of data search and retrieval will be one of the cornerstones of community building on the FHR Web site. We want users to be able to explore the rich qualitative data provided by community members and generate their own questions for further exploration and discussion. The community-building function will be aimed at fostering social networking within the membership of the site. Users will be encouraged to create in-depth profiles and personally controlled communication spaces. The profiles should increase the level of trust within the community by providing a way for users to display and authenticate their credentials. Moreover, we hope to provide tools that encourage users to interact with the larger community. Such tools should include a private messaging system, blog, and public commenting on each user's blog and first-hand report, which will be under the control of that user, who owns his or her own communication space.

Accommodating Informants and Researchers

The FHR Web site will be designed to accommodate the needs of two macro categories of users: informants and researchers. A third user role is really a non-role: guests. Visitors to the site who are not members of the community will be severely restricted with regard to what information they can access; this restriction is necessary to maintain a high level of trust within the community. Non-members will have access to public-facing information such as the administrator's news/blog, the site FAQ, a public discussion board, links and resources, and the sign-up/login interface.

Informants are those who contribute to the site's knowledge base by writing first-hand reports. The PI collects data from informants initially through the structured questionnaire that generates the first-hand report. Informants will not have administrative access to the Web site. They will interact with the Web site in two roles: as members and as moderators.

Members are the core community. They request full access to the Web site's database of first-hand reports, a request that must be approved by either an administrator or a moderator. In exchange for full access to the first-hand reports of other members, new members must agree to share their own story of technology use and/or adoption related to single sourcing and/or content management. If they have no experiential story to share about those technologies, they must, at the very least, provide a real-identity profile and agree to the informed consent, copyright, and usage policy of the site (to which members with first-hand reports must also agree).

We refer to the researcher role as the PI, as though it were a single person, but on any given project the PI might well be two or more persons on a research team, working collectively to share the PI's functions. The PI is defined by two distinct roles: researcher and administrator. In the role of researcher, the PI is responsible for collecting and interpreting data. As an administrator, the PI controls all access to the Web site's content management system through the admin control panel. The PI assigns roles and sets the precise details of what functionality a user may access on the site; these decisions affect the user's views of the site. The front-end view is the general presentational view for all users. The back-end view is a customized view that allows users to edit or change elements of the Web site from the permission levels set by the PI (individual or team).

Making Participatory Research Practical

The role of user advocate is central to the professional identity of technical communicators. In like manner, the role of "practitioner advocate" is central to the way that many researchers in our field view the studies they undertake and the various modes of communication that are critical to their ultimate mission, which is to inform practice and to help build the profession's body of knowledge by writing articles, books, and textbooks, and by giving presentations, leading workshops, and teaching courses.

Just as technical communicators over the past two decades have increasingly aspired to involve users in the design and evaluation of their information products, so too have researchers aspired to an ideal of participatory research. In a landmark article examining this ideal, Blakeslee and colleagues (1996) reviewed perspectives in our field about what constitutes validity in qualitative research. Their discussion assumes familiarity with the issues, so we provide a brief, high-level overview as background.

In scientific research, validity is shown if the researchers can marshal the arguments, from experimental methods and the analysis of results, to persuade fellow scientists that an experiment has indeed provided a reliable test of the hypotheses, and the results have produced relevant new information to help answer the questions that motivated the study. The nature of qualitative investigation, however, is thoroughly interpretive and focused on subjective observations and opinions. Qualitative research represents a different paradigm of knowledge making, and so its quality must be evaluated using different criteria.

Over the past three decades, thought leaders in qualitative research have proposed a host of concepts by which the quality of qualitative studies may be measured (see Seale (1999) for an engaging discussion). Lincoln and Guba (1985) first proposed the term *trustworthiness* to sum up the essential characteristic of good- quality research within qualitative traditions, and they broke the concept into four components that could be shown empirically to some extent: credibility, dependability, transferability, and confirmability. They later added to trustworthiness the concept of *authenticity* (Guba and Lincoln 1989): according to Seale (1999), this means being consistent with the constructivist (that is, relativistic) view that undergirds the qualitative paradigm, while at the same time offering a standard by which one research-based view might be considered more worthy of belief than any other.

> Authenticity, they say, is demonstrated if researchers can show that they have represented a range of different realities (fairness). Research should also help members develop "more sophisticated" understandings of the phenomenon being studied (ontological authenticity), be shown to have helped members appreciate the viewpoints of people other than themselves (educative authenticity), to have stimulated some form of action (catalytic authenticity), and to have empowered members to act (tactical authenticity).
>
> (Seale 1999, 468–69)

The views of Blakeslee and colleagues (1996) are consistent with the quality criterion of authenticity, although they do not use that term. They keep their discussion within the literature of technical communication in arguing for a similarly participative ideal in evaluating the validity of qualitative research. Contrasting their stance with the views of several other scholars in technical communication and in composition studies, they state, "[W]e need to view validity as being more than a matter of determining whether, in fact, we are measuring what we think or say we are measuring, which is how many scholars continue to define validity" (1996, 128).

Paraphrasing the views of Kirsch (1992, 257), Blakeslee and colleagues agreed that researchers doing qualitative studies in technical communication "should open up our research agendas to our participants, listen to their stories, and allow them to actively participate, as much as possible, in the design, development, and reporting of our research" (1996, 132). They acknowledge the difficulties of implementing that vision of participatory research.

Traditional methods of qualitative research rely mainly on one-to-one communication between informants and researchers—some form of interviewing. In many studies, each informant is interviewed only once. In most studies, the opportunities for informants to dialog with the researchers about their findings and conclusions are greatly limited or nonexistent. Rarely do informants in a qualitative study get the chance to exchange views with other informants about the study and what the researchers plan to publish about it. Even in focus groups, the participants typically leave the moderated discussion unable to predict what generalizations the researchers will write to sum up the many opinions expressed by a dozen or so people over the course of an hour or longer. The participants will never get a chance to talk about those findings and the implications drawn from them.

The barriers to implementing participatory research are mundane, practical constraints and not attitudes: researchers do not usually have the time, the resources, and the means to incorporate as much dialog with informants into their research as they would wish. By the same token, many informants would not necessarily be willing to take the time and effort that would be required of them if researchers solicited more input and feedback.

Our concept for an FHR Web site removes most of these barriers to participatory research. Table 10.1 summarizes why we believe this is so by comparing traditional interview-based research to the FHR Web site in terms of methodology, what gets published, and what informants get in exchange for their participation.

Development and Implementation Progress

Our conceptual model for the FHR Web site is the roadmap we used to launch the development process, with Thacker serving as lead developer assisted by his friend and business partner Patrick Kim, who is a Web programmer. Dayton served as both the client for the project and as a consultant in user-centered design.

Initially, Thacker and Kim planned to build the site using the Java Enterprise 2 (J2EE) platform. Their goal, apart from building a site enabling Dayton to carry out his research, was to create an open-source Web application that could help

Table 10.1 Interview-Based Qualitative Research Versus the First-Hand Reports Web Site

Traditional interview-based research	First-hand reports Web Site
Methodology: Principal investigator or research team conducts interviews, reduces qualitative data through analysis to distill generalizations and reach conclusions.	**Methodology:** Principal investigator or research team (PI/RT) structures prompts for first-hand reports, moderates draft reports, and solicits additional details and clearer explanations.
• Research report summarizes and interprets the information collected, but the raw data are not accessible to anyone outside the research team, and in many cases only a lone researcher has access.	PI/RT may also add public comments, moderate discussions, maintain a blog to "think out loud" about themes, issues, ambiguities, which can prompt further discussion in comments on the blog entries. At some point, PI/RT reduces data through analysis to distill generalizations and reach conclusions.
• Informants typically are not given a chance to review and respond to the research report's representation of the information they provided; if they are given the chance, their perspectives may not be adequately represented or even included in the final report.	• Research report summarizes and interprets the information collected, but the report is linked to an online knowledge base that can be examined by others and also restudied later by the PI/RT.
• Raw data remain inaccessible to anyone outside the research team; in effect, the data vanish, replaced by the published report's generalizations illustrated by selected quotations.	Informants—members of the Web site—may discuss the report on the site and debate alternative interpretations.
• This methodology rarely produces longitudinal studies—follow-up interviews with the same informants over time. (Time and expense of conducting interviews and continued cooperation of informants create major impediments to longitudinal studies.)	• Longitudinal studies are not only possible, but they become a relatively convenient and therefore compelling option—over relatively short time frames, at least (less than a year). If participants and institutional review boards are amenable to the possibility, some FHR sites could have an even longer life cycle.
What gets published: A summary containing a small, highly selective fraction of the data collected.	**What gets published:** A summary containing a small, highly selective fraction of the data collected.
• Others do not have access to the data interpreted by the researcher(s).	• Others can examine the same source information and conduct their own analyses.
• The published interpretation cannot be challenged by re-analyzing the same data.	• The published interpretation of the data can be challenged by others offering alternative interpretations based on the same data.
What informants get in return for their participation: The gratitude of the researcher(s); possibly some insights into the activities/attitudes under study as a result of the interaction with the researcher(s) and/or because of the report's analysis and findings.	**What informants get in return for their participation:** Informants will be motivated mainly by self-interest, finding value in the Web site community if it becomes a continually expanding source of useful information. Those who also enjoy the social interaction on the site will value it the most.

others build FHR Web sites. Thacker and Kim envisioned a commercial, off-the-shelf product that could be easily configured by anyone with a modest degree of technical skill. Users with adequate technical know-how could choose to alter the underlying code of the application to improve upon the basic FHR model. Access to this product would come with an explicit agreement that any innovation would have to be shared by making the revised code available on the Web, along with a similar open-source licensing agreement. Ultimately, anyone would be free to adopt and further develop any changes to the interaction design and functionality that improved the FHR site builder.

That vision proved overly ambitious. Constraints of cost, time, and technical knowledge brought the project to a crawl. At Dayton's suggestion, Thacker decided to evaluate open-source social networking platforms that might enable he and Kim to achieve most of the functionality envisioned in the FHR model but at far less cost in time and effort.

At first, Thacker and Kim looked into the feasibility of combining separate off-the-shelf applications (Wiki, blogs, discussion boards) within an open-source Web CMS such as Drupal. However, the time needed to build out the FHR model in that way was still more of a challenge than two unpaid part-time developers with full-time "day jobs" could manage. Their decision not to experiment with an open-source Web CMS was validated by a recent conference paper from a marketing firm that built a prototype research Web site with social networking features using Drupal (Johnson and colleagues 2007). Although the marketing firm reported success in building their Web 2.0 site for testing novel ways to elicit people's reactions to products, they described technical challenges that went far beyond what Thacker and Kim could manage.

The continuous innovation that is a hallmark of Web technology companies presented an unanticipated alternative: Web sites where anyone who can read and click their way around an interface can create a social networking site in a matter of minutes. At Dayton's suggestion, Thacker tried out the two such Web service providers whose social networking platforms seemed to offer most of the functionality essential to building out the FHR model: KickApps.com and Ning.com.

In 2009, both companies (and a number of others) were offering "software as a service" (SaaS) platforms designed to let anyone build a customized social networking site, including such standard features as blogs, chat, member-to-member messaging, discussion forums, comments, ratings, and embedded video. Anyone wanting to create their own specialized Facebook-like site for a group could obtain the necessary Web applications from these companies; the site could be hosted and maintained by the same companies, initially for free.

After evaluating both the KickApps and Ning platforms, Thacker concluded that they were remarkably similar in what they offered, but that Ning was easier to use. In addition, KickApps instituted a usage-based pricing plan for all users after a 30-day free trial. Ning, in contrast, retained its free-to-use option for sites that did not exceed storage and bandwidth limits—more than enough to serve Dayton's needs, for a Web site community of fewer than 50 people. Table 10.2

Table 10.2 FHR Web Site Features Compared to those Available in a Ning.com Web Site

Features with our rating of Ning's ability to deliver	FHR model assuming custom build	What Ning.com offers
Whole-site search Poor (significant tradeoff)	Granular, highly customizable search that allows researchers to explore site content by selecting any variety of pre-defined tags, conditions.	Limited search feature; administrator can only search by name, tag, and topic. The site offers no customization for its search feature.
Forum Adequate, with possibly important tradeoff	Simple threaded discussion board; allows multiple levels of user access. Guests would have access to at least some discussions.	Basic discussion board. The user-interface is a bit confusing, and the threading system is counterintuitive. User access is restricted to members only.
Security Not provided, but not needed (by us)	Encryption of data transmitted to/from the site is not a concern for our initial project; however, this security could be essential for research involving sensitive subjects.	Password protected. No data encryption option.
Wiki Not provided, but not needed	Fully integrated wiki to offer user-generated guidance and instructions for using the site.	No option for this, but ease-of-use and Ning.com's user assistance makes the wiki we envisioned unnecessary.
User interface Excellent	Designed for our specific research purpose; bare-bones, with custom-designed features.	Limited customizable interface allows you to add pre-packaged features.

Administrator tools Excellent with caveat	Customized tools that allow administrator(s) to control all aspects of the site including user access, reports, and user interface. Easy, flexible content export options.	Limited export features for data mining; robust control of user access; user-interface controls; messaging and user analytics.
Profile Excellent with caveat	Required personal information set up by the administrator with option for disguised identity. Profile space to be similar to that of popular social networking sites.	Customizable look and feel; administrator can require a limited set of personal identifiers such as a photo, name, location, age, and occupation. Disguised identity users possible but more difficult to set up, control.
Blog Excellent	Envisioned to be the main portal for data collection.	Robust blogging feature that integrates new media and commenting.
Group space Excellent added feature in Ning	Group space was not envisioned in the original model. We relied on the discussion board to serve this purpose.	Similar features to the profile page; integrates a group discussion, commenting, and notes.
New media Excellent added feature in Ning	We did not envision allowing users to embed new media or third-party applications. These features could make the site "stickier" if enough community members use them.	Each point of interaction (blog, discussion board, group space, commenting) allows for the integration of external links, embedded audio and video (podcasts, YouTube).
User-installed third-party applications Excellent added feature in Ning	Our design did not intend for users to install their own third-party applications.	Allows for user-installed third-party applications; limited to preselected applications such as Box.net (Filesharing), and Twitter (Text Broadcasting).

summarizes Thacker's assessment of the features that Ning offered (as of summer 2009) compared to the most important features in the FHR Web site conceptual model.

The comparative analysis summarized in Table 10.2 shows that a Ning Web site provides most of the key features of our FHR model. Currently, Ning does not support encryption of transmitted data, which would be essential for an FHR site dealing with sensitive information. That security feature is not needed, however, for our research Web site about technical communicators' experiences and opinions regarding single sourcing and content management. We are also optimistic that future versions of Ning will offer the option to use secure protocols for transmitting data, perhaps for an added fee.

For our project, the most noteworthy tradeoff of using Ning at present is the lack of a powerful search function coupled with limited export options. We envisioned a site with a robust query-builder and search engine that would make it easy for researchers to mine all the site's text, or narrowly targeted subsets of it, applying granular filters with tags and key words. In addition, we planned for researchers to be able to export content from the site in formats that would make it easy to import texts into any number of content analysis software programs.

We believe that the lack of robust search and export features in Ning has to be kept in perspective: our vision of the ideal search tools for researchers was unrealistic, given our limited programming resources and time available for developing them. If Ning can make it a relatively simple matter to create an FHR Web site for collecting qualitative data, we think researchers will happily accept the tradeoff of having to use their usual digital tools for exploring, organizing, and analyzing those data.

The only other noteworthy tradeoff in using Ning, as noted in Table 10.2, is inconsequential. Our FHR model called for allowing guests to view at least some discussions and even to participate in the discussions. That does not appear to be an option with Ning sites at this time. Our vision of the FHR site has evolved, and we now see little to be gained in allowing guests to have access to any content on the site while the research phase of the project is under way. We anticipate that the research phase of our first FHR site may be short-lived, with most of the activity occurring within the first three months. Once the research team has collected the first-hand reports and all the ensuing discussions, we plan to let the site community decide the future of the site. The options would need to be spelled out at the outset of the project because our university's institutional review board (IRB) will require that we describe how we plan to store and ultimately dispose of all the information we will have collected on the site during the life of the project.

We are now at the phase of building a prototype FHR site using Ning, so that we can prepare a proposal to our university's IRB. Thus, we are more aware than ever that crucial details need to be hammered out with regard to managing the FHR Web site, particularly in these three areas: (1) confidentiality and security; (2) ethical and legal requirements; and (3) expansion and/or transfer of the site (see Table 10.3). These potential threats can be mitigated through effective

Table 10.3 Threats to Community Trust and Policies and Procedures to Manage Them

Confidentiality and security	Ethical and legal requirements	Site life cycle concerns
Issue: Personal and demographic information in profiles and postings could leave informants vulnerable to outsiders who have little or no interest in the welfare of the community.	**Issue:** Institutions of higher education require rigid adherence to federal protocols for protection of human participants in research; permission from Institutional Review Board (IRB) must be obtained for all research activity involving human participants.	**Issue:** What becomes of the FHR site once the researchers have gathered what they consider to be sufficient content?
Potential threats:		**Potential threats:**
• Marketers—use open discussion forums to spam users with marketing messages or posting mainly to sell their product or service.	**Potential threats:**	• Will community members be allowed to vote whether to: (1) keep the site going; or (2) let researchers archive all content and delete the site from the Internet server?
• Data Farmers—firms that aggregate personal data culled from Web sites. Information is sold for marketing and other purposes (including scams and identity theft).	• Research involving human participants must meet or exceed the standards outlined by federal regulations: Title 45 of the Code of Federal Regulations, Part 46.	• Community members may have conflicts regarding what to do with discussion board threads to which multiple members contributed.
• Special Interest Groups—promote their issues and steer discussions. May include consultants who seed discussions to promote products and services in which they have a vested interest.	• Procedures for gathering, storing, and using data must be reviewed by IRB(s) of all universities represented on the research team.	• Community members may disagree about assignment of administrator rights once the research team is ready to transfer site ownership.
• Trolls—Internet jargon for forum participants who purposefully post fallacious or inflammatory messages.	• Best practices have yet to be proposed for dealing with data collected from informants on an FHR Web site.	

Table 10.3 Continued

Policies and procedures to manage threats to community trust

• Restrict site access to trusted participants; no guests allowed. • Authentication process. • Terms of service agreement, with strict enforcement if members violate confidentiality rules. • Code site to hide it from search engines.	• FHR Web site members must affirm that they understand and agree with all the provisions of an informed consent form that will have to include the site's terms of service agreement. • University researchers will need to obtain evidence of participants' informed consent that is acceptable to the IRB(s).	• If the community decides to keep the site going, each member will need to document a decision as to whether all or certain parts of what they contributed to the research site should continue to appear on the community-owned site. • All confidential information from research site must be erased before new administrators take over, with access to all site data.

communication from the research/administrative team running the site, starting with the development of clear, accessible, and reasonable policies and procedures.

Summary and Conclusion

Shortly after the turn of the 21st century, first-generation Internet technologies began to be applied and combined in novel ways that converged with improvements in Web technology under the rubric Web 2.0. More recently, the terms *social media* and *social networking* have become attached to a wide range of Web 2.0 sites that feature interactive multimedia entertainment, personal instant publishing with easy content aggregation and sharing, and asynchronous socializing through comments and discussions. We believe that integrating basic Web 2.0 technologies to facilitate a small online community's extended discussion of a limited topic would produce compelling benefits for qualitative research in technical communication. We have argued that such Web sites offer a means to realize, much more consistently and fully, the vision of participatory research that scholars in our field have been advocating for over a decade. To advance that collective ideal, we described our conceptual model for a first-hand reports Web site. We then related the obstacles that made programing the site a difficult, long-term challenge for us, even though our team included technical expertise that few researchers in technical communication possess. As an alternative, we investigated the feasibility of using an open-source CMS, but then we discovered that we could implement most of the features we wanted in an FHR site by using Ning, a free Web tool for building social networking sites. We summarized and discussed the benefits and tradeoffs of using Ning for our project. Finally, we identified several kinds of threats to community trust posed by an FHR Web site and suggested protective policies and procedures that must be presented in persuasive detail, first to institutional review boards, and then to those volunteering to join an FHR Web site's community.

Acknowledgment

*This article was originally published in *Technical Communication* (2008), 55: 383–91.

References

Allen, J., and S. Southard 1995. "Strategies for Research in Technical Communication: Purpose and Study Design." Paper presented at the 42nd annual conference of the Society for Technical Communication, Washington, DC.

Blakeslee, A. M., C. M. Cole, and T. Conefrey. 1996. "Evaluating Qualitative Inquiry in Technical and Scientific Communication: Toward a Practical and Dialogic Validity." *Technical Communication Quarterly* 5: 125–49.

Dayton, D. 2001. "Electronic Editing in Technical Communication: Practices, Attitudes, and Impacts." PhD dissertation, Texas Tech University.

Fuchs-Kittowski, F., and A. Köhler. 2005. "Wiki Communities in the Context of Work Processes." In *WikiSym '05: Proceedings of the 2005 International Symposium on Wikis*, ed. Dirk Riehle, 33–39. New York, NY: The Association for Computing Machinery.

Guba, E. G., and Y. S. Lincoln. 1989. *Fourth Generation Evaluation*. Newbury Park, CA: Sage.

Gurak, L. J., and C. M. Silker. 2002. "Technical Communication Research in Cyberspace." In *Research in Technical Communication*, ed. L. J. Gurak and M. M. Lay, 229–48. Westport, CT: Greenwood Publishing Group.

Hart-Davidson, W. 2007. "Web 2.0: What Technical Communicators Should Know." *Intercom* 54 (September–October): 8–12.

Hayhoe, G. F. 2006. "Who We Are, Where We Are, What We Do: The Relevance of Research." *Technical Communication* 53: 393–94.

Johnson, A. J., J. Miller, and G. Davies. 2007. "Exploring the Practicalities of Developing Web 2.0 Applications in Online Research." In *The Challenges of a Changing World: Proceedings of the Association for Survey Computing 5th International Conference*, ed. M. Trotman and colleagues, 45–54. Berkeley, CA: Association for Survey Computing.

Kastman, L., and L. Gurak. 1999. "Conducting Technical Communication Research Via the Internet: Guidelines for Privacy, Permissions, and Ownership in Educational Research." *Technical Communication* 46: 460–69.

Kirsch, G. 1992. "Methodological Pluralism: Epistemological Issues." In *Methods and Methodology in Composition Research*, ed. G. Kirsch and P. Sullivan, 247–49. Carbondale, IL: Southern Illinois University Press.

Lincoln, Y. S., and E. G. Guba. 1985. *Naturalistic Inquiry*. Beverly Hills, CA: Sage.

O'Neil, T. 2005. "What is Web 2.0?" http://www.oreillynet.com/pub/a/oreilly/tim/news/2005/09/30/what-is-web-20.html.

Rainey, K. T. 1999. "Doctoral Research in Technical, Scientific, and Business Communication, 1989–1998." *Technical Communication* 46: 501–31.

Rheingold, H. 1993. *The Virtual Community: Homesteading on the Electronic Frontier*. Reading, MA: Addison-Wesley.

Seale, C. 1999. "Quality in Qualitative Research." *Qualitative Inquiry* 5: 465–78.

Smith, F. R. 1992. "The Continuing Importance of Research in Technical Communication." *Technical Communication* 39: 521–23.

Thacker, C. A. 2007. "The Prospects and Challenges of Using Web 2.0 Technologies for Qualitative Research in Professional Writing." Master's thesis, Towson University.

———, and D. Dayton. 2008. "Using Web 2.0 to Conduct Qualitative Research: A Conceptual Model." *Technical Communication* 55: 383–91.

Treese, W. 2006. "Web 2.0: Is it Really Different?" *netWorker* 10: 15–17.

Chapter 11

Heuristic Web Site Evaluation

Exploring the Effects of Guidelines on Experts' Detection of Usability Problems*

Marieke Welle Donker-Kuijer, Menno de Jong, and Leo Lentz

Editors' Introduction

Investigating the usability of communication artifacts provides fertile ground for qualitative research, but the underlying methods of usability studies can also be analyzed using qualitative techniques. In "Heuristic Web Site Evaluation: Exploring the Effects of Guidelines on Experts' Detection of Usability Problems," Welle Donker-Kuijer, de Jong, and Lentz investigate the role that heuristics play in expert evaluations of Web usability and the ways that experts use those heuristics. The result is a mixed study, a quantitative and qualitative analysis in which the authors examine the contributions that heuristics make to expert evaluation by comparing the results of such evaluations made with and without the use of heuristics.

They examined three questions in their study of how 16 communication professionals assessed the navigation and comprehensibility features of a municipal Web site:

- Do heuristics simply reflect knowledge of the field, such that experts have already internalized the points that the heuristics make explicit?
- Do guided and unguided expert evaluations differ in the type and number of annotations that experts make?
- Does the use of high-level or low-level heuristics make a difference in the type and number of annotations made by the experts?

The authors' mixed-methods design means that their approach and terminology is quite different than the studies highlighted in other articles in this collection. They first had their participants conduct an evaluation without using heuristics, and then they asked them to conduct a second evaluation using heuristics. This approach is called a within-subjects or within-group design. In a traditional between-subjects (or between-groups) design, you would have two groups of participants and one "treatment"; in a within-subjects design, you have one group of research participants and two treatments. This allows you to compare the results with and without the treatment. Quantitative researchers like within-subjects designs because you can use more subjects with each treatment, and this increases the statistical "power" of the results. The within-subjects design also reduces the statistical error variance between the two treatments, because you know that the groups receiving each treatment are identical (whereas in a between-subjects design, even with random assignment into the groups you could end up with notable differences in the makeup of the groups).

Once the expert evaluations were complete, the experts' annotations about features of the site were coded according to the comprehension and navigation heuristics that were used for the half of the evaluations that were guided. Participants were also interviewed using a structured interview protocol. The results of the study provide valuable information for practitioners.

- They confirmed the validity of the heuristics used—69 percent of the comments on navigation and 86 percent of the comments on comprehensibility corresponded to the heuristic guidelines.
- They found that the guided evaluations produced significantly fewer comments than the unguided evaluations, and that the guided evaluations also produced significantly fewer negative comments.
- Finally, they found no differences in the number of comments between high-level and low-level guidelines in the number of comments, the number of positive versus negative comments, or the various guideline categories.

Obviously, the counting of various phenomena revealed through this study is the basic quantitative technique used, while the content analysis and coding of the annotations was the principal qualitative method.

Interestingly, none of Giammona's nine themes directly addresses the concerns that Welle Donker-Kuijer and her colleagues focus on—except, indirectly, how future practitioners should be educated. That is the case because this article examines technical communication techniques, rather than the themes that are central to Giammona's study: the profession and its significance, or technical communicators themselves. Nevertheless, we welcome the validation of a technique commonly employed in usability studies, as well as the implication that guidelines commonly used with that technique have been internalized by experts in the field.

Introduction

When it comes to evaluating Web sites and other communication means, qualitative research strategies are the dominant approach. Rather than assessing the overall usability of a Web site or document, usability evaluators want to know in detail where usability problems occur and how these problems may be solved. Qualitative, troubleshooting methods help to uncover these problems, either by carefully examining the usage process or by urging participants to make detailed and very specific comments (de Jong and Schellens 1997). A distinction can be made between expert-focused and user-focused methods (Schriver 1989). The majority of the methodological research thus far has focused on the validity of user-focused evaluation methods and on the comparison of expert-focused and user-focused approaches (de Jong and Schellens 2000). Within the domain of expert-focused evaluation, various approaches have been developed to enhance the quality of expert evaluation, one of which is the use of heuristics. This article describes an empirical study combining quantitative and qualitative analyses to explore the pros and cons of heuristics as a qualitative approach of Web site evaluation.

Heuristic evaluation has become a popular method among Web site designers and usability professionals for assessing the quality of Web sites (Vredenburg and colleagues 2002). In a heuristic evaluation, experts systematically review a Web site by judging its compliance with recognized usability principles or guidelines—that is, the heuristics (Nielsen 1994). Originally developed for the evaluation of software, it is nowadays applied to all kinds of IT media ranging from Web sites to virtual reality applications. Various sets of heuristics have been developed for many aspects of Web site quality, such as accessibility, usability, navigation, and comprehensibility (de Jong and van der Geest 2000). It is generally assumed that heuristics facilitate the evaluation process and enhance experts' ability to detect usability problems in a Web site.

The evaluation process may be facilitated in various ways. Specific guidelines may complement the experts' own knowledge about the design of effective Web sites or may serve as mnemonic devices. Furthermore, the complete set of heuristics used may raise experts' awareness to focus primarily on the needs of potential users or to evaluate particular aspects of a Web site. For instance, experts working with heuristics about visual design may be expected to be more sensitive to visual presentation issues and see more problems in this area than experts in an unguided evaluation process. Heuristics may also support experts to evaluate a Web site systematically by offering a structured framework. A final advantage of heuristics is that they may be helpful in experts' communication about the evaluation results (van der Geest and Spyridakis 2000). Nevertheless, little is known about the actual contribution of heuristics to the process of expert evaluation and about the way that experts incorporate the use of heuristics into their evaluation process. Despite the potential advantages mentioned above, heuristics also provide experts with the difficulty of having to switch between the heuristics and the Web site, thus complicating an already complex task.

Empirical research into the contribution of heuristics to the quality of expert evaluation has been limited. Studies by Sutcliffe (2002) and Paddison and Englefield (2004) found that experts are not always satisfied with the results of their heuristic evaluations. Sutcliffe also found that experts judged some of the heuristic items to be ambiguous when they had to be used in an actual evaluation. A recent study by Tao (2008) showed that information system professionals recognize and know many of the Web design guidelines available, but perceive difficulties in applying them to a specific Web site. Experts in a study by Hvannberg and colleagues (2007) mentioned the heuristic guidelines they had to work with as both a facilitator (seven times) and a hindrance (six times) in detecting usability problems.

Only two studies explicitly addressed process characteristics of heuristic evaluation. Faulkner (2006) observed that experts do not necessarily use heuristics to identify usability problems in a Web site; some experts in her study primarily used the heuristics to label problems they had found by relying solely on their own expertise. This might be indicative for the task complexity of evaluating a Web site when switching between heuristics and Web site is involved. The contribution of the heuristics to the detection of usability problems would be limited, and the requirement to label all problems afterward would possibly even have a negative effect on the number of usability problems identified.

Apart from that, and on a more detailed level, Faulkner (2006) showed that the procedure of a heuristic evaluation may affect its effectiveness (in terms of the number of usability problems found). One half of the participants in her study worked for 40 minutes without breaks, whereas the other half worked in four 10-minute blocks divided by five-minute breaks. Although some participants found the breaks distracting, the participants with breaks proved to be more productive in identifying usability problems.

However, regardless of the set of heuristics used or whether or not the participants took breaks, work experience seemed to be the most important predictor of the number of problems detected, as was also found in studies by Nielsen (1992) and Saroyan (1993). A study by Hvannberg and colleagues (2007) focused on the effects of the medium of reporting problems (paper and pencil versus a Web-based registration tool) on the detection of usability problems and did not find significant differences between the two alternatives.

Several studies compared the results of heuristic evaluation with those of user-focused evaluation approaches, such as think-aloud usability testing (Bailey and colleagues 1992; Desurvire 1994; Fu and colleagues 2002; Hvannberg and colleagues 2007; Jeffries and colleagues 1991). Typically, these studies focused on the overlap in problems detected, as well as on the total number and types of problems found. The results of these studies are mixed: in some cases, heuristic evaluation proved to be an effective way of detecting user problems; in other cases, heuristics only enabled experts to predict small portions of the (severe) usability problems.

Two studies compared evaluation results of experts working with and without heuristics, again with mixed results. Bastien and colleagues (1999) found that experts using one set of heuristics (the so-called Ergonomic Criteria) performed better than both experts using another set of heuristics (the ISO/DIS 9241–10 dialog principles) and experts without any guidelines. Apparently, the effects of heuristics depend on the specific list of heuristic guidelines used. In the other study, by Connell and Hammond (1999), no differences were found between heuristic and unguided evaluations.

Surprisingly, no in-depth comparisons were made between unguided expert evaluations and heuristic evaluations. Furthermore, the empirical literature seems to neglect the potentially important distinction between high-level and low-level heuristics (de Jong and van der Geest 2000; Wright 1985). This distinction is comparable to the distinction between goal and action rules in the field of safety science (Hale and Swuste 1998). In the case of high-level heuristics, experts are given a limited number of more or less general guidelines, which are formulated as design aims rather than as specific design specifications. The guidelines define the goal, without specifying how it should be achieved. An example is the advice to "work to ensure that users will view and notice links" (Farkas and Farkas 2000). High-level guidelines rely strongly on experts' professional knowledge to assess which design options are most suitable to achieve these aims. In the case of low-level heuristics, experts are given a larger set of detailed guidelines, which are formulated as design specifications rather than as design aims. The guidelines define a concrete action or a required state of the Web site. Low-level guidelines rely less strongly on experts' professional judgments about suitable design options, but instead are more likely to prescribe the desired action. "Well-established cues such as underlining and the raised 'button' appearance should be used to indicate links. Do not use these cues for other purposes," is an example of a low-level guideline (Farkas and Farkas 2000).

In this article, we present the results of a quantitative and qualitative study in which the contribution of heuristics was examined through a detailed comparison of experts' evaluation results, both with and without heuristics. The study involves a within-subjects design in which 16 participants first conducted an unguided evaluation, and, after that, used a set of heuristics to evaluate a municipal Web site. One half of the participants worked with high-level heuristics and the other half with low-level heuristics. Our analysis focuses on the annotations (the problem detections and positive remarks) made by the experts. Specifically, the following questions are addressed:

- What does the unguided expert evaluation tell us about the validity of the heuristics?
- Are there any differences between heuristic and unguided expert evaluation regarding the number and types of annotations made?
- Do high-level and low-level heuristics have different effects on the annotations made by experts?

Our study contributes to the knowledge about heuristic evaluation in several respects. The results from the unguided evaluation are used to check whether the heuristics concerned actually cover all relevant aspects of navigation and comprehensibility. If experts implicitly use criteria in their unguided evaluations that are covered by the criteria in the heuristics, the heuristics can be said to reflect the knowledge in the field. A comparison of the numbers and types of annotations made in the unguided and the heuristic evaluation will shed light on the added value of heuristics for expert evaluations

The comparison of the effects of high-level and low-level heuristics is a first attempt to check whether this potentially important design feature of heuristics actually affects their usefulness. On the one hand, high-level heuristics may be more easily incorporated in the evaluation process (because they are easier to memorize and because it is easier to gain an overview of the complete set of heuristics) and may facilitate the detection of a broader range of usability problems (because of their goal instead of action orientation). On the other hand, low-level heuristics provide more specific cues for detecting usability problems.

Materials and Methods

Participants

Sixteen communication professionals participated in this study. Considering Saroyan's (1993) finding that differences in experts' background and perspective may lead to process and outcome differences, we kept the background of the participating experts as similar as possible. Experts were defined as communication professionals with a master's degree in communication studies at the University of Twente and with at least one year of professional experience of designing and/or maintaining Web sites. Participants were approached through the university's alumni network. As an incentive, they received a gift voucher and a summary of the results of the study.

Participants' professional experience ranged from one to seven years (mean = 3.5 years). Most communication professionals worked as communication officers in commercial and noncommercial organizations (including municipalities), with responsibility for one or more Web sites. Four communication professionals were responsible for Web sites as part of another function (for example, a consultant in a small firm taking on the responsibility for a company Web site). Two communication professionals ran a commercial usability laboratory. The participants' age ranged from 25 to 36 years (mean = 28.5 years). Seven men and nine women participated.

Procedure

The communication professionals evaluated two parts of a municipal Web site. During the evaluation, they had to think aloud and record their positive and negative comments using *Infocus*, a software program for evaluating Web sites developed by Utrecht University. *Infocus* works as a normal Web browser but also offers the experts the opportunity to make screenshots; annotate them with boxes, lines, and arrows; and add explanatory text if they notice a problem or positive feature in the Web site (see Figure 11.1). It is also possible for the research coordinator to offer evaluators a list of annotation categories to facilitate later analysis of the annotations.

First, the communication professionals performed an unguided evaluation for 25 minutes to assess their normal evaluation style. This evaluation was unguided in the sense that the communication professionals did not receive any (external) guidance as to how to perform the evaluation. The only instruction they received was to pay special attention to the navigation and comprehensibility of the Web site. In this session, they had to assign every annotation to the single category "General." After the unguided evaluation, they were presented with either high-level or low-level heuristics on the navigation and comprehensibility of Web sites (see Figure 11.2 and the subsection "Heuristics" for examples). They were first asked to read the heuristics and give a first impression about their usefulness. After that, the communication professionals evaluated a second part of the Web site using the heuristics (again, for 25 minutes). In this session, they had to assign a heuristic category to each annotation. This evaluation was followed by a structured interview on their experiences and the annotations they had produced. In this article, we used the interviews mainly to solve unclear annotations. In a separate article, we will use the observation, think-aloud, and interview data to analyze further the process characteristics of experts using heuristics.

The entire session lasted for between two and three hours. All communication professionals except two were asked to stop when they were still busy evaluating. Two of the communication professionals finished evaluating a section within the time: one during the unguided evaluation (18 annotations) and the other during the heuristic evaluation (8 annotations). The sessions took place in quiet offices or (conference) rooms at the workplace, at home, or at the university.

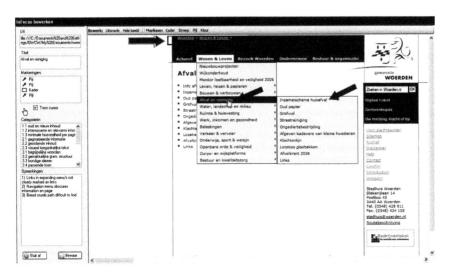

Figure 11.1 Example of an Annotation Screen in *Infocus*

Note: On the left side of the screen, one sees boxes that display the URL and title ("Titel") of the page, the markers used ("Markeringen"), the coding categories for the annotations, which in this case are from the items in the heuristics ("Categorieen"), and the text comments about the Web page ("Opmerkingen"). On the bottom left, the buttons for cancel ("Sluit af") and save ("Bewaar") are placed. The top bar contains buttons for different markers ("Markeer"), such as box ("Kader"), line ("Streep"), arrow ("Pijl"), and the color of these markers ("Kleur"). Note the arrows toward the unmarked links in the expanding navigation menu and the box around the bread crumb path.

Excerpt from Spyridakis (2000):
3.1 Use words that readers can easily and accurately understand.
Effective text features:
* Concrete words.
* Words that appear frequently in the language.
* Short words (fewer syllables).
* Pronounceable words.
* Link labels that create clear context for the linked page.
* Words that readers are familiar with (the audience's vocabulary set).

Excerpt from high-level heuristics:
3.1 Use words that readers can easily and accurately understand.

Excerpt from low-level heuristics:
3.1 Comprehensible words
* Concrete words.
* Words that appear frequently in the language.
* Short words (fewer syllables).
* Pronounceable words.
* Link labels that create clear context for the linked page.
* Words that readers are familiar with (the audience's vocabulary set).

Figure 11.2 Example of Guidance Regarding Comprehensibility

From Spyridakis (2000) and the corresponding comprehension item in the high-level and low-level heuristics.

Web Site

The Web site used in this study was the municipal Web site of Woerden. The target group for this Web site is very broad, consisting of the citizens and resident organizations of the municipality, and visitors interested in coming to the municipality for personal or professional reasons. The Web site contains information about all aspects of life in the municipality, ranging from policy information to passport applications; from municipal taxes to sightseeing spots. Annually, the Dutch government evaluates all Web sites by municipalities, provinces, boards of public works, and departments to see whether they conform to criteria of transparency, service quality, participation, and accessibility. The Web site of Woerden ranked 215th in the 2006 ranking for municipalities (total 485 places), which makes it an average-quality municipal Web site. To ensure that the Web site remained the same during the research period, an offline copy was used.

Both sections used were part of the "Dwelling and Living" domain of the Web site. One section focused on "Garbage and Cleaning" and the other on "Education, Sports, and Well-Being." These two sections were chosen because of their relative length—needed for the time duration of the evaluation—and their similar number of subpages. In addition, the topics were general enough for the experts not to need specialized expertise about the content. The order of the Web site parts alternated:

one half of the communication professionals conducted their unguided evaluation using the "Garbage and Cleaning" section, and the other half started with "Education, Sports, and Well-Being."

Heuristics

The heuristics used were developed by combining two sets of research-based guidelines published in the autumn 2000 special issue of *Technical Communication*. The first set offers advice on designing Web site navigation (Farkas and Farkas 2000), and the second focuses on the comprehensibility of Web pages (Spyridakis 2000). Of the comprehensibility heuristics, only the first three sections were used; to maintain a strong focus on comprehensibility, the sections on credibility and globalization were eliminated. Both sets of heuristics consist of general criteria formulated in one or two sentences, supplemented with several "key points" or "effective text features."

The sentences can be seen as high-level heuristics, whereas the key points or effective text features are examples of low-level heuristics. To create a high-level set of heuristics, only the high-level sentences were used. In the low-level set of heuristics, the sentences were summarized in a short headline, and all the key points were presented underneath. The high-level heuristics consisted of one page. The low-level heuristics consisted of four pages. Examples of high-level and low-level heuristics are given in Figure 11.2. The examples illustrate why the length of the two heuristics differed so much: in low-level heuristics, many more items are needed to cover the same content as one item in high-level heuristics.

Analysis

The evaluation sessions were recorded and analyzed with the usability software program *Morae*, and a Webcam. *Morae* records the desktop activity, audio, and Webcam video, and synchronizes all into a single file, which can be further analyzed. We marked every change of page to assess which pages the communication professionals had seen, as well as the start of all *Infocus* annotations to assess when communication professionals decided to make an annotation. In addition, we used the Infocus output to analyze the content and nature of the annotations.

Although the communication professionals were asked to write down only one comment per annotation, some annotations did contain two or more comments. Therefore, all annotations were first checked for the occurrence of multiple comments. Two annotations were set aside, because they did not contain any comment on the Web site. This resulted in a set of 466 comments that formed the dataset for further analysis.

The first step in our data analysis involved the coding of the annotations made by the communication professionals. All comments were independently categorized by the first two authors. The coding scheme (Table 11.1) was initially based on the heuristics that the communication professionals used.

Table 11.1 Overview of Heuristic Coding Scheme for Annotations

Level 1	Level 2	Level 3
Comprehension (C total)	Comprehension general (C general)	
	Selection and presentation of information (C1)	Presentation that facilitates orientation (C1.1)
		Selection of relevant and interesting information (C1.2)
		Limited amounts of information per page (C1.3)
	Organization of content on page (C2)	Grouping of content (C2.1)
		Logical order of content (C2.2)
		Visually accessible and scannable content (C2.3)
	Style and language (C3)	Comprehensible words (C3.1)
		Comprehensible syntax (C3.2)
		Conciseness (C3.3)
		Appropriate tone of voice (C3.4)
Navigation (N total)	Navigation general (N general)	
	Design of effective links (N1)	Links that are recognizable as links (N1.1)
		Noticeable links (N1.2)
		Clear link destinations (N1.3)
	Management of large numbers of links (N2)	Effective breadth/depth ratio in hierarchy (N2.1)
		Combination of primary and secondary links (N2.2)
		Appropriate converging of hierarchy branches (N2.3)
		Interface design that reveals underlying structure (N2.4)
		Effective breadth/depth ratio in hierarchy (N2.1)
	Provision of orientation information (N3)	Orientation information on home page (N3.1)
		Orientation information on lower pages (N3.2)
	Augmentation of link to link navigation (N4)	Sitemaps (N4.1)
		Search facilities (N4.2)
		Link to homepage on each page (N4.3)
Other (O total)		

The categories in level 3 were directly based on the heuristics. The categories Comprehension general, Navigation general, and Other were added, based on a first analysis of the annotations.

During the coding process, this scheme was deemed too limited, and three extra categories were added. These categories were "Comprehension general," "Navigation general," and "Other." The categories "Comprehension general" and "Navigation general" were added because we encountered annotations that clearly had to do with comprehension or navigation issues but were not covered by items in the heuristics. "Image supports text. Image gives more variation on the screen" is an example of such an annotation.

The category "Other" was added to account for possible annotations that did not fit in with the navigation and comprehension heuristics at all—for example, "It may be a good idea to also offer the possibility to e-mail about complaints." All comments were also coded as either positive or negative. We used Cohen's kappa, a test that would assess the consistency of the work performed by the two coders. Cohen's kappa was 0.69 for the heuristic coding and 0.91 for the positive/negative coding, which indicates a satisfactory to almost perfect intercoder reliability (Landis and Koch 1977).

We ran a few statistical tests on the data to test whether differences in evaluator behavior were significant. We wanted to know, for example, whether the types of annotations produced during unguided evaluations are significantly different from the types of annotations produced during heuristic-based evaluations, and we wanted to know whether the annotations produced during evaluations using high-level heuristics are significantly different from the types of annotations produced during evaluations using low-level heuristics. Having examined similarities and differences between these different evaluation approaches, we moved to the heart of our study, which is the in-depth analysis of the content of annotations produced by the participating communication professionals. The statistical tests allowed us to consider whether different approaches are significantly different; the qualitative analysis allowed us to consider the nature of the differences and to delve into the specific results produced in the three different evaluation approaches that we considered.

Results

We will first discuss the validity of the heuristics and then go into the comparison between unguided and heuristic evaluation. After that, we will address the comparison between high-level and low-level heuristics and describe in more detail the comments in both conditions on one particular Web page. Differences were tested using nonparametric tests. These statistical tests are most appropriate when the data come from a small sample and are not normally distributed, as was the case in this study.

Validity of the Heuristics

An important criterion for the validity of heuristics is that they reflect state-of-the-art knowledge about effective Web site design (de Jong and van der Geest 2000).

The unguided evaluation data from our study can be used to check whether the heuristics indeed cover all (or most) relevant aspects of navigation and comprehensibility. To assess the heuristics' validity, we examined which percentage of all navigation and comprehension comments made by the communication professionals were covered by the specific guidelines in the two heuristics.

Of all comments regarding navigation, 69 percent had a clear relation with the guidelines offered in the heuristics. This means that the content of the annotations was very similar to the content, and sometimes even the wording, of an element in the heuristics. An example of this was the annotation, "It is not clear to me what I can find behind this link. Is there more information? Or something else?," which is very similar to the N1.3 item of the navigation heuristics ("Be sure that all links clearly indicate their destinations").

Of all comments regarding comprehension, 86 percent corresponded to the guidelines in the heuristics. An example is the annotation, "Try to make the text a bit more scannable by separating the list from the other text," which even uses some of the wording of the C2.3 item of the comprehension heuristics— "Use organizational cues to make text visually accessible and scannable (easily skimmed or quickly read through at a top level); and to facilitate search tasks, comprehension, and recall. Do not distract readers with unnecessary cues."

Combined, the heuristics covered 78 percent of all comprehension and navigation annotations made by the communication professionals. These figures confirm that the content of the heuristics matches the expert knowledge of communication professionals about navigation and comprehension and thereby confirm the relevance of these heuristics for experts who want to evaluate Web sites on these issues. At the same time, however, they question the novelty value of the heuristics: given the fact that communication professionals more or less naturally adopted many of the same evaluation criteria as those comprised in the heuristics, the heuristics' actual contribution to the detection of user problems cannot be expected to be very strong. If experts are already inclined to look at the visibility of links, getting advice from the heuristics to look at this issue will not lead to additional problem detections.

An important omission in the navigation heuristics involved possible disorientation after users had clicked a certain link. The heuristics cover the clarity of the destination of links but do not sufficiently cover the requirements of the destination page. For instance, communication professionals were concerned about users' disorientation when a link opened a new screen, which seemed to replace the original Web site. They also criticized the fact that users landed smack dab in the middle of the destination place and were forced to start looking for the desired information all over again. An omission in the comprehension heuristics involved the possible contribution of images to help the users visualize and understand the information offered. One example is: "Now that I've seen page X with a photo, it might be handy to add a picture of the blue garbage container on this page, so that everybody knows what is meant by that." This and similar annotations might warrant an extra subcategory in the heuristic "Visual support for information."

Comparison Between Unguided and Heuristic Evaluation

In total, the communication professionals wrote down 466 annotations: 269 in the unguided evaluation and 197 in the heuristic evaluation. The communication professionals differed in the number of comments they made within the same time frame. The mean number of comments in the unguided evaluation was 16.8 (range, 6 to 31), and the mean number of problems in the heuristic evaluation was 12.3 (range, 6 to 28). The trend is toward a lower number of comments in the heuristic evaluation (Wilcoxon signed-rank test, $Z = 2.93$, $P < 0.005$; this test result indicates that it is unlikely that the trend is accidental or random but rather that it is statistically significant). (Note: Like the familiar t-test, the Wilcoxon signed-rank test compares the differences between measurements, but without making assumptions as to how the measurements are distributed.) The *Morae* footage shows that this difference may be attributed to the time communication professionals needed to find the appropriate heuristics to categorize their comments. This categorizing time seems to have slowed them down. In general, this may mean that experts performing a heuristic evaluation for the first time need more time to complete it.

Subdivision in Positive and Negative Annotations Between Evaluation Conditions

Heuristic evaluation may yield both positive and negative comments on a Web site, because it also directs evaluators' attention to areas without problems ("There are no problems with language use" translates into "The language use is good"). In this study, it is interesting to see that only those communication professionals who created positive annotations in the unguided evaluation also created positive annotations in the second round. The number of negative annotations was significantly lower in the heuristic evaluation (10.3) than in the unguided evaluation (13.9; Wilcoxon signed-ranks test, $Z = 3.22$, $P < 0.01$). There were no differences in the number of positive annotations between the unguided and heuristic evaluation (Wilcoxon signed-ranks test, $Z = 1.01$, not significant).

Subdivision in Heuristic Categories Between Evaluation Conditions

If we look at the level 1 coding—that is, the subdivision over the three categories Comprehension, Navigation, and Other—significant differences were found between the unguided and heuristic evaluation in the number of annotations within the categories Navigation and Other. The number of "Other" annotations fell from 3.3 in the unguided condition to 1.4 in the heuristic condition (Wilcoxon signed-ranks test, $Z = 2.05$, $P < 0.05$). Surprisingly, however, the mean number of "Navigation" annotations also dropped from 5.5 to 3.7 in the heuristic evaluation (Wilcoxon signed-ranks test, $Z = 2.11$, $P < 0.05$). The number of annotations in the category "Comprehension" did not differ between the two evaluation conditions (Wilcoxon signed-ranks test, $Z = 0.83$, not significant).

On the second level, there were no differences in the number of "Comprehension" annotations per criterion between the two evaluation conditions. Within the "Navigation" annotations, however, one significant difference was found: the number of navigation problems that did not correspond to the specific navigation criteria included in the heuristics dropped from 1.8 in the unguided evaluation to 0.7 in the heuristic evaluation (Wilcoxon signed-ranks test, $Z = 2.48$, $P < 0.05$). For the other second-level "Navigation" annotations, no differences between the evaluation modes were found.

The number of annotations in all the third-level categories did not differ between the unguided and heuristic evaluations, except for one category. These annotations involved the heuristic C1.2, "Selection of relevant and interesting information." In the heuristic evaluation, the communication professionals created significantly less annotations in this category than in the unguided evaluation: 2.8 in the unguided evaluation versus 1.5 in the heuristic evaluation (Wilcoxon signed-ranks test, $Z = 1.99$, $P < 0.05$).

If we look at the content of these C1.2 annotations, they are usually very specific, asking for more information regarding the topic of the page. An example is: "Now that I have found the information [about the city dump], I miss information about whether the garbage needs to be presented/packaged in a special way or not." In the heuristic evaluation, we find several more general annotations that copy the wording of this heuristic but that do not go into details. Examples are: "I miss relevant information," and "Residents find this interesting and useful information." In the second case, the lack of specificity is not a problem, but a designer needing to remedy the first annotation will have problems deciding what needs to be added.

Comparison of High-Level and Low-Level Heuristics

No differences were found between the high-level and low-level heuristics in the number of annotations, the subdivision in positive and negative annotations, and the subdivision in the different heuristic categories.

In-Depth Analysis of Annotations

The general trend seen in the previous section was that, under the influence of the heuristics, the annotations were increasingly focused on navigation and comprehension issues covered by these heuristics. To see the effect in terms of actual annotations, we will now take a closer look at the annotations regarding a page that all communication professionals evaluated: the "Sports Policy" Web page. This page is a part of the "Education, Sports, and Well-Being" section and is listed at the top of the overview page and the navigation menu, so all communication professionals were likely to see this page early in their evaluations. It contains a medium amount of content and has some secondary links to related information, both in the text and at the bottom. General secondary links, such as contact, sitemap, accessibility, and the link to the English-language site, are listed

Figure 11.3 Screen Shot of "Sports Policy" Web Page in Infocus Browser

Note: The arrows in the top bar are backward and forward buttons that function just like similar buttons in other browsers. The central white box contains the URL of the current page and can be used to navigate to other pages by typing in an address. The camera/annotation button ("Bewerk") is placed on the right. Clicking on this button opens the annotation screen as seen in Figure 11.1. The camera is symbolic for the screen shot Infocus makes of the current Web page and that can subsequently be annotated in the special annotation screen.

on the right-hand side of the page (see Figure 11.3). Most of the heuristics could be applied to this page.

In all, the communication professionals made 29 annotations in the unguided evaluation and 39 annotations in the heuristic evaluation. Similar to the general trend of increasing specificity, the number of "Other" annotations dropped from 9 to 2, and the number of general "Comprehension" (C-general) and "Navigation" (N-general) annotations dropped from 3 to 0. In contrast with this decrease in general annotations, the number of annotations that corresponded directly to specific heuristics rose considerably. The number of annotations that had to do with the selection and presentation of content (C1) rose from 6 to 10 and the annotations about style and language (C3) increased from 1 to 10. There was also an increase in the number of navigation annotations that referred to the design of effective links (N1), which went from 3 to 7.

The drop in the number of "Other" annotations between the two conditions was consistent with the general picture of this study, but the content of these annotations also became more specific. The subjects of the "Other" annotations in the unguided evaluation showed a wider variation. They often regarded the general secondary links on the right-hand side of the screen ("The text button for partially sighted people is too small") and aspects of the layout ("Clear, readable, no strange colors in text or background" and "The text itself is also relatively small"). In contrast, the only two "Other" annotations in the heuristic evaluation were: "Well-organized site, not crowded, no banners, etc.," and "It would be convenient to open this type of downloads in a new window, so that when you close the download, you do not close the site of woerden.nl."

The content of these two annotations is much closer to the subject of the heuristics. The heuristics mention that links on a Web page should be noticeable and that the text should be visually accessible and scannable. Both aims can be achieved with well-structured pages that are not overcrowded. Regarding the second annotation, the heuristics pay attention to issues such as clear indications of the destination of external and internal links. Saying that downloads should open in a different window is not a big leap from that guideline.

In the content of the "Comprehension" annotations, the influence of the heuristics can also be seen. Six annotations in the unguided evaluation dealt with the selection and presentation of the content (C1). Comments were made about the amount of text on the page (three annotations), its relevance to the user (two annotations), and the desirability of publishing the address of and directions to the town hall on every page of the site (one annotation). Interestingly, one communication professional thought that there was "quite a lot of text to read," whereas two communication professionals said, about the same page, it was "short: on one page, I like that." The two communication professionals commenting on the relevance of the text both wondered about "the added value" and "the relevance" of the information "for the citizen."

In the heuristic evaluation, similar concerns about the length of the text (three annotations) and the relevance for the citizen (four annotations) were raised. As in the unguided evaluation, the communication professionals in the heuristic evaluation did not agree about the length of the text. One commented, "Short page, not a long-winded story," whereas number 2 thought that "this is about the maximum amount of information I would put on a Web page," and number 3 said, "This story looks like it is too long. I would shorten it or present a list in between." The communication professionals making annotations about the relevance showed a stronger agreement in their doubts regarding the relevance of the information for citizens of Woerden.

In addition to the annotations about length and relevance, three communication professionals were not satisfied with the content on the page and asked for more information about specific subjects ("What are the consequences? What are the changes?").

The picture of increased specificity in the annotations is even clearer when we look at the annotations that deal with style and language (C3). In the unguided evaluation, only one communication professional commented that there was "maybe a superfluous word" on the Web page. In the heuristic evaluation, however, the communication professionals saw many more problems with the style and language. There were problems with "difficult words" (four annotations) and "difficult syntax" (two annotations); the text contained "unnecessary details" (one annotation), and the communication professionals "would choose another tone to address inhabitants; it is now much too official" (three annotations). These examples show that the communication professionals suddenly seemed much more focused on the selection and presentation of information and the style and language used. The use of the same terms in both the annotations and the heuristics are another indication of this influence.

If we look at the annotations regarding the design of effective links (N1), two of the communication professionals in the unguided evaluation commented on the quality of the secondary links on the right-hand side of the screen. Their annotations read:

> Certainly the navigation at the right side raises some questions because of general terms like "digital counter" and "municipal guidebook" (does not refer to content, more to services).

> Introduction, is meant for English people. I don't think it's very logical. I would make it more eyecatching that this is the English version.

On the other hand, another communication professional stated that there was "structure in the information: links are clear." In the heuristic evaluation, the annotations were a bit more specific: five annotations regarded a link that did not state its destination. The complaints ranged from "It is not clear this link starts a dialog for downloading," to "Preferably indicate the size of the file that can be downloaded"; and from "Is this a site from the municipality or a commercial site or the like; where will I end up or what does this site mention?" to "Use link labels." In contrast, judging by the annotations "The links are clear" and "Here, links are clearly accentuated," two communication professionals were positive about the quality of the links.

On the basis of the annotations given, it seems the communication professionals were already aware of the necessity of having recognizable links with a clear, unambiguous name but that the heuristics reinforced this awareness and alerted them to the possibility of adding more destination information than just an informative link name. Under the influence of the heuristics, they were more specific in their annotations. Also, as was the case with the comprehension annotations, the influence of the navigation heuristics was visible in the use of the same terminology, such as link labels and download dialog.

To summarize, this analysis of the exact wording of annotations has shown how the general trends from the more general analyses can also be seen in the content of very specific annotations. The content of the "Other" annotations had more bearing on the subject of the heuristics, and the formulation of the comprehension and navigation annotations was more specific, sometimes even using the same terminology as the heuristics.

Discussion

The results of our study confirm the validity of the heuristics on navigation (Farkas and Farkas 2000) and comprehensibility (Spyridakis 2000): both heuristics strongly reflect the state-of-the-art knowledge that the communication professionals in our study brought to the unguided evaluation task. At the same time, however, our results raise questions about the practical usefulness of the heuristics in this particular setting. A remarkable result is the decrease in the number of annotations between the unguided and heuristic evaluation, because the communication professionals needed time to find the appropriate heuristic to label the annotation.

This does not necessarily correspond to a structural disadvantage of heuristics but underlines two unfavorable aspects of the heuristic evaluation under study, which must be considered by organizations and professionals who think of adopting heuristic evaluation. First, the practical usefulness of heuristics depends on their validity and novelty value, and there is often a tension between the two criteria. The heuristics used in this study proved to reflect strongly the knowledge in the field about navigation and comprehensibility, but had relatively little novelty value for this particular group of communication professionals. The usefulness of heuristics can only be assessed by considering the prior knowledge and evaluation practices that communication professionals already have. Second, the practical usefulness of heuristics will probably be enhanced when communication professionals become more experienced with them. Heuristics will probably be more beneficial when organizations or experts have adopted them as a standard evaluation procedure for many Web sites and/or when they are introduced in an educational program.

Another important effect of the heuristics is that they appeared to focus experts' attention to navigation and comprehensibility. The heuristics seem to cause communication professionals to limit their attention to issues that fit within the framework of the heuristics. Two results are indicative for this phenomenon. One indication is a drop in the category "Other" annotations between the unguided and the heuristic evaluation. The use of the heuristics did not lead to the detection of more navigation and comprehension problems but to a decrease of problem detections that did not correspond to "Navigation" and "Comprehension." The heuristics narrowed the experts' attention for user problems, or, more positively formulated, gave them a clearer focus. Another, more unfavorable indication is the decrease of navigation annotations that did not correspond to the specific navigation guidelines. Working with the heuristics narrowed the communication professionals' views on the various aspects of user-friendly navigation.

The distinction between high-level and low-level heuristics did not seem to have any impact on the annotations of the communication professionals in this study. We can only speculate on the reasons for this finding. This might be because of the short period between the introduction and the use of the heuristics. Maybe with a longer training period or a longer duration of the evaluation, the influence of the type of heuristics will be more pronounced.

Limitations of the Study

This study has two limitations. First, the experts had to use the heuristics immediately after receiving them. They had difficulty in finding a modus operandi to integrate the heuristics into their evaluation process. Maybe with more time, the use of heuristics would become more natural for them, and they would have the chance to internalize (parts of the) heuristics. This might speed up the evaluation process and lead to more, and more diverse, annotations. In future research, we will therefore focus on more structural ways of using heuristics to evaluate Web sites. Second, the assignment to think aloud may have affected the communication professionals' evaluation process in both conditions. It is not likely that the cognitive load on the professionals was too high, because the evaluation of Web sites was not a very novel or complex task for them. However, the assignment to think aloud probably slowed down the evaluation process and might have urged participants to work more systematically than they would have done in normal evaluation settings.

Practical Implications

Practitioners who consider conducting a heuristic evaluation of a Web site for the first time need to be aware that this may take up more time and energy than an unguided evaluation from their own expertise. In addition, the subject of the heuristics needs to be chosen carefully, because their focus will be mostly limited to this subject. However, if professionals find a modus operandi for incorporating the heuristics in their work process, heuristic evaluation may be a valuable method. Such a modus operandi could be a combination of the unguided and heuristic evaluation, in which each page is first scanned for problems that "jump to the eye" and subsequently evaluated according to the (order of the) heuristics. In addition, after the evaluation is completed, the annotations of both unguided and heuristic evaluations can be categorized according to the heuristics to achieve a more structured discussion of the results.

Acknowledgment

*This article was originally published in *Technical Communication* (2008), 55: 392–404.

References

Bailey, R. W., R. W. Allan, and P. Raiello. 1992. "Usability Testing Vs. Heuristic Evaluation: A Head-To-Head Comparison." *Proceedings* of the Human Factors and Ergonomics Society, 36th Annual Meeting, 409–13. Santa Monica, CA: HFES.

Bastien, J.M.C., D. L. Scapin, and C. Leulier. 1999. "The Ergonomic Criteria and the ISO/DIS 9241–10 Dialogue Principles: A Pilot Comparison in an Evaluation Task." *Interacting with Computers* 11: 299–322.

Connell, I. W., and N. V. Hammond. 1999. "Comparing Usability Evaluation Principles with Heuristics: Problem Instances Versus Problem Types." In *Human–Computer Interaction—INTERACT '99*, ed. M. A. Sasse and C. Johnson. Amsterdam, Netherlands: IOS Press.

de Jong, M., and P. J. Schellens. 1997. "Reader-Focused Text Evaluation: An Overview of Goals and Methods." *Journal of Business and Technical Communication* 11: 402–32.

——, and ——. 2000. "Toward a Document Evaluation Methodology. What Does Research Tell Us About the Validity and Reliability of Evaluation Methods?" *IEEE Transactions on Professional Communication* 43: 242–60.

——, and T. van der Geest. 2000. "Characterizing Web Heuristics." *Technical Communication* 47: 311–26.

Desurvire, H. W. 1994. "Faster, Cheaper!! Are Usability Inspection Methods as Effective as Empirical Testing?" In *Usability Inspection Methods*, ed. J. Nielsen and R. L. Mack. New York, NY: John Wiley.

Farkas, D. K., and J. B. Farkas. 2000. "Guidelines for Designing Web Navigation." *Technical Communication* 47: 341–58.

Faulkner, L. L. 2006. "Structured Software Usability Evaluation: An Experiment in Evaluation Design." PhD dissertation, University of Texas at Austin.

Fu, L., G. Salvendy, and L .Turley. 2002. "Effectiveness of User Testing and Heuristic Evaluation as a Function of Performance Classification." *Behaviour & Information Technology* 21: 137–43.

Hale, A. R., and P. Swuste. 1998. "Safety Rules: Procedural Freedom or Action Constraint?" *Safety Science* 29: 163–77.

Hvannberg, E. T., E. L.-C. Law, and M. K. Lárusdóttir. 2007. "Heuristic Evaluation: Comparing Ways of Finding and Reporting Usability Problems." *Interacting With Computers* 19: 225–40.

Jeffries, R. M., J. R. Miller, C. Wharton, and K. Uyeda. 1991. "User Interface Evaluation in the Real World: A Comparison of Four Techniques." In *Proceedings* of the SIGCHI Conference on Human Factors in Computing Systems: Reaching Through Technology, ed. S.P. Robertson, G.M. Olson, and J.S. Olson. New York, NY: ACM.

Landis, J. R., and G. G. Koch. 1977. "The Measurement of Observer Agreement for Categorical Data." *Biometrics* 33: 159–74.

Nielsen, J. 1992. "Finding Usability Problems Through Heuristic Evaluation." In *Proceedings* of the SIGCHI Conference on Human Factors in Computing Systems, ed. P. Bauersfeld, J. Bennet, and G. Lynch. New York, NY: ACM.

——. 1994. 'Heuristic Evaluation." In *Usability Inspection Methods*, ed. J. Nielsen and R. L. Mack. New York, NY: John Wiley.

Paddison, C., and P. Englefield. 2004. "Applying Heuristics to Accessibility Inspections." *Interacting With Computers* 16: 507–21.

Saroyan, A. 1993. "Differences in Expert Practice: A Case from Formative Evaluation." *Instructional Science* 21: 451–72.

Schriver, K. A. 1989. "Evaluating Text Quality: The Continuum from Text-Focused to Reader-Focused Methods." *IEEE Transactions on Professional Communication* 32: 238–55.

Spyridakis, J. H. 2000. "Guidelines for Authoring Comprehensible Web Pages and Evaluating their Success." *Technical Communication* 47: 359–82.

Sutcliffe, A. 2002. "Assessing the Reliability of Heuristic Evaluation for Website Attractiveness and Usability." In *Proceedings* of the 35th Annual Hawaii International Conference on System Sciences (HICSS'02), vol. 5. Washington, DC: IEEE Computer Society.

Tao, Y.-H. 2008. "Information System Professionals' Knowledge and Application Gaps Toward Web Design Guidelines." *Computers in Human Behavior* 24: 956–68.

van der Geest, T., and J. H. Spyridakis. 2000. "Developing Heuristics for Web Communication: An Introduction to this Special Issue." *Technical Communication* 47: 301– 10.

Vredenburg, K., J.-Y. Mao, P. W. Smith, and T. Carey. 2002. "A Survey of User-Centered Design Practice." In *Proceedings* of the SIGCHI Conference on Human Factors in Computing Systems. Minneapolis, MN: ACM.

Wright, P. 1985. "Editing: Policies and Processes." In *Designing Usable Texts*, ed. T. M. Duffy and R. Waller, 63–96. Orlando, FL: Academic Press.

Chapter 12

Making and Acting

Ethnographic Development
of a Case Study Approach*

*Thomas Vosecky, Marika Seigel,
and Charles Wallace*

Editors' Introduction

Vosecky, Seigel, and Wallace provide an interesting account of an innovative use of qualitative research techniques to create classroom simulations of the complex working environment of technical communicators. The methods described in their article could easily be adapted to other situations in which technical communicators need to achieve a detailed understanding of the workplace realities of customers, users, and readers.

Vosecky and colleagues note that technical communicators inhabit workplaces where interdisciplinary teams must reach decisions about how to balance the needs of customers and the needs of their employers. Many of the skills needed to succeed at these endeavors are not governed by general rules that apply to all social situations but rather are based on contextual knowledge that varies from situation to situation. To account for this complexity the authors distinguish between *techné* (or making) and *phronesis* (acting), where the former is essentially rule-based technical or rational activity that is relatively easy to teach, while the latter is more concerned with how human values are intertwined with social action. They are saying that although we can teach new technical communicators how to make documents, we cannot teach them how to act effectively and appropriately in all workplace social environments and situations. Instead, we have to invent ways that allow new technical communicators to learn for themselves how their actions produce different outcomes in the complex workplaces where they will one day be employed. As they comment, "Working *with* others, rather than just *for* them, is becoming the valued action of the workplace."

Vosecky and colleagues offer a solution to this dilemma that makes innovative use of qualitative methods. Although it may be impossible to duplicate the rich social environments of the workplace for students, it is possible to create simulations. The authors describe how they used qualitative data-gathering and analysis methods to create detailed and realistic case studies of the communication challenges that arise in software development teams, and how they use these case studies in their education of technical communicators.

The authors use a case study design that relies upon ethnographic data-gathering methods, and their project was approved by an institutional review board. They used their data to construct hyperlinked case studies that include text, audio recordings, and videos (you can see their recorded case studies at www.speaksoft.mtu.edu/cases). Their research participants were senior computer science students who were carrying out their workplace practicums, as well as the various clients and stakeholders with whom the students interacted. They gathered data through document analysis, participant observation (which included the creation of interaction diagrams), audio recordings, video recordings, field notes, and semi-structured interviews. They used the data to create detailed chronological descriptions of the case, as well as more succinct thematic accounts (and they created these latter accounts by using the coding and theming techniques of grounded theory).

The authors also explain that they decided to present the cases as "realist tales," referring to van Maanen's (1988) excellent monograph on the alternative ways in which qualitative researchers construct their research reports. They wanted their case studies to offer a close and realistic simulation of workplace experience so students can in effect become participants in the actions that are presented. It is easy to see why they selected qualitative methods, which avoid the reductionism inherent in quantitative statistical analysis, and instead promote the development of thick descriptions of contextual situations.

Vosecky and colleagues touch on several of the themes discussed by Giammona. Their emphasis on the complex social environment of technical communication workplaces sheds light on our evolving identity and role; their focus on end-to-end development projects also shows one way in which technical communicators work within interdisciplinary teams to create new innovations. Most importantly, though, they offer insights into one way in which we can educate future practitioners to become architects of both information environments and social environments.

Reference

van Maanen, J. (1988). *Tales of the Field: On Writing Ethnography*. Chicago, IL: The University of Chicago Press.

Speaking of Software . . . Again

In a 2006 issue of *Technical Communication*, Brady and colleagues reported on the interdisciplinary project, "Speaking of Software: Integrating Communication and Documentation Techniques into an Undergraduate Software Engineering Curriculum." As you may recall, this is a project funded by the National Science Foundation (NSF), intended to "increase opportunities for all those participating to view communication as a rhetorical act and to integrate the theories and pedagogies of technical communication and software development" (Brady and colleagues 2006, 318). To reach this goal, project participants developed case studies to be used in software engineering courses to teach the rhetorical complexities of communicating with stakeholders—from the client to the user—in the software development process. These case studies are available online at http://www.speaksoft.mtu.edu/cases/. At the time that this first article was written, one case study had been developed but had not yet been used in the classroom. Now, three years into the project, we can report on some of the results of using these case studies in the classroom, and offer unexpected insights that the qualitative methods used to develop them have yielded.

The literature of software engineering presents the need for capturing recurrent difficulties in the process of software development. In other words, groups of individuals often create a massive and valuable knowledge base concerning the process of project design, developing effective means of dealing with the ambiguities of the situation as they go. Our ethnography-based method captures that knowledge from the lived experience of groups and individuals. Incorporating this knowledge and experience into pedagogical tools, our cases have the potential to instill that experience in others through their reflection on and imitation of the concrete examples of situations that can arise in any workplace.

Our initial goal was to develop case studies for software engineering classroom applications. This goal was complicated by the proprietary nature of the processes and products found in the workplace. This remains particularly true in software development, where the need for secrecy often outweighs pedagogical goals and prevents students and faculty from sharing their workplace experiences. To avoid this problem, we turned to the students themselves. Although this may be considered a convenience sample in some respects, we believe it is a valuable and representative one. The software engineering program requires a "Senior Design Project" capstone course, where computer science (CS) students work on real, practical software projects for outside clients and other stakeholders. An added benefit is that our method captures and presents the real experiences of fellow students, which students can extrapolate to situations they are likely to find in entry-level positions. Students appreciate these compelling stories about problematic communications that are familiar to them. Here we focus on one particular project, development of control software for a ship-mounted crane, known as the Seabase Project.

Brady and colleagues (2006) emphasized the possibilities that the Speaking of Software project provides for interdisciplinary collaboration, particularly for outlining what technical communication might "give back" to other disciplines in ways that might increase the visibility of the field in both academia and industry. In this article, we want to focus on another related aspect of the project. The case studies we developed using qualitative research methods shed new light on the dilemma recognized by rhetoricians since Aristotle: that phronesis—the practical wisdom of acting with others for the common good—cannot be taught in the same way that making, or techné, has traditionally been taught.

Our first purpose here is to review the concepts of techné (making) and phronesis (acting), and explore them through one ethnographically based

Speaking of Software case study. The initial use of the Seabase case study was in a technical communication course made up of students majoring in technical communication, business, engineering, software engineering, and computer science. Second, we report on two trials, showing how these concepts are "rediscovered" by students when the cases are ultimately deployed in the classroom. Finally, we present our method of using ethnography, not just for research but to train (Miller 1991). We believe that if the concepts of making and acting are not understood and intentionally incorporated at the beginning of development of the cases, they will be present in only the most tangential of ways, unnecessarily complicating the discovery process of the students.

Because these concepts are critical to our pedagogical goals of helping students to create documents (the making, or techné aspect) that are ethical and promote the common good inherent in any situation (the acting, or phronesis aspect), we seek to share our method of developing these rich, compelling cases, and show how our goals have been accomplished through their use.

Why Teach "Making" and Learn "Acting"? (and How?)

For students who pursue degrees outside the liberal arts, the technical communication classroom is perhaps the best hope for acquiring the communication skills they will need in the workplace. These dedicated writing courses first appeared in engineering schools as a response to a perceived lack of writing ability on the part of their graduates. Basically prescriptive, they followed a "forms" approach that taught the correct way to make memos, letters, and reports. Most educators and practitioners today would probably consider this approach to making documents a "degenerate" techné, even if they might not use that precise terminology, proposed by Wild (1941). A degenerate techné would imply that the person—the technical communicator, in this case—does not understand why she is doing what she is doing—merely that she must do it. Making (techné) in its proper application is a "reasoned state of a capacity to make," which includes knowing the "why" of what is done (Aristotle 1947a, NE 1140a8). Atwill (1993, 1998, and with Lauer 1995) has called for a production-based approach to teaching writing, seeking to revitalize techné.

In 1979, Miller called for a more humanistic approach, one that included considerations of civic responsibility and social action—phronesis—as opposed to the more instrumental methods found in the textbooks of the day. Dunne (1993) also sought an approach based more on acting specifically to counter the behaviorist emphasis on making (and quantifiable results) that he saw gaining strength in the school system. Phronesis is a "reasoned state of a capacity to act" and has as its goal good action, in public, with others (Aristotle 1947a, NE 1140b5). Concerned with the particular situation and gained through experience (Aristotle 1947c, 1142a14), it is a wisdom acquired through habituation (augmented, in a way we will explore, with "training" through imitation). Although Aristotle admits that in situations where practical purposes are concerned, experience is "in no way inferior" to techné (Aristotle 1947b, Meta 981a14), it remains that "men [or women] of mere experience" cannot directly teach their understanding (Aristotle 1947b, Meta 981b9) because each situation is unique and not amenable to reduction to a set of rules or procedures.

The importance of the distinction between making and acting is not limited to the theoretical realm of the academic classroom; it has implications and applications on the job as well. Brady (2007) has discussed the implications of our teaching, showing how what we teach and what students learn is taken to the workplace. Once on the job, the technical communicators she studied used their knowledge to make sense of "existing information" and negotiate more effectively with the stakeholders. Brady's respondents recognized this negotiation as a social process: being in the workplace moved their knowledge from the mechanistic applications of "making" that they had initially used in the classroom to "acting," or working with others in the "rough and tumble of public spaces" (2007, 59). The noted professor of philosophy Dreyfus has accounted for this movement with a series of stages, relying on Heidegger's *Being and Time* (Dreyfus 2006). The student, or new employee, begins as a novice, moves through competence to expert status, and on to mastery. Although Dreyfus's noncognitivist approach to acquiring new skills raises certain difficulties for us, discussion of these difficulties is beyond the scope of this paper.

Miller (1991) notes the deficiencies of inexperienced workers: although they possess the theory and the know-how, they lack the experience and common sense to put that knowledge into practice. This resembles the difference that Johnson-Eilola (1996) pointed to between a functional and a conceptual understanding of technology or genre. For example, those entering a technical and professional communication course or a workplace as novices will frequently rely on prepackaged templates to create new (to them) genres such as resumés, memos, or business letters. In Johnson-Eilola's words, these templates are "instructing users in functional but not conceptual aspects of technologies" (1996, 179). Novice students with little or no experience in resumé design, for example, plug in information as instructed by a word processing application's "wizard," often making resumés that "look" right but paying little attention to their ultimate end: getting hired. Created with only functional skill, they prominently feature summer employment at fast-food chains in chronological work histories and leave out internships, course projects, and other valuable material that actually shows that they are qualified for the job at hand. After receiving instruction about the conceptual—which for us means rhetorical—aspects of resumé design, they may begin to acquire competence, using their full, "reasoned capacity to make" to tailor the design and content of their resumé to particular employers. They will only master the resumé when they have used it in real situations—at career fairs and during job interviews, for example. Having entered this stage, the (now not-so-new) student or employee can encompass the total context, both task-related and social—not just making a resumé, but acting with it. By now the individual will have moved toward acquiring the practical wisdom that is necessary to engage with others about the resumé (or any other task in the classroom or the workplace), modify it on the fly, and even know when to break completely with best practices.

Dreyfus proposes an additional stage, where the individual who has mastered a skill couples that mastery with creativity to transform the world. Although Dreyfus's examples—Galileo, Martin Luther King Jr, Larry Bird, and Henry Ford—are, by his own admission, dramatic, Brady (2007) provides examples that are more mundane. For instance, "Frances" mentions involving clients in the process of creating documents, seeing them more as partners than just as recipients of the final product. "Billie" concurs: "You can't say 'this is just the way it's going to be'" to the client (Brady 2007, 57). Compared with the past, when the mantra of the engineer was "we can make whatever they throw over the wall," a transformation is under way. User testing was perhaps the beginning. Working with others, rather than just for them, is becoming the valued action of the workplace. Technical communication practitioners, like Frances and Billie, have changed the world of documentation, developing and implementing new best practices that include, rather than ignore, the client and the user.

The workplace does not always welcome this kind of change, especially from new employees. When doing business, sometimes the status quo is enough. It is not always necessary to reinvent the wheel when the old wheel works just fine. Although it would be counterproductive to stifle continuously the very knowledge and creativity sought from the new hire, mundane jobs sometimes call for mundane solutions. Understanding and explaining tasks in terms of making and acting provides a resolution to this dilemma. Some tasks require one, some the other, and some both. It is experience that determines which is called for in any given situation.

Why Use Case Studies to Educate?

In the classroom, unfortunately, there is neither time nor availability for "real-world" making that can accommodate all students and prepare them for the world of work. Projects where the students work for a client are a start, but often are less than adequate as simulations of real experience: they are often seen as "superficial" by the students and confounded by misdirection on the part of the "clients" (Blakeslee 1997, 2001; Freedman and colleagues 1994). In internships or co-ops, students may find that three or four years of work in their major did not adequately prepare them for real-world tasks. They may become frustrated and disillusioned. The aim of a case study approach is to expose the students to the workplace situations, ethical dilemmas, and political realities they may face by simulating that "real-world" experience.

Aristotle's *Poetics* suggests that well-crafted literature can be "an imitation . . . of action and life" (1947c, 1450a17–18). Good case studies that capture the essence of a situation do the same. Nussbaum, in *The Fragility of Goodness* (2001), has explicated this overlooked aspect of the *Poetics*, noting that Aristotle suggests that "imitation is natural" and that it is through imitation that we learn (1947c,

1448b, 5–9). When students reflect on and discuss the case, they bring their knowledge to discursive consciousness, reinforcing their learning. They will have acquired the craft, the techné, of what they make. That knowledge will become, as Brady (2007) found, a guide that will allow them to cope better with the various social contexts and practices of the workplace.

Why Our Case Studies?

As noted above, Freedman and colleagues reported that students sometimes see case study simulations as "artificial" (1994). Thomas (1995) suggested that some of that feeling may come from unrealistic expectations placed on the students by having them adopt the role of a bank president, for example; we are asking them to play roles too far from their existing or imminent experience. Our cases present the experiences of students as they complete their senior design projects, imitating the kinds of "entry-level" positions they will find when they enter the workforce. This approach has the dual benefits of holding their interest while in school and later of easing their transition to the workplace. We have enlivened our case studies even more by responding to user reports of what works and what does not. We have had some success: one reader of a case, for instance, said he "couldn't put it down; it read like a novel." Comments like these confirm that case studies can be improved by including some literary techniques, such as those suggested in the *Poetics*.

Exploring the benefits of client-based projects, Spinuzzi (2004) and Blakeslee (2001) have both written on workplace collaborations. For Spinuzzi, any workplace will do; the importance is to bring those experiences back to the classroom, share them with others, and reflect on them (2004). Blakeslee finds that the kind of feedback the students receive on their performance is critical yet varies widely depending on the client or supervisor (2001). Our cases simulate the ambiguity and goal-driven nature of the workplace. Their use in the classroom asks the students to make their analysis and interpretation of the situations presented explicit, fostering learning about how they might act were they actually in that situation.

Developing the Cases

Applied Ethnography

Moss describes ethnography as "a qualitative research method that allows a researcher to gain a comprehensive view of the social interactions, behaviors, and beliefs of a community or social group" (1992, 155). We applied proven ethnographic methods to capture and present views of real software development settings, making them available for future study. However, because of our limited resources, our work should not be confused with classic ethnographic studies. We could not perform the years of fieldwork required for such endeavors, nor did we try to record much beyond the development sessions proper.

Using a qualitative case study approach, we sought to focus particularly on the "important aspects or variables" of individuals working in small groups (Lauer and Asher 1988, 23). From those identifications, we developed case studies to be explored and analyzed by other students. Our goal was to simulate the workplace situations they will need to be able to cope with as they pursue careers in technical communication, computer science, or other fields.

Collecting Data

Any development process produces a massive amount of data, and the Seabase Project was no exception. The CS senior design students generated various versions of the code itself, along with meeting minutes, e-mail, internal documentation, and weekly progress and final summary reports for their clients. The CS students also reflected daily on their efforts and consolidated that information and their documented code into one-page progress reports submitted to the senior design instructor. Accessing this material was the first step of our case development. On the senior design projects, ad hoc e-mail lists was the method of choice for the project teams to communicate among themselves and with others. By the simple measure of being included on these lists, our research team of graduate students and faculty was able to view, and more importantly to save, this original material. The senior design instructor, client, and CS students themselves provided additional material.

The first case study we developed was primarily archival in nature. Observing standard qualitative research practice (Agar 1996; Kirsch and Sullivan 1992; Lauer and Asher 1988), graduate students from the Computer Science Department and the Rhetoric and Technical Communication program in the Humanities Department collected the written material from the senior design students and attended and recorded selected meetings. This case (Seabase) contains more than 100 pages of text, half an hour of audio excerpted from the meetings, and six video files.

For subsequent cases, the graduate students were present at the majority of the students' meetings, including meetings with clients, acting as participant observers. The graduate students recorded the conversations with digital audio equipment, diagrammed the rooms (including how people moved about and where they positioned themselves in relation both to the room and each other), and wrote field notes. These field notes included fine details of the interactions that might not be apparent on the audio recordings, so as to aid the researchers in their later recall, reconstruction, and analysis of the situations (Emerson and colleagues 1995). Furthermore, the field notes attempted to capture and explain what the researcher thought of key events and incidents and noted the nonverbal reactions of all participants. Key events and incidents included major changes in the direction of the project, sometimes caused by unexpected changes in the client's requirements and sometimes caused by factors outside the stakeholders' control. Incorporating these descriptions of the setting and details of the action (both visual and oral) captured the ambiance of the encounters and brought life to the cases, providing future users of the cases with that "you are there" feeling.

These later cases tend to be richer than the first in that they make much more use of audio recordings. One in particular contains more than five hours of audio excerpts, in addition to 148 pages of text and 14 video files, some with animation. (Note that these figures represent primarily the contribution of the senior design teams. They do not include the original code with which they were furnished, which itself totals more than 140 pages of single-spaced text.)

As the projects wound down, the graduate students conducted semi-structured interviews with the senior design students and their clients. These interviews helped to triangulate results and interpretations, to support early findings, and to "earn the confidence of the reader that the researchers have 'gotten it right'" (Hesse-Biber and Leavy 2005, 66). We have on file signed and IRB-approved informed consent forms from all participants covering the observations, interviews, and any subsequent use of the material collected.

Constructing the Cases

The first step in constructing the case studies was to arrange the entire corpus of original material into a chronological summary account, including hyperlinks to the original documents. The graduate students parsed it into modules, with each module covering (usually) one week of the semester. To assist the students in identifying and examining the issues presented in the cases, we also prepared a set of questions for each module designed to provoke inquiry into the events. As an instructional aid, we developed some password-protected background material for the instructor, giving an "insider view" of the situations, where the CS project teams were headed, and suggestions for classroom use.

For ease of use by both instructors and students, we also created a thematic version. Rather than moving chronologically through the account, we identified some themes that allow the users to focus on certain aspects of the case, potentially reducing the time commitment considerably. The entire case is still available, but the themes provide preselected entry points that can be followed independently of the case as a whole, if desired.

These versions were prepared using the grounded theory method of Strauss and Corbin (1998). Reading through the chronological account, the graduate students started with a detailed, line-by-line analysis, moving between the chronological account and the descriptions available in the transcripts. This analysis generated initial categories, focused around inherent meaning and details, and identified central issues. These were labeled as emergent themes, and used the students' original terms whenever possible.

For example, one theme was related to the difficulty the senior design team working on the Seabase Project experienced when required to learn and work with an unfamiliar engineering-specific programming language, Matlab. The various occasions when the students had problems were abstracted from the case and listed separately in chronological order, using the heading "MATLAB." Other problem areas were also identified and grouped under their own thematic headings.

By looking across themes and comparing the content of interactions and the times when they occurred, an analysis can begin to demonstrate some explanatory and predictive power. For example, on occasion questions that arose in the MATLAB theme (when the students were working on their own) were not satisfactorily answered in the "client interactions" theme (when they received no clear direction from the client, who was quite familiar with the Matlab program). This frustrated the CS students and impeded their progress on the project. These observations help to explain why the project did not stay on schedule. This method of comparing across themes also provides some predictive power. It is likely that any future interactions between the students and their client would repeat those of the past if no new approaches were tried.

Presenting the Cases

We chose to present our cases in the "realist tale" style because this is "by far the most prominent, familiar, prevalent, and recognized form of ethnographic writing" (van Maanen 1988, 45). Those familiar with documentaries will recognize how this style can expose the mundane, but nonetheless important, details of a situation. Like any documentary, these details gather to make a point about life. Beyond that, the details can inspire a sense of intimacy in the user: they feel they are in the presence of the participants; a part of the action, involved. To this end we use quotations, recordings, and the students' own documentation and reflections to represent the participants' point of view authentically. We consulted with the participants for input on the accuracy of those representations to insure we remained true to their impressions of the process of dealing with stakeholders, tempered by our analysis and pedagogical goals.

Bowing to the exigencies of the modern world and the modern student, we have moved beyond the typical, paper-based presentation. Our case studies (which can be found at www.speaksoft.mtu.edu/cases) combine text, audio, and video material in multimedia packages, where students in a software engineering or technical communication course can read text, hear audio clips, or watch animations. The "Seabase" case, for example, is made up of 14 modules, each of which corresponds to roughly one week of the semester. Each module contains a narrative detailing what happened that week, in which links to original documents that the team produced and to audio files are embedded. By embedding hypertext links to the original documents and recordings rather than including them in an unwieldy appendix, this electronic format enhances usability, accessibility, and portability. In addition, it adds to the reality of the case. For example, excerpts from an e-mail might be included in the module, but the document that came with it as an attachment is left as a separate file. To access the information, the student must open that attachment, as did participants with the original e-mail. This action expands the reader's role from passive observer to active participant, raising interest in the case by making it seem more real.

Our cases preserve the plain language used by the students and retain specific references to people, programs, and equipment. Using the vocabulary of the application domain (mechanical engineering, in this instance) imparts contextual information with which the CS students struggled but that subsequent users might find insignificant. Encouraging this kind of constructive questioning—even of something usually seen as transparent and mundane, such as language—elicits important details that might otherwise be missed (Sutcliffe 2003).

To simulate the problems of dealing with the issues of real clients, we designed the presentation of the material so that information concerning project requirements is imparted in stages, much as often occurs in the real world (this presentation is described in more detail below). In some cases, we present examples of communication failure, helping students to reflect on what went wrong, and develop alternatives they might pursue to avoid these difficulties themselves (Gale 1993).

Making and Acting: Theorizing One Case

The importance of the concepts of making (techné) and acting (phronesis) can be demonstrated by looking for their presence in the cases. Often, individuals will have difficulty separating what they make from how they act, but this ability can be critical to success in the classroom and, by extension, in the workplace. The *Challenger* disaster is a case in point. The engineers involved became so focused on making the flight happen that they overlooked whether that making would be acting for the greater good. The following examples demonstrate that difference and how case studies can bring these concepts to light, and make them useful as heuristics to guide what we make and how we act, increasing effectiveness in both areas.

In the "Seabase" case study, senior design students (called the "CS team") worked to develop software to control a ship-mounted crane. The crane would be used to transfer loads from one ship to another while at sea and would be controlled by a joystick, much as in a video game. The CS team would write a program that would compensate for the rolling of the ocean, making the crane easier for the operator to use. The collaborative project with the Mechanical Engineering Faculty and students was at times overwhelming. The CS Team had to learn to program in a new language, Matlab, and to navigate the culture of mechanical engineers as well. As new members of an ongoing project, they had to make both their controller and their place in an established—yet, to them, foreign—work environment.

Making the controller seemed straightforward to the CS Team. As one member described the situation when interviewed at the end of the project:

Question: What product are you making?

Answer: We're supposed to port code, existing C-code into a "Matlab" format. Basically the code is supposed to stabilize a payload for the crane. It was pretty vague in the beginning. We thought we were going to actually be designing the code, whereas we end up to find out we're just like porting some existing code.

We can see from his answer that at the beginning of the project, this self-described "technical guy" was concerned about the scope of the task: designing the code seemed formidable; merely porting it well within their "capacity to make." However, complicating the task was the "acting" aspect—working with others in an unfamiliar situation. This excerpt from the case, shown in Figure 12.1, captures the difficulty.

As an example of how we presented the case, the hyperlink to an e-mail from one project advisor in mechanical engineering is available only after following the hyperlink to the meeting minutes shown in the excerpt. To get to this information, the undergraduate students analyzing the case must dig into and ferret out the information, much like the CS team had to do at the time. This adds to the effectiveness of the case as a simulation of real life, where not everything is presented clearly and immediately. The e-mail highlights an all-too-common problem when working with others. The two advisors, in their roles of clients, hold differing opinions on what should be the most difficult part of the project. Advisor Hank Taylor feels "the crane [controller] is the 'biggest, nastiest' part" and a side project to design a GUI is "the easiest part." However, one week before, advisor Nancy Smith indicated the opposite: "The GUI design is a good project for the CS team," and "working on only the crane controller would be 'too simple.'" The questions posed to the student users of the case (excerpted in Figure 12.2) ask them to reflect on and offer solutions to this disparity.

This exemplifies the dilemma of working with others while trying to promote a common good. The arena of acting is too dependent on the particulars of the situation; it cannot be codified into a set of rules, a "how-to" manual, as can making a product. The multitude of variables involved helps explain the difficulty of teaching phronesis directly. Aristotle's "capacity to act" and do what is appropriate

---

Module C Story

On Wednesday of the fourth week of the semester (Sept. 22) the leaders of the three crane project teams meet with project advisors Hank Taylor and Nancy Smith. They decide that since the "point of meeting is to get regular coordination of the teams, they will continue the meeting of team leaders on Wednesday from 12–1 on". Representing the CS Team are JoAnn, Ken and Bob; Matt and Ben come for the crane builders; and Jon is there to talk about the platform.

Minutes of Sept. 22 Crane Team Leaders meeting

The items on JoAnn's summary of the meeting are:

- The CS Team will work on crane, not on the platform, this term.
- In a discussion of scope of the CS Team's part, Hank says the crane part is the "biggest, nastiest part" and he thinks the GUI for the platform will take about an hour and is the easiest part.

Figure 12.1 Excerpt from Module C Story, Seabase Case Study

Module C Questions

1. How would you characterize the interactions among Hank, Nancy, and the team members?
2. It's interesting that Hank says that the "crane part" is going to be "the biggest, nastiest part", and that the GUI design will be easiest. On the other hand, Nancy seems to be saying the opposite: the controller will not be very difficult, and the GUI will be more challenging.
   * Why might they have such different opinions?
   * How can the CS team resolve this difference?

Figure 12.2 Module C Questions, Seabase Case Study

in any situation can only be gained by experiencing a variety of situations and reflecting on them.

Never would we argue that teaching "making" is easy—we know it is not. It is just that teaching directly the aspects of life that comprise "acting" is impossible: what rules, what universals are there to rely on and pass on to students when it comes to human interactions? It is only through experience and imitation that we learn how to act well. The richness and immediacy of our cases provide a path to simulating the experience of the workplace. In responding to the excerpt above, students must put themselves in the place of the CS team and learn how to deal with the particulars of the situation. It is here that we find the value of our ethnographically developed case studies and our theorizing of them through the concepts of making and acting.

Simulating Workplace Actions in the Classroom

We have successfully used our case studies in both technical communication "service" courses where the students come from a variety of majors, and in software engineering classrooms where a much more homogeneous audience receives the cases. The depth of the cases helps to simulate the complexity of working on a collaborative software project, and their breadth helps to hold the interest of students from other disciplines. Although our initial focus was on improving communication skills, here we look deeper to see how the concepts of making and acting have facilitated that endeavor. The examples we offer here draw on our first use of the Seabase case study in technical communication classes, where the heterogeneity of the audience more accurately reflects a typical workplace than does the homogeneity of a software engineering class. This interdisciplinary service course typically attracts students majoring in technical communication, business, engineering, and software engineering and computer science. Here we report specifically on two trials, summer 2006 (a 2.5-week unit out of 7 weeks) and fall 2006 (a 4-week unit out of 14 weeks). We were fortunate in that each interdisciplinary team working on the case included a software engineering or computer science student. The overall goal of the unit was to analyze the multiple communications efforts of the CS team (including e-mails, meeting minutes, timelines, requirements documents, risk documents, reports, and presentations) as they developed the software for the crane controller. Specific assignments included writing memos and final reports and preparing for in-class discussions.

Making/Techné

In one exercise, the technical communication students considered and discussed the purpose(s) and audience(s) for a given set of meeting minutes. They rewrote the minutes to improve their usefulness as a record of the meeting and as a report to other stakeholders. This production can be read as techné, the reasoned capacity to make. For both the CS team and the technical communication class members, creating these documents contributed to their communication skills. The skills of the CS team, unfortunately, never seemed to progress much past Dreyfus's novice stage. Their reliance on rules and procedures led to confusion and, ultimately, failure.

The technical communication students, on the other hand, by vicariously experiencing and reflecting on the efforts of the CS team, began to move beyond competence. Their analysis of the situation led to alternative, more appropriate ways to create the documents. They began to master the art of technical communication. Like Aristotle, we value the masters of their crafts "because they know the reasons of the things which are done. . . . Thus the master craftsmen are superior in wisdom, not because they can do things, but because they possess a theory and know the causes" (1947b, Meta 981b1).

The students also show a discursive awareness of their skill, in that not only can they find and repair inadequacies in the documents, but they can also explain why they are important. Wild begins his definition of techné as "any act that can give a rational account of itself, explaining why it does what it is doing" (1941, 256). In compiling a final report, some students noted that the CS team's timeline was missing dates: "The biggest problem is that there are no dates at all on the timeline," they wrote. "The team has not worked out starting times, durations, and, most importantly, deadlines." The technical communication students recognized that the timeline was probably written in a rote manner merely to fill a requirement of the course, with no apparent realization by the CS team that it had a purpose, or an audience that would really use it. In our terms, the technical communication students recognized the CS team as novices using a "degenerate" techné (if that). Furthermore, the technical communication students showed their more advanced skills by producing a revised timeline with dates and deadlines, and laying out who on the CS team should have been responsible for which tasks.

Acting/Phronesis

Their discussions and analysis also explored how the inadequacy of the communicative efforts on the part of the CS team contributed to the action (more usually inaction) of those involved in the Seabase Project and, eventually, to its outcome. Going beyond just an analysis of the communication-related causes that contributed to the failure of the Seabase Project, the technical communication students wrote recommendation reports that were crafted to help the next generation of CS senior design classes avoid the pitfalls the Seabase Project encountered when making their controller. The technical communication students modeled behaviors in an attempt to advance the knowledge of future CS students. They became teachers, an ability Aristotle considers a sign of knowledge (1947b, Meta 981b8). Taking the point further, Aristotle describes this acting for the human good as the realm of phronesis (1947a, NE 1140b20). To echo Dreyfus's claim, the technical communication students are potentially able to transform a portion of the CS world by demonstrating virtuous behavior that the CS students could then imitate. As Miller hoped, our technical communication offerings have "ceased being a technical skills course and instead become 'practical,' in the most valuable sense of that word" (1991, 71).

Conclusion

The examples we provide of our experience using the Seabase case in a technical communication classroom show that the students have learned valuable lessons about communication and about working on interdisciplinary teams. One lesson concerns the importance of valuing the skills of others; that is, of appreciating their capacity to make. The technical communication students learned this when they recognized that their own skills were frequently unappreciated, an intrateam conflict that Myers and Larson (2005) suggest is rooted in relations of power. In evaluating the situation, one technical communication student explained her frustration with working on a multidisciplinary team analyzing the Seabase project: "I did not like how my team functioned. Skills possessed by some were overlooked or not valued. The function of my team was to 'please the instructor' and not to do a good job working on the assignment."

Another valuable lesson exemplifies the possibility of the Isocratean ideal of learning to act well with others through imitation. In a case still under development, the senior design project team needed root access to the departmental computers; something the system administrators were unwilling to grant. Although the situation was frustrating to his team, one member drew on his workplace experience in explaining the inherent difficulties of acting in the interests of "the good" that are sometimes conflicting, depending on from which side of the situation the good is seen:

> They want to keep the machines [computers] secure, and that's reasonable. That's actually what I do in my co-op internship position, I'm an information security intern, and we handle thousands of requests each month asking for access.

He was able to appreciate the dilemma from both sides and showed an insightful understanding of the workplace and the situation-bound nature of acting in it. He was also able to communicate that understanding not only to his team but through the case study, to future students. The advantage of ethnographically developed case studies like ours is that they capture the realities of those situations and present them to students as compelling examples of real dilemmas, simulations they can learn from through imitation, analysis, and reflection.

Like Brady (2007), we believe that these lessons are taken from the classroom and applied when students get "on the job." We believe this is facilitated by the reality of our cases, made possible through the method we used to develop them. By relying on the tested procedures of ethnography, we have faithfully recorded and reported the real experiences of real students working with real clients. Students particularly appreciated the inclusion of the actual documents as created by the CS team and the conversations that surrounded their creation. Too often, the material in case studies is reduced to a minimum, rendering it sterile and incomplete. By using ethnographic methods, our cases overcome this problem.

The students in our technical communication courses became aware of how documentation can contribute significantly to a project's success or failure. In the Seabase Project, the inclusion of the actual e-mails, minutes, reports, and so forth created by the CS team allowed the students to experience first-hand the repercussions of lacking both knowledge of and experience in applying the principles of audience, purpose, and context.

The challenge of working with people from other disciplines is common. Management, designers, engineers, production, sales, and technical communication personnel must be able to reach consensus if the goals of the company or client are to be achieved. Simulating the experience of real situations, and encouraging reflection on and discussion of those situations, helps to prepare students majoring in technical communication for those challenges and deepens the appreciation of communication skills for those from other fields. For students graduating with a degree in scientific, professional, and/or technical communication, this experience is crucial as they join the workforce and become practitioners.

Acknowledgment

*This article was originally published in *Technical Communication* (2008), 55: 405–14.

References

Agar, M. 1996. *The Professional Stranger*, 2nd edn. San Diego, CA: Academic Press.
Aristotle. 1947a. *Nicomachean Ethics*. Trans. W. D. Ross. In *Introduction to Aristotle*, ed. R. McKeon. New York, NY: Random House.
——. 1947b. *Metaphysics*. Trans. W. D. Ross. In *Introduction to Aristotle*, ed. R. McKeon. New York, NY: Random House.
——. 1947c. *Poetics*. Trans. W. D. Ross. In *Introduction to Aristotle*, ed. R. McKeon. New York, NY: Random House.

Atwill, J. M. 1993. "Instituting the Art of Rhetoric: Theory, Practice, and Productive Knowledge in Interpretations of Aristotle's *Rhetoric*." In *Rethinking the History of Rhetoric: Multidisciplinary Essays on the Rhetorical Tradition*, ed. T. Poulakis. Boulder, CO: Westview Press.

———. 1998. *Rhetoric Reclaimed: Aristotle and the Liberal Arts Tradition*. Ithaca, NY: Cornell University Press.

———, and J. Lauer. 1995. "Refiguring Rhetoric as an Art: Aristotle's Concept of *Techné*." In *Discourse Studies in Honor of James L. Kinneavy*, ed. R. J. Gabin. Potomac, NJ: Scripta Humanistica.

Blakeslee, A. M. 1997. "Activity, Context, Interaction and Authority: Learning to Write Scientific Papers in Situ." *Journal of Business and Technical Communication* 11: 125–69.

———. 2001. "Bridging the Workplace and the Academy: Teaching Professional Genres through Workplace Collaborations." *Technical Communication Quarterly* 10: 169–92.

Brady, A. 2007. "What We Teach and What They Use: Teaching and Learning in Scientific and Technical Communication Programs And Beyond." *Journal of Business and Technical Communication* 21: 37–61.

———, R. R. Johnson, and C. Wallace. 2006. "The Intersecting Futures of Technical Communication and Software Engineering: Forging an Alliance of Interdisciplinary Work." *Technical Communication* 53 :317–25.

Dreyfus, H. 2006. "Can There be a Better Source of Meaning than Everyday Practices? Reinterpreting Division I of *Being and Time* in the Light of Division II." In *Heidegger's Being And Time: Critical Essays*, ed. R. Polt. Lanham, MD: Roman and Littlefield.

Dunne, J. 1993. *Back to the Rough Ground: Practical Judgment and The Lure of Technique*. Notre Dame, IN: University of Notre Dame Press.

Emerson, R. M., R. I. Fretz, and L. L Shaw. 1995. *Writing Ethnographic Fieldnotes*. Chicago, IL: University of Chicago Press.

Freedman, A., C. Adam, and G. Smart. 1994. "Wearing Suits to Class: Simulating Genres and Simulations as Genre." *Written Communication* 11: 193–226.

Gale, F. C. 1993. "Teaching Professional Writing Rhetorically: The Unified Case Method." *Journal of Business and Technical Communication* 7: 256–66.

Hesse-Biber, S. Nagy, and P. Leavy. 2005. "Validity in Qualitative Research." In *The Practice of Qualitative Research*, 62–68. Thousand Oaks, CA: Sage.

Johnson-Eilola, J. 1996. "Relocating the Value of Work: Technical Communication in a Post-Industrial Age." In *Central Works in Technical Communication*, ed. J. Johnson-Eiola and S. Selber. Oxford, UK: Oxford University Press.

Kirsch, G., and P. Sullivan. 1992. *Methods and Methodology in Composition Research*. Carbondale, IL: Southern Illinois University Press.

Lauer, J. M., and W. Asher. 1988. *Composition Research/Empirical Designs*. New York, NY: Oxford University Press.

Miller, C. R. 1979. "A Humanistic Rationale for Technical Writing." *College English* 40: 610–17.

Miller, T. P. 1991. "Treating Professional Writing as Social Praxis." *Journal Of Advanced Composition* 11: 57–72.

Moss, B. J. 1992. "Ethnography and Composition: Studying Language at Home." In *Methods and Methodology in Composition Research*, ed. G. Kirsch and P. Sullivan. Carbondale, IL: Southern Illinois University Press.

Myers, L. L., and R. S. Larson. 2005. "Preparing Students for Early Work Conflicts." *Business Communication Quarterly* 68: 306–17.

Nussbaum, M. 2001. *The Fragility of Goodness: Luck and Ethics in Greek Tragedy and Philosophy*. Cambridge, UK: Cambridge University Press.

Spinuzzi, C. 2004. "Pseudotransactionality." In *Teaching Technical Communication*, ed. J. M. Dubinsky. Boston, MA: Bedford/St. Martin's Press.

Strauss, A., and J. Corbin. 1998. *Basics of Qualitative Research: Techniques and Procedures for Developing Grounded Theory*, 2nd edn. Thousand Oaks, CA: Sage.

Sutcliffe, A. 2003. *Scenario-Based Requirements Engineering*. Paper presented at the IEEE International Conference on Requirements Engineering, September 8–12, Monterey Bay, CA.

Thomas, S. G. 1995. "Preparing Business Students More Effectively for Real-World Communication." *Journal of Business and Technical Communication* 9: 461–74.

van Maanen, J. 1988. *Tales of the Field: On Writing Ethnography*. Chicago, IL: University of Chicago Press.

Wild, J. 1941. "Plato's Theory of Techné: A Phenomenological Interpretation." *Philosophy and Phenomenological Research* 1: 255–93.

Chapter 13

Integrating Experts and Non-Experts in Mathematical Sciences Research Teams

A Qualitative Approach

Linda Phillips Driskill and Julie Zeleznik Watts

Editors' Introduction

Linda Driskill and Julie Watts's "Integrating Experts and Non-Experts in Mathematical Sciences Research Teams: A Qualitative Approach" is a case study, like a number of the other articles in this book. Unlike all the others, however, it deals with the work of mathematicians, statisticians, and applied and computational mathematicians, not technical communicators. It is also different in that the research it reports was conducted over more than two years, a much longer period than any of the other studies that we have included here. Such longitudinal studies tend to have greater credibility in the research community because they are based on larger bodies of data and often reveal changes in social, communication, and business patterns over time.

The case study conducted by Driskill and Watts is also unusual because the population of interest represented a relatively large group of four different populations: undergraduate students, graduate students, postdoctoral fellows, and university faculty—all participants in small research groups supported by a National Science Foundation (NSF) grant to transform the culture and structure of mathematics education in the United States. A second NSF grant funded Driskill and Watts's research into how communication methods and patterns might change over time in those research groups.

So this article is notable for yet another reason. Not only is it a longitudinal study of individuals representing four disciplines at four different levels of disciplinary maturity rather than technical communicators, but it is also a report of research funded by the NSF, which seldom funds studies in our field.

The techniques used in the study included: surveys of student participants; semi-structured interviews of faculty, postdocs, and graduate students, conducted at intervals between fall 2003 and summer 2005; and regular observations of the interactions of each research group over the two years. In addition, this article draws on insights gained throughout the 2003 to 2008 grant period, as well as drawing on a few comments from exit interviews in 2008.

As Driskill and Watts reviewed the data they were collecting, they described and interpreted the behaviors and communication practices they were observing using commonly accepted theoretical models such as organizational theory, activity theory, genre theory, community of practice theory, and critical discourse analysis.

What Driskill and Watts discovered as a result of their research was that groups of novices and experts working together on research projects communicated in ways quite different from those seen in the traditional classroom. The novices learned that even those with considerably more knowledge were themselves sometimes novices, and that conversations that included "asking questions that 'might or might not sound stupid' were a badge of being a mathematical scientist." Similarly, the need to "explain in English," so that novices could understand without losing status, helped the experts learn to become more effective communicators by addressing more diverse audiences about their areas of expertise in language the non-experts can understand.

Obviously, this article does not directly address any of Giammona's themes, which concern full-time technical communicators, but much of what Driskill and Watts convey in their fascinating article will remind us that all technical professionals spend a significant portion of their time communicating specialized information to others. By focusing on communication in the realm of mathematics, these two researchers are simultaneously expanding the body of knowledge in our field.

For example, their study concerns the way in which knowledge is shared between distinct social groups (experts and novices within a practice), and underscores the importance of interaction and community for the creation, sharing, and use of knowledge. The creation and transmission of knowledge is of fundamental concern for technical communicators, and this article may help us to understand why the early qualitative researchers of workplace writing came to see writing as a social activity, and why so many researchers and commentators (including Giammona) insist that interpersonal skills are an essential component of the technical communicator toolkit.

Converging political, social, and technological trends have created a striking challenge for technical communicators and scholars: how to understand communication in groups that include both experts and non-experts. Linear transmission models of communication that separate writers and readers or speakers and listeners do not represent communication in such groups well. These situations, we contend, demand a rethinking of the nature of learning, communication, and organizational membership. Specifically, we need to understand how a group structured to include both experts and non-experts can work and learn together by developing appropriate genres of communication that accomplish group goals and foster identification with the group.

Nearly 10 years ago, the National Science Foundation promoted university research communities of experts and non-experts as well as close linkages between K-12, undergraduate, and graduate programs in mathematical sciences. The qualitative study presented here illustrates how learning and identity construction in one university program occurred. As individuals with different levels of knowledge and experience pursued research questions together, they changed their face-to-face interaction, developed new genres and conventions, and created knowledge artifacts. We offer the case as an example of how new structures and missions lead to new roles and genres. This article also explains how we adapted models from activity theory, genre theory, and communities of practice theory in order to account for their actions. The case and the models show how rhetorical analysis can illuminate the processes of learning and identity formation when communication networks include individuals of widely varying prior knowledge and experience.

A Case of Innovation and Knowledge Creation in a University Setting

By the late 1990s, United States mathematics higher education was widely perceived to be in crisis. In 1999, United States mathematics departments graduated 26 percent fewer majors than in 1986 (United States National Science Foundation [NSF] science resource statistics). The high proportion of advanced mathematics degrees awarded to non-United States citizens from the mid-1980s onward had motivated the NSF to find ways to reduce dependence on foreign students by attracting and retaining more United States citizens as majors and graduate students. In September 1997 a special emphasis panel reported to the NSF's Division of Mathematical Sciences (DMS) on a program to address these problems. Its report strongly endorsed the concept of vertical integration, which they described as "constructing undergraduate, graduate and postdoctoral programs to be mutually supportive" (Lewis 1997).

However, this proposed solution to the pipeline problem introduced other challenges. First, it proposed a change that would admit people into research activities who would formerly have been considered unqualified. Second, it proposed establishing a culture of collaboration and mentoring where a culture of individualism and competitiveness had dominated. Third, it proposed that a vertically integrated program's activities co-exist with other activities and university structures that reinforced and enacted the old cultural distinctions. Thus, participants found themselves in a problematic new structure with a revised mission that combined research and learning.

Mathematical sciences (statistics, mathematics, and computational and applied mathematics) have typically separated research activities from instruction, limiting participants' involvement in research to those who had completed at least several graduate-level courses. Moreover, unlike many natural sciences, mathematics research has been the province of individual scholars. As one mathematician casually expressed it, "[Mathematics] research is not a team sport" (Interview notes 2008). However, other natural sciences such as biochemistry, chemistry, and physics had long since developed positions for novices in research projects. The

NSF's VIGRE program urged mathematical sciences departments to do the same through vertical integration and mentoring.

This demand for vertical integration and mentoring challenged the traditional culture and structure of mathematical sciences education. After the University of California at Berkeley lost its own VIGRE grant in 2002, the chair of the mathematics department, Calvin Moore, downplayed the need for mentoring, as reported by Mackenzie: "'One of our goals is to cultivate self-reliance. Berkeley is a tough place. Berkeley is not a warm and fuzzy place. Students react to this atmosphere: Some thrive, and others don't,' he said" (Mackenzie 2002, 1389). Any fault lay with the students; the culture would remain as it was.

This individualist culture common in mathematics and the segregation of research activities had tended to create a lack of knowledge about career options among both undergraduates and graduate students. In the special interest panel report, mentoring appeared to be a kind of career counseling for mathematicians, conceived of as information about varied and widely available jobs that were unknown to both graduate students and undergraduates. This emphasis is understandable, given the panel's concern with the deficit in workforce numbers.

The panel had also taken a utilitarian or vocational skills view of com-munication that was fundamentally based on an implicit transfer model: "The panel endorses the need to improve the communication skills of undergraduate students. In particular, they should develop the ability to speak and write about mathematics, and to communicate ideas to their peers and to non-mathematicians. Enhanced communication skills will give mathematical science majors even greater employment opportunities" (Lewis 1997). The panel did not acknowledge the powerful role that communication might have within a program, in terms of mentoring, classroom talk, oral presentations, research meetings, and synthesizing knowledge, yet that was where most of the changes they proposed would be played out as values and established practices were renegotiated.

The Rice University VIGRE Program

Rice University received a VIGRE grant in 2003, which was renewed in 2008 for another five years. The Rice VIGRE Program offered to achieve NSF objectives by creating small research-and-learning groups that had four to twelve members (Forman 2003). Composed of postdocs, faculty, undergraduates, and graduate students, the groups were called PFUGs, based on an analogy with musical fugues, a form in which multiple independent melodic lines interact harmoniously. The faculty, postdocs, and graduate students mostly came from three departments: statistics (STAT), mathematics (MATH), and computational and applied mathematics (CAAM). However, undergraduates from a wide range of majors over the five-year period, from physics to engineering to music, signed on. An undergraduate course and a graduate seminar related to the research topic was to be attached to each group.

In 2003 the three original PFUGs were interdisciplinary in their choice of research topics, which included computational finance, computational algebraic geometry, and biomathematics; and, to a more limited extent, they were interdisciplinary in their membership. The mixing of disciplinary topics would be expected to create additional procedural and communication complexities (Klein 1990, 1996; Klein and Porter 1990; Kent-Drury 2000). Klein notes that com-munication is the heart of interdisciplinarity, calling for multiple kinds of negotiation, including negotiation of the nature and definition of the problem, the types of methods to be used, the ways data or results will be interpreted, and the ways results will be communicated to complex audiences (1996, 224).

The Rice University VIGRE Program Communication Study

At the same time that the Rice University VIGRE Program began, the NSF funded a qualitative communication study to describe the changing patterns of communication in these small groups (Driskill and Zeleznik 2003). Following typical qualitative designs (Marshall and Rossman 2006; Koretz 1992; Campbell 1999; Lampert 2000), we used surveys, participant interviews, observations, and document collection to facilitate triangulation, which would be especially important since the total number of participants would be small. While we collected data we also extended and developed models (described later) to use in analysis and interpretation.

Technical communication researchers have found qualitative methods valuable for revealing the complex contexts in which communication occurs. For example, researchers have relied on qualitative approaches to explore students' experiences (Artemeva 2005, 2008; Flynn and colleagues 1991; Freedman and Adam 2000; Winsor 1996, 2001), disciplines (Berkenkotter and Huckin 1995); professionals' practices (Bazerman and Paradis 1991; Orlikowski and Yates 1994; Rymer 1989; Paré 2001; Schryer, Lingard, and Spafford 2007) and types of reports written in specific courses (Herrington 1985; Mehlenbacher 1992). Qualitative approaches include several theories and methods in addition to those associated strictly with anthropology and sociology: activity theory, rhetorical genre studies, sociolinguis-tics, critical discourse analysis, community of practice theory, and others. Scholars in the field, which is now well established, have consolidated their theories and research approaches (Artemeva 2008; Paré and Smart 1994) and have recently engaged in strategic reflection and planning (Artemeva and Freedman 2006).

Rejecting a naïve view of social categories as neutral "givens," qualitative rhetorical studies scholars have analyzed the way that language works to create power relationships and validate social positions (Bazerman 1994, 2002; Blyler and Thralls 1993; Longo 1998). Drawing on Foucault, Longo argues that technical discourse can be analyzed to discover how, within cultural contexts, "institutional, political, economic, and/or social relationships, pressures, and tensions" can influence struggles for knowledge legitimation. Clark (2007), like several other scholars, adopts Bourdieu's distinction (1987) between an individual's *habitus* (a set of deeply internalized precepts or ideas) and a *field* (a structured system of positions and allowed activities) that he or she encounters. The field determines what range of agency an individual may have and what cultural, economic, and social "capital" an individual may accumulate in order to advance to a higher position.

These approaches are relevant to VIGRE because the NSF's program alters the *field* in which individuals participate. VIGRE: (a) redefines the scope of the educational process as a chain of continuously viewable career stages; (b) declares undergraduates eligible to participate in research; and (c) combines research and instruction. Communication is the means by which group transactions occur and mathematical issues are negotiated, as well as the place that these processes can be viewed. The NSF's critical goal presents a special research opportunity inasmuch as qualitative methods can illuminate the role of communication in carrying out these transactions and helping students become members of mathematical sciences communities. This article focuses on the start-up years of the Rice University VIGRE Program (2003 to 2005), and portions of a larger qualitative study (to be reported elsewhere) that continued through 2008.

Given the possible conflicts inherent in the NSF approach, we wanted to know how communication functions might contribute to research productivity, team formation, members' identities, and cultural changes. If VIGRE were to produce a new mathematical sciences workforce, how would novices become members? What expectations did faculty and students have about this new program? Would the definition of their mathematical sciences research activity change? How would familiar communication patterns such as lectures and seminar discussions change as members with differing levels of knowledge and experience joined in? Which genres and patterns of communication would affect members' sense of identity as mathematical scientists?

We used qualitative methods to study these changes; especially changes that resulted from the new organizational structure, interdisciplinary research topics, group membership, and communication practices. Like many others, we found it useful to combine high-level theories and models with ones that could describe specific exchanges and events. Our models synthesize theories and work from several fields. They are grounded in activity theory (Leont'ev 1978; Engeström 1987; Russell 1997a), genre theory (Bazerman 2004; Miller 1984; Russell 1997b; Orlikowski and Yates 1994), critical discourse analysis theory in linguistics (Halliday 1978, 1994; Rogers 2004; Gee 1999, 2004), and the social theory of learning that underlies work by community of practice scholars (Lave 1988; Lave and Wenger 1991; Wenger 1998; Wenger, McDermott, and Snyder 2002). The community of practice literature examines learning in the context of activities leading to changes in participant identity, membership in groups, task performance, and construction of meaning. Like Bereiter (2002), we believe that the knowledge creation function of the community as well as the knowledge conservation function must be appreciated. We will explain the relevant models in later sections.

Methods

Study Approval

The study design submitted in the proposal was developed according to best practices in qualitative studies (Denzin and Lincoln 1998) and approved by the university's Internal Review Board before it was sent to the funding agency.

Survey

We designed a survey to learn the rank, experiences, majors, and expectations of students who signed up for one of the three VIGRE research groups the first year because there were no comparable surveys available. A 28-question survey instrument (Appendix 13.1) was completed by 31 of the 47 student participants for a 66 percent return rate (15 of 26 undergraduates completed surveys; 16 of 21 graduate students completed surveys). Accepted strategies for designing, distributing, and analyzing survey results were followed (Lauer and Asher 1988). The survey documented participants' experience with vertically integrated groups, their intentions and career plans, and their expectations for participating in VIGRE. It also asked why students had chosen to join the program and their relevant experiences. Students had the option to complete the survey electronically or on paper. Initially we planned to distribute surveys to participating postdocs and faculty; however, we interviewed these participants instead because of the substantial leadership roles they were given.

Interviews

The three principal investigators (PIs) of the Rice VIGRE program, nine additional faculty, four postdocs and four graduate students were interviewed individually during September 2003 and spring 2005, and follow-up interviews were conducted during February 2004 and summer 2005 (see the sample survey guides in Appendix 13.2 and Appendix 13.3). Exit interviews were conducted with faculty and postdocs in spring 2008, but these are not the focus of this article.

The semi-structured interviews allowed participants to discuss their roles in the research and teaching groups and provided critical details concerning their expectations for group success (Creswell and Miller 2000). Specifically, the interviews included the following topics: background and collaborative research experience; expectations about participation; perceptions about VIGRE; and expectations for student learning and department curriculum. Interviews were audio-recorded, transcribed, and analyzed (see list of codes, Appendix 13.4). The codes were developed on the basis of recurring themes that emerged in answer to our principal questions. We analyzed interviews to identify participants' responses to these recurring themes, as well as observation logs and meeting notes. These included cultural predispositions, undergraduate and graduate student participation in research, undergraduate research roles, graduate student teaching and roles, postdoc teaching and roles, faculty teaching and roles, mentoring, recruitment and retention strategies, interdisciplinarity, and evidence of curricular change.

A few excerpts from spring 2008 interviews have been included in this article because they contained remarks related to the start-up period experience. At that time an additional round of interviews was conducted and audio-recorded with a sample of faculty, postdocs, graduate students, and undergraduates from the 14 PFUGs. Eight faculty members, five graduate students, seven undergraduates, and two postdocs were interviewed (total 22), each for about one hour. Several PFUG meetings were recorded as well. Special attention was given to updating information on the three groups that had been in existence since the beginning of the program. Undergraduates who left the program after only one semester were included, as well as graduating seniors who had participated in three or more semesters and could reflect on a longer period. New as well as experienced faculty were selected for interviews. Faculty were identified by name since it would be very easy to discover their identities from Web site records, but we gave undergraduates and graduate students fictional names since they were assured at the beginning of the communication study that their real names would not be used. All interview tapes were transcribed and coded to facilitate analysis of themes and concepts and assess the quality of participants' experiences over time. Aspects of the program that benefited or frustrated participants early in the program were noted as part of the assessment.

Observations

During the 2003 to 2004 academic year, we regularly observed the VIGRE PFUGs' seminars and planning meetings. We observed and recorded pedagogy, participants' communication strategies, and collaboration. The principal investigators conducted most of the observations. In 2004 the addition of five more PFUGs made it necessary to recruit additional colleagues in the engineering communication group to observe some of the new PFUGs and interview (and audio-record) the leaders of those PFUGs. Observers wrote observation notes at least once every three weeks. The observers met weekly to discuss what was going on in the PFUGs. Handwritten field notes were transcribed and analyzed. We closely described interactions among VIGRE participants during seminar and group meetings and noted the types of questions participants asked of one another and the discussions that ensued. We triangulated our observations with the interview transcripts and student surveys to help validate our analysis and identify genres of communication that appeared to be emerging in the various groups.

Web Sites, Syllabi, and Printed Materials

During the period 2003 to 2005 we saved online materials such as course Web sites (saved weekly), syllabi, some e-mails, and other program materials. However, none of these materials is included in this article.

The Conditions in 2003

In 2003 each of the three participating mathematical sciences disciplines valued some problems more than others (as Fisher's early description of colleagues in his own field of mathematics in 1978 predicted). They preferred some methods of investigation over others, and they were influenced by different disciplinary communities beyond the university or the department—such as professional organizations, colleagues at other institutions, and government funding organizations. We identified the following departmental cultural predispositions toward research problems by reviewing the 2003–2005 interviews and observational notes. We kept in mind theories of organizational culture summarized in Allaire and Firsirotu (1984), but applied especially the research distinction offered by Martin (1992) that organizational culture may be studied in one of three ways: with an integrative approach (one culture identified as present throughout the whole organization); with differentiation approaches (contrasting cultures present in different parts of an organization); and with fragmentation approaches (discrete values across an organization but not comprising a whole). Our study used a differentiation approach, noting both the overarching university culture and the disciplinary cultures that departments shared in part with national disciplinary organizations. The results were as follows.

Mathematicians were interested in abstract, patterned, logical relations. They prized logical problem solving, time-consuming work, and elegant, rigorous, self-contained projects. They reported their most memorable experiences as independent, even solitary work, focusing on "why."

The statisticians focused primarily on actual events, changes, causes, and consequences, especially the "who, what, and when." Their mathematical science revealed patterns in data that suggested the possibility for power over events through prediction and prevention. A wide variety of problems and situations from voting and weather to medical outcomes interested them. They transformed these problems into statistical representations and developed new techniques for dealing with the data.

The computational and applied mathematicians specialized in "how." They developed power through mathematical modeling and analyzed numerical algorithms and mathematical software. They explored inverse problems, discrete and continuous optimization, computational neuroscience, partial differential equations, constrained optimization, and large-scale numerical linear algebra.

Members of disciplines enact their values in their processes and structures. Not surprisingly, the cultural values attached to different research questions and methods affected how easily undergraduates (and sometimes graduates, too) could be brought into research projects. Disciplines that used some experimental methods or computer programs known to undergraduates could more easily offer roles for novices. Bringing undergraduates into a faculty member's ongoing research in advanced mathematics was sometimes more difficult, but sometimes possible.

We observed that vertical integration was indeed achieved in the PFUGs that developed during the first two years. PFUG membership was drawn from all four ranks in most of the PFUGs (Driskill and Zeleznik 2006). Students' responses to questionnaires at the end of the first and second semester agreed with observation notes showing that people of multiple ranks worked together. For example, consider the following undergraduates' responses to the question about what they had liked about being in the biosciences PFUG:

> I liked interacting and learning from all the CAAM/STAT people. I also really liked being able to interact w/ grad students, post docs, and professors.
>
> The excitement of both students and professors in learning this material. Lots of energy in each PFUG.
>
> Exposure to broad range of topics and variety of presenters (students, professors).
>
> I enjoyed working with many different types of people.
>
> A Plus: People of varying levels of background work together.
>
> (December 2003)

Already at the end of spring 2005, the number of PFUGs had grown from three to nine, and one of these had four subgroups. By 2008 there were 14 PFUGs, an astonishing increase. During the start-up period (2003–2005), individuals' roles varied from PFUG to PFUG, but by May 2005 they became *regularized* or "fixed for now," rather than *regulated* (by faculty requirement or rules) as particular types of relationships became more familiar and seemed workable (an important distinction, as noted by Schryer, Lingard, and Spafford in 2007). The VIGRE PIs were not inclined to prescribe what PFUG leaders did. As one co-PI explained, "[the management team] spent a good bit of time discussing how to organize this project without micromanaging the PFUG directors and to transfer our vision for how the VIGRE activity should proceed. . . . That's a bit challenging" (Ensor interview, September 22, 2003). However, since graduate students as well as undergraduates eventually earned their diplomas and moved on, roles always fluctuated and could vary when an individual with an unusual *"habitus"* (Bourdieu's term) enabled a student to take on responsibilities and powers not formerly exercised in the group. The co-PI elaborated, "One challenge will be how to integrate new people into the group as we continue forward and how to let people transition out of the group without losing their efforts toward the success of the whole group. That's probably the biggest challenge." Her comments were confirmed in several interviews with faculty in 2008.

Interdisciplinary Groups Motivate Cultural Adaptation as Well as Ruptures

Significantly, when brought together in interdisciplinary collaborations, faculty members tended to suspend expression of their disciplinary cultural preferences and to relax authority according to an etiquette of hospitable collegiality, and in this more tolerant atmosphere graduate students and undergraduates as well as faculty from other departments had more opportunities to speak up without anxiety. In Bourdieu's terminology, VIGRE had offered more positions that students could occupy than traditional mathematical sciences classrooms offered. An example from the interdisciplinary biomathematics PFUG is explained later.

However, preferences for particular types of problems remained pressing, as might be expected (Huber 1990). A group that favored a statistical approach broke off from the larger biology-related interdisciplinary group during the first year so that it could, one faculty member remarked at the time, "deal with real data, not just endless stochastic models" (Observational notes, September 3, 2004). This comment was consistent with other statisticians' comments about their preferences for addressing actual situations and "counting the bodies." When asked in 2008 about this earlier comment, the faculty member grinned and said, "I don't recall that, but it sounds like something I would say" (Interview notes, April 29, 2008). This kind of congruency was used to affirm cultural categories we identified.

A second example shows how cultural preferences are related to activities (and why interdisciplinary poses a social challenge). These excerpts are from transcribed shorthand notes of a mathematician's beginning-of-semester lecture to prospective recruits and returning PFUG members:

> Pure mathematicians model information to say WHY things are happening. Statisticians find all the similar circumstances and PREDICT what will be found in the future. They do not explain what principles cause the systems to work the way they do. There is no explanatory power in statistics. Mathematicians try to explain WHY you're seeing what you're seeing. Mathematicians think in this powerful way.
>
> . . . CAAM (computational and applied mathematics) uses computers to describe things; mathematicians use their work, but it isn't the way they [mathematicians] think about things initially.
>
> (Transcription of nearly word-for-word shorthand lecture notes, January 19, 2005)

The values implicit in this description function as underlying "rules" for decision making and carrying out research: mathematicians seek to explain the "why" of observed results. Therefore, they have certain types of methods; they think in particular ways. The phrasing hints that the mathematician thinks mathematical work is somewhat superior to the approaches of other disciplines: Mathematicians have explanatory power that statisticians don't have; computational and applied mathematicians use computers to describe things but when mathematicians use computers their purposes are explanatory, not descriptive. It is mathematicians' way of thinking that he emphasizes as particularly desirable and exciting.

Applying Activity Theory and Genre Theory in Analyzing Interdisciplinary Work

We anticipated that a new organizational structure (vertically integrated groups) and a combination of purposes (to conduct research simultaneously with providing education) would result in a new definition of a PFUG's activities that differed from ordinary mathematical sciences lectures or seminar-style activities. Most mathematical sciences courses at Rice before 2003 had been one of these two types. We used a framework from activity theory to focus our attention on the underlying relationships between a PFUG's rules, tools, and division of labor, as shown in a typical activity model, illustrated in Figure 13.1.

Engeström's model is especially apt for analyzing the processes of cultural change that were likely to occur as the structure of groups changed and individuals who would not have interacted much in the past became members of the PFUGs. Since individuals from all three departments would now work together, participants would also have to negotiate changes in rules and division of labor. Communication practices also had to be reconciled. Figure 13.2, an application of the activity theory model to the situation in the research and learning groups, illustrates the factors affecting communication.

When an activity involves language production or communication, the "rules and tools" of the community include the system of communication types or genres such as lectures, seminar presentations, proposals, dissertations, articles, questions or criticisms, and so on. Figure 13.2 displays several aspects of communication that had to be negotiated. As we will explain later, some of the modifications in activity were peculiar to a specific PFUG; others spread across several PFUGs, especially new conventions of formal oral presentations and poster design.

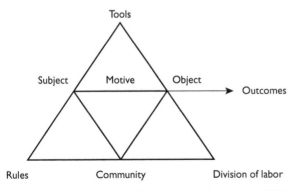

Figure 13.1 Activity System Components (Engeström 1987)

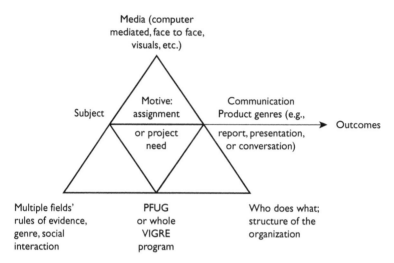

Media (computer
mediated, face to face,
visuals, etc.)

Subject

Motive:
assignment

or project
need

Communication
Product genres (e.g.,
report, presentation,
or conversation)

Outcomes

Multiple fields'
rules of evidence,
genre, social
interaction

PFUG
or whole
VIGRE
program

Who does what;
structure of the
organization

Figure 13.2 Model of Integrated VIGRE PFUG Genre System and Activity System

Some of the changed communication "rules and tools" pertained to the settings (computer-equipped classrooms, laboratories, and laptops owned by students and used in meetings), but the use of these facilities changed as the purpose of the groups changed. For example, instead of using the computer projection equipment to display a faculty member's lecture notes or PowerPoint, the computer equipment in a PFUG meeting might be used part of the time for displaying a Web site that a pair of students had located, or to search for an article on a topic or problem that had come up in discussion. In a typical classroom, all present would have looked at the faculty member's PowerPoint display, but in the informal PFUG sessions some groups might be working in conversation while others looked at the Web site displayed and another one or two might be talking at a dry-erase board. In short, the genre of classroom talk changed in most of the PFUGs.

These rules and tools can be described as belonging to the larger disciplinary community of professional publications, but they may also be influenced by the university, which has its own rules for, say, proposal approval forms, and by the technologies available (Russell 1997a; Russell and Bazerman 1998). In this respect, changes in PFUG communication were related to larger systems of genres and organizations, as Yates and Orlikowski have described (2007).

Genre theory, on the other hand, has methods for analyzing what counts as a genre and explanations for how aspects of a genre change over time to enable groups to deal with recurring situations (Russell and Bazerman 1998; Yates and Orlikowski 1992, 2002, 2007). These concepts add ways to explain dynamics and changes over time; they extend the range and specificity of activity theory. Current genre theorists argue that each time someone creates a document or presentation that has conventional aspects, there is also the possibility of modifying one or more conventions to satisfy the situation's conditions (see essays in Coe, Lingard, and Teslenko 2001). Thus rhetoricians see genres as "stable for now" arrangements (Schryer's phrase, 1993) that are also subject to change as the groups using these forms of communication transact their work (Bazerman 2004; Schryer 1993; Schryer, Lingard, and Spafford 2007).

The genre conventions of the various disciplines from which faculty members came were not the same. Most of the PFUGs were working on interdisciplinary topics, such as computational finance or biomathematics. They therefore had to deal with the activity structures and genre systems manifest in other fields' ways of investigating problems and publications, as shown in Figure 13.2. Although faculty had some genres in common, such as the published article, the argument structures and types of evidence used within these genres differed. For example, faculty in computational and applied mathematics published articles and their graduate students wrote dissertations in which the results from computer programs such as Matlab were important evidence. In contrast, the mathematicians published articles in journals that usually required proofs and arguments based on equations. When they did use computer models or computed results, these served as illustrations, not proof. In addition to differences in evidence and argument structure, genres sometimes differed in length (which is major if one field uses much lengthier arguments and many more diagrams than the other), as did more minor features: abstract formats, conventions for labeling figures, and so on. As will be explained later, these differences were addressed as PFUG members prepared presentations, posters, and reports together.

New and Adapted Genres for New Activities

The co-PIs of the Rice VIGRE program hoped to provide participants with a vertically integrated research and teaching environment in which faculty, postdocs, graduate students, and undergraduates could work and learn together. Overall, VIGRE participants, such as this statistics professor, perceived the structure to be an aid: "I understand the PFUG mostly as a tool or a vehicle or a framework in which we can help students, and even postdocs, to learn about research and to contribute to research" (Interview notes, August 29, 2003). However, the general framework had few specific guidelines. Creating or modifying genres was one of the ways that PFUGs worked out what would characterize the new kinds of activities and mission created by the VIGRE program. Vertical integration—the combination of undergraduates, graduate students, faculty, and postdocs in groups—appeared to contribute to the formation of new rules and ways of interacting, as suggested by figures 13.1 and 13.2, and these were oral and visual rules, not just written conventions.

The integration of research and learning into a single activity system was perhaps the most complicated innovation in the NSF-sponsored program. VIGRE's mandate to integrate teaching and research for this vertically integrated population was a major modification of the activity systems in the three departments. Each PFUG had an undergraduate course and a graduate seminar associated with it, but since research as well as instruction had been combined, the undergraduate "course" strained participants' tolerance as it strove to accomplish merged purposes. Similarly, the VIGRE graduate seminar was supposed to introduce advanced work related to the PFUG's topic, which differed from graduate seminars that students felt belonged to their particular areas. The graduate students said the seminars ate up time, a scarce resource, and competed for attention, distracting them from their chosen specialties.

Furthermore, the VIGRE program had to be integrated into the ongoing structure of graduation requirements, major requirements, semester-length limits, course schedules, and college life. To encourage students to "try out" VIGRE, undergraduates' participation in a PFUG was allowed "variable credit" of one to three credit hours. Therefore, participation—even for one credit—was, in some sense, to be "a course." Students felt the need to have a sense of completion, material covered, and techniques learned, and they had quite different ideas about how much work should be done for one, two, or three credits. On the other hand, research does not run on a railway schedule; it is rarely possible to provide a tightly organized syllabus for the progress of research activities. These differing expectations and conflicts challenged both faculty and students. Although some of the students had worked on research or project teams in other classes, the vertical integration made the team membership strikingly different from the team experience described as optimal by Kleid (2004).

At the beginning (2003), the conflicts were not fully appreciated. One faculty member assigned summer students to write up "white papers" that would be the required text for their peers in the fall VIGRE undergraduate "course," but the students were not at all sure what a "white paper" was—a genre problem. Their initial drafts varied widely; some resembled personal essays, others encyclopedia entries without clear arguments, others research reports. Another PFUG leader tried to incorporate undergraduates in a seminar-style weekly discussion of articles to prepare students to undertake research projects in the spring. Members were supposed to present and lead discussions of current research articles. Undergraduates' lack of experience and knowledge caused some to flounder, some to leave; gradually graduate students and postdocs realized the importance of mentoring the undergraduates who were presenting and began helping them to prepare. Gradually the genre "leading a seminar" was renegotiated so that a "seminar presentation" became a summary of an article's main points and evidence followed by a great deal of questioning and discussion among the group without the presenter's "direction" of the discussion.

Faculty members present did not completely direct the discussion either, although their questions or issues might take more group time than others'. On the plus side, where the PFUG subject was interdisciplinary, the undergraduates who were majoring in a related topic also began to function as mentors for mathematical sciences graduate students. More collaboration flourished as people recognized others' potential to contribute. Over the first two years, VIGRE "seminars" took on a different set of roles, conversational moves, and conventions than were found in typical graduate seminars, as will be discussed below.

Because research topics were unlikely to be the subject of complete books, students' textbook expectations could not be met easily. Occasionally a book could be found to serve as the foundational reading. Many groups chose to read recently published articles and to have members try out the methods or algorithms in the articles to evaluate their usefulness for the group's own concerns. Faculty offered on-the-spot summaries of background to scaffold members' learning; everyone encouraged the asking of questions, no matter how elementary. The least successful approach was to have undergraduates "sit in" on graduate student seminar discussions and then drop by the faculty members' office to hear and discuss explanations of the articles. One of the most successful tactics was for the faculty member to begin working a problem from an article in class and invite under-graduates to carry forward with it before the next meeting. The students could not always finish the problem, but they were expected to report on how far they had got with it and to be able to discuss their attempts at the next class.

New Genres That Foster Identification as Mathematical Scientists Communication

We found that participants in the PFUGs frequently used three types of com-munication in ways that appeared to accelerate individual and group learning and to foster identification with the group and the field: (1) one-on-one conversations (in particular, between participants who typically would not discuss research); (2) formally scheduled presentations; and (3) situationally motivated presentations that enabled participants to disseminate research progress occurring in the broader mathematical sciences field and to deliver information and report on work they or the research group had generated. We contend that the vertical integration of the PFUGs along with the small member-size of these teams (5 to 15 participants per PFUG) affected the activity system and encouraged participants to use these communication strategies.

One-on-One Conversations: Peer and Inter-Level

One-on-one conversations were the backbone of intra-PFUG communication. Obviously, all VIGRE participants used one-on-one conversations between peers (for example, an undergraduate speaking to another undergraduate). However, during seminars and group meetings we also regularly observed one-on-one conversations between participants who typically would not have opportunities to discuss research. For instance, in contrast with typical classes, we noticed how frequently postdocs talked with undergraduates, graduate students talked with undergraduates, and members of all ranks talked with faculty. This type of communication was the key to the mentoring we observed.

Formally Scheduled Presentations

Faculty, postdocs, graduate students, and undergraduates all delivered formal presentations to their PFUGs and, in the summer, to the "VIGRE interval," the Wednesday afternoon colloquiums in which each PFUG presented once to all the rest, including faculty.

Formally scheduled presentations were usually announced at the beginning of the semester, typically appeared on the syllabus of courses if syllabi existed, and were announced in class. These communication forms followed the classroom formats students were familiar with and described activities of achievement and recognition. For these, participants tended to spend several days or even weeks preparing for their talk. The content of the talks "belonged" to the students; they identified with the work they presented. These presentations had predetermined objectives (for example, to present preliminary research results), which were usually stipulated by the faculty, and predetermined time limits (for example, 30 minutes) within which participants were asked to speak.

Situationally Motivated Presentations

Faculty often requested presentations that were not on the syllabus or asked for volunteers to deliver presentations. At other times, PFUG members volunteered to share results or to describe a research article that they had found. Many VIGRE meetings were structured to include these more impromptu presentations. Overall, both formal and situationally motivated presentations helped VIGRE participants report progress and contributed information that was needed by the entire group.

The Biomathematics PFUG Example

Mathematics professor Robin Forman asked participants in his Developmental Biology PFUG to deliver both formally scheduled as well as situationally motivated presentations. This PFUG included approximately nine members—one faculty, two postdocs, three graduate students, and three undergraduates. Each week, participants met for 90 minutes. The following scenarios illustrate the two types of presentations that characterized this PFUG as well as the Rice VIGRE program overall.

During an unscheduled presentation in the developmental biology seminar (February 10, 2004), Bert, a mathematics undergraduate, offered to review the article he had read for that day. Bert spent about 20 minutes presenting the main points of the article at the board in front of the class. During his talk, he also responded to 16 questions asked by faculty, postdocs, and graduate students. These questions helped Bert further explain the finer points of the article and enabled the participants to speculate about how this scholarship impacted on the PFUG's research trajectory (Observation notes, February 10, 2004).

The next scenario, also from a developmental biology seminar (October 28, 2003), shows the collaboration between an undergraduate and a graduate student as they deliver a formal presentation about their work with the Gillespie algorithm. Barbara, a mathematics/statistics/economics undergraduate, and Ross, a mathematics graduate student, presented their talk, "Minor Generalization of the Gillespie Algorithm and a Program to Implement It," to the members of the developmental biology PFUG. This pair, along with an additional graduate student who did not participate in the presentation, collaborated outside of the seminar for several weeks on the work for this talk.

Barbara and Ross split their presentation time so that each of them explained different aspects of their work in front of the class at the board and with transparencies. When the pair was interrupted with questions from the audience, both Barbara and Ross answered the questions about their work (Observation notes, October 28, 2003).

A New PFUG Genre

Often these presentations also provided learning opportunities not anticipated by faculty. For example, during Barbara and Ross's presentation, Professor Forman interrupted Ross and instructed him to "explain in English what your equation says" (October 28, 2003). At first, explaining the equation was difficult for Ross; however, Barbara jumped in and they were able to respond. This exchange led to a new requirement for students presenting their work (or the work of others): they were expected to explain equations clearly and accurately so that their audience did not need to struggle over them.

Barbara and Ross's interaction and the PFUG leader's direction illustrate how PFUGs developed their own PFUG-specific genres of presentations and documents. Whereas in departmental colloquia faculty or visiting scholars are expected to "talk over the heads" of junior members in the audience, the commitment to joint research created new requirements of clarity and comprehensibility that also reflected an ethical commitment to mutual understanding. Outside of the university, "explain it in English" is a colloquial expression for demanding a translation in familiar terms. However, in the PFUG it became a shorthand for a requirement that meant much more than simplified talk. It became understood as a particular process of chronologically organized, simple-to-complex unpacking of new equations and materials, replete with definitions. Although easier to understand, these explanations might be longer due to the inclusion of "for instance" illustrations, definitions, and connections to work already studied or performed.

In summary, activity theory and its high-level model of significant components directed our attention to ways that the changes in structure (proposed by the program leaders) and purpose (determined by the NSF) were being negotiated through communication genres as they contributed to the identities of both groups and individuals and created knowledge outcomes that Bereiter (2002) calls "conceptual products of group activity." Genre theory prompted us to take a look at systems of communication types and the transactions they negotiated. Genres of communication and mathematical reasoning produced research results as well as roles and subject positions for speakers and writers.

Applying Community of Practice Theory to Understand the Development of Group and Individual Identities

Community of practice theory was developed initially from studies of apprenticeships in European industry (Lave and Wenger 1991), and was elaborated in studies in large institutions and multinational companies (Wenger 1998). In these descriptions, a community of practice and an operating unit, such as the claims processing department of an insurance company, are described as one and the same. However, this formulation left out the ways that some units' practices are controlled by the expertise of other groups. In *Cultivating Communities of Practice* (2002), Wenger, McDermott, and Snyder develop a model to explain how communities of practice (CoP) might differ from the departments and operations units in which the expertise about practice is applied. Communities of practice share, conserve, and develop new knowledge; operations units apply that knowledge. Members of operations units who also belong to a community of practice elsewhere in the organization then take their experience back to the community of practice, where it is refined, compared with former knowledge constructions, stored, and disseminated to operations units in the future. They describe an ideal organization as "tightly knit" (2002, 18). At least some individuals will hold multimemberships in both a community of practice knowledge group (say, statistics) and in a department or project group engaged in problem solving (say, operations management). Thus, their model for learning separates the conservation of existing knowledge and the application and testing of these tenets. Their model appears in Figure 13.3.

The ordinary technical company differs from this idealized multimembership structure by housing conservation or stewardship in research and development (R&D) and having a training department transmit the knowledge to operations staffs, much as medical research results are disseminated to practitioners. Those in operations do not become members of the research and development community, just as many nurses do not find themselves accepted as members of research communities. This situation is far from the community of practice ideal.

The Rice program did *not* follow the industry model, either in an ideal multimembership configuration or in the R&D department configuration. The vertically integrated PFUGs *combined* the stewardship of available knowledge, application and testing of that knowledge, and development of new knowledge all in a single group, the vertically integrated PFUG. The PFUGs collapsed the distinctions in the multimembership model as they involved faculty, postdocs, graduate students, and undergraduates in one structure. Furthermore, they were not constrained to practices approved by company policy, unlike the claims

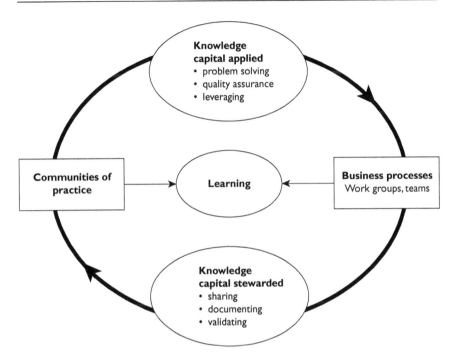

Figure 13.3 Wenger, McDermott, and Snyder, *Cultivating Communities of Practice* (2002, 19)

processors of Wenger's 1998 community of practice. An adapted model was needed to capture the PFUGs' situation.

In some of the PFUGs, some hierarchical distinctions persisted as individuals were sometimes slotted into tasks suited for their knowledge level, but most of the PFUGs adopted an egalitarian attitude that allowed and encouraged individual students to speak up and participate freely. Our observational notes recorded the contrast between the more abrupt openings to lecture courses (for example, "All right, let's get started. Last time we were on the radius of convergence. The next section is on polar coordinates.") and the relatively social PFUG sessions where people from more than one field were present. In those groups, at least five minutes would be spent in conversational pleasantries—"How's it going?" "How far did you and Tom get with that algorithm?" or "What did you guys find out about . . .?" to students as they arrived, and a discussion that grew out of those gambits. If the faculty (up to four might be members) did not arrive first, the conversation started anyway unless a faculty member was scheduled to bring in something that was needed before the group could proceed. Asking questions in lecture, usually an infrequent occurrence unless initiated by faculty, was expected in PFUG meetings and took a bit of getting used to.

In the following excerpt from an interview in 2008, a senior described how students had been induced to accept new identities expressed in communication as they became seasoned members of the community of practice in their PFUG a couple of years earlier:

> And I spent that first semester in VIGRE really not knowing. I felt out of my element because I felt like everyone around me knew what they were doing and kind of had it all together. And Dan [Cole, the faculty member] was real good about kind of, you know, prodding and coaching and so were the other guys. . . . They're about to graduate. That's why they're confident with the material, . . . confident about spouting out an idea that may or may not be stupid, you know [laughs]. And they said—they kept telling me, "Over time, you know, you'll open up. It's okay."
>
> (Interview notes, April 28, 2008)

After two semesters in the PFUG and a summer spent in the VIGRE program with daily progress presentations and helping give a team presentation to all the other PFUGs, the student had stepped up to a leadership role himself:

> And then this year Mike Nelson and I, we were essentially in the same place that the senior guys were in whenever we showed up, you know. So, we had some nervous new recruits not very confident and not sure when or whether they should speak up. And, you know, it was kind of fun 'cos we got to engage in that role of, "No, you're doing fine. Really, we all sound like idiots [laughs]. So, I mean, no one's going to judge you. That's just part of it." It's been a lot of fun. I hope to continue with it after I graduate.
>
> (Interview notes, April 28, 2008)

The Reifying Consequences of Communication

We argue that statements uttered during sessions that refer to the speaker as subject or object ("I" or "we," "me" or "us"), as well as stories participants told outside of the work group, function in a reifying way, expressing to others and storing in the speaker's memory a constructed identity connected to disciplinary membership and meaning. This interview excerpt illustrates how the community of practice approach focuses on learning in the context of activities leading to changes in participant identity, membership in groups, task performance, and construction of meaning.

In the PFUG meetings, new norms for discursive interaction gradually developed that resembled neither a mathematics seminar nor a lecture. The types of statements we are about to describe can also occur in lectures or seminars, but in those settings they generally serve a different purpose and constitute a different proportion of the discourse. By fall 2004, a "discussion about research articles" had distinctive rhetorical traits. Figure 13.4 shows our adaptation of Figure 13.3, the community of practice model shown earlier (originally Figure 1-1 in Wenger, McDermott, and Snyder 2002, 19), to depict one version of the NSF's proposed combination of a research and instructional situation. Figure 13.4 reflects PFUGs' combined research and learning mission and therefore collapses the stewardship of past knowledge function and the application and creation of knowledge function. It illustrates how two types of communication—which we label *forward-looking* and *retrospective*—appear to have helped participants to fuse research and learning in the several Rice VIGRE groups.

In our theoretical model, *forward-looking communication* occurs when a faculty participant (a PFUG leader or another faculty member, and many PFUGs had two to four faculty members) explains established concepts or the group discusses recent publications so that the group can undertake new research activities or change activities. In doing so, for example, students may develop understanding, if not mastery, of routine problem-solving techniques, plan their work by creating a narrative of what they would do or expect to do, work out the equations or algorithms used in a new article, and propose equations or codes to try or test. They may also learn to evaluate alternatives. The unfamiliarity of the material caused the PFUGs to place a high priority on asking questions without regard to a risk of appearing "stupid." All of this tended to move the discussion into the future tense or a subjunctive mood: "If we could make the algorithm on the Yale site—and it's only for two dimensions—so it would handle three dimensions, maybe Toby and Zheng will be able to use it in the stuff they're doing" (Observation notes, October 12, 2004).

Forward-looking communication certainly occurs in regular mathematics lectures and seminars, but it is seldom directed toward research. Much more common is a faculty comment such as, "You'll need to keep this in mind on next Monday's problem set," or "You will definitely use this next semester when you get to PDEs (partial differential equations)." In a PFUG, forward-looking communication was generally aimed at the immediate future, and it occurred much more frequently because the purpose of discussing what was already known or had recently been published was to change or evaluate the PFUGs' current or future work.

Retrospective communication occurred when VIGRE participants (including faculty, postdocs, graduate students, and undergraduates) interpreted the research tasks they and/or their group were performing or had already performed and the results they or their group found; in doing so, participants constructed a retrospective or reified account of activities. For example, in completing a particular research activity, participants might switch to communication that discussed acts as they performed them:

Faculty: "How did that work you and Zheng did turn out?"
Toby (looking at his open laptop): "I'm not getting what we expected here."
(Observation log notes, November 16, 2004)

Or, they might narrate what a group or individual had accomplished thus far.

As the design of Figure 13.4 attempts to indicate, both retrospective and forward-looking communication are part of a recursive process. Participants switch

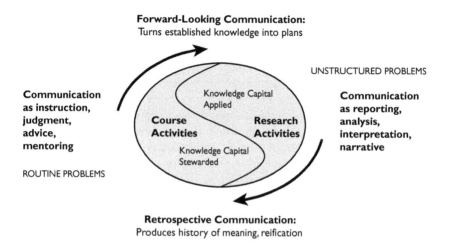

Forward-Looking Communication:
Turns established knowledge into plans

UNSTRUCTURED PROBLEMS

Communication
as instruction,
judgment,
advice,
mentoring

ROUTINE PROBLEMS

Knowledge Capital
Applied

Course
Activities

Research
Activities

Knowledge Capital
Stewarded

Communication
as reporting,
analysis,
interpretation,
narrative

Retrospective Communication:
Produces history of meaning, reification

Figure 13.4 Communication Processes in a Research/Learning Community of Practice
Affecting Identity and Development of Expertise

back and forth between these framing discourses over time in a discussion. Some PFUGs, especially those which were trying to accommodate the sense of a "course" by prescribing a list of readings, would talk about the assigned readings first and then switch to a discussion of how the article related to the overall research work and finally to a progress report on work done. Other PFUGs began with what had been accomplished during the interval and then looked at new readings or articles that people had found and considered how those might be used in the future, if at all. The distribution and patterning of tenses created the texture of the overall genre of "meeting" in each PFUG.

These two types of communication affected students' sense of their identities as mathematical scientists. The linguistic artifacts generated through retrospective communication comprised a history of practice. As students reported on, questioned, and argued about published work, they took subject positions in relation to the mathematical sciences disciplines. They became characters in their own stories. As characters entitled to subject positions in these artifacts, members had acquired identities that they ascribed to themselves as they told stories. In these they were "reified," represented as "things," mathematicians who played a certain part. As they increased the group's store of knowledge and experience, they also developed expertise. The comments quoted above about being encouraged to speak up and participate are but one instance of this reification of the student's experience and past speech. Reification appears to have been one of the most powerful factors in constructing participants' identification as mathematical scientists, regardless of rank (undergraduate to postdoc). And subject positions and reification involve emotions and feelings, too, we should add. However, unlike the relatively slow changes in an established community of practice's activities, the PFUGs' combined research and learning missions accelerated changes both in identities and in practices.

We note that, partly as a result of the policy to offer PFUG leaders broad latitude, new postdocs and graduate students who were given charge of a PFUG had a more difficult time devising and adapting genres and communication patterns. They were the ones who were more likely to attempt to continue a seminar or lecture format when participants were dissatisfied; although they praised the outcome as having improved and broadened the range of their teaching approaches, they expressed frustration with their initial semester of teaching. The seasoned faculty who pioneered the PFUGs mostly had an easier time. Orientation and training as well as mentoring by experienced faculty now can facilitate new leaders' transitions.

Changes in communication types were instantiated in changes in linguistic forms. Both in the intra-PFUG presentations about what individual members had been doing and in discussions about published articles, the discourse tended to move from past tense to present and future tense. In both types, the speaker describes himself or herself as the subject of the sentence, but in retrospective communication the action is expressed in the past tense whereas in forward-looking communication the anticipated action is expressed in the future tense. This regular fluctuation contrasts sharply with discourse in a typical lecture. It has been our experience that lecture-hall students seldom raise their hands and deliver a past-tense commentary in which they represent themselves in the subject position. We did not attend many lectures outside of PFUG meetings, but we believe that our general impression is fairly accurate.

Communities of practice seek to create mutually understood knowledge about their ways of working and problem-solving plans. These communities work best when every member is able to share as much of his or her knowledge as possible and when the group refines those contributions as they relate to their shared goal-directed activities. As the Rice University VIGRE Program began in 2003 to 2005, the PFUG research and learning teams adapted their communication genres and conventions to complete their research projects and efficiently increase every member's level of knowledge.

Conclusions

The introduction of people of all ranks into learning and research groups modified the traditional power structure and the purposes of their activities. In traditional classrooms, the organizational structure and roles guarantee only faculty members (usually one per course) the right to speak and prescribe limited exchange genres such as the prepared lecture and instructor-initiated questions as the dominant forms of interaction. Learning is aimed at recall of currently approved disciplinary knowledge and application of standard problem-solving techniques in routine problems.

We contend that the vertical integration of the PFUGs, along with the small team size (5 to 15 participants), affected the activity system and encouraged participants to adapt familiar genres of classroom talk or to develop new conventions that suited the new purposes of their activities. In the case of the Rice University VIGRE Program in 2003 to 2005, vertical integration was achieved, and groups developed new genres or significantly modified their existing genres of communication to achieve the NSF's purposes and implement the Program's activities. One-on-one conversations about research occurred between people who typically would not have discussed how to conduct research. Both formally scheduled presentations and situationally motivated presentations reported individuals' experiences and shared knowledge among the teams. Formal summer presentations by each team to all the other teams helped members to grasp the broad range of mathematical sciences research.

The new communication genres affected participants' sense of individual identities as well as group identities. As individuals or pairs of students reported to their groups, their identities were reified in their narratives. They expressed these identities in retrospective talk, PowerPoint presentations, written reports, and poster shows. They developed a context of disciplinary discourse through forward-looking talk that connected past work to their present or future work, and this too affected their identities by placing them in an intellectual tradition and history of action. Specific conventions, such as asking questions that "might or might not sound stupid" were a badge of being a mathematical scientist, one who was willing to risk stature as an expert in order to persistently pursue inquiry. "Explaining in English" was another finely honed convention suited to effective communication in a group that contained both experts and novices without sacrificing anyone's status.

Activity theory, genre theory, and community of practice theory were helpful in accounting for participants' actions. Activity theory directed attention to changes in the configuration of factors that affected situations. Genre theory helped identify the ways that typical classroom genres such as lecture and seminar were changed to carry out the new research and learning mission and the new roles of participants who varied in their expertise. Community of practice theory helped to explain how individuals and groups achieved identities as mathematical scientists and as collaborative rather than individual researchers. They made research (at least some kinds of mathematical research) into a team sport.

The case and theoretical models show how rhetorical analysis can illuminate the processes of learning and identity formation when communication networks include individuals of widely varying prior knowledge and experience. Future studies should be done in other technical communication situations to understand the activities of participants who are now connected by interactive communication technologies and new organizational structures. Traditional experts will benefit from seeing themselves as members of a broader community of practice. Former "recipients" will gain a voice and contribute to the transformation of technical into local knowledge. Overcoming the limits of one-way models can facilitate better decision making and benefit people in multiple industries and systems worldwide as new communication technologies create global communities of stakeholders. Technical communication scholars and practitioners have a vital opportunity to explore and empower such communities through the application of activity theory, genre theory, and community of practice theory adapted to the challenges of particular industries.

Appendix 13.1: VIGRE Survey (Graduate Students and Undergraduates)

Fall 2003/spring 2004

Name:_____

Directions. Please respond to the following items as they relate to your academic work. Circle the letter that most closely describes your characteristics. For those questions that require a written response, please write directly on the survey form.

I. BACKGROUND

1. Indicate your major or field of study:
   A = Computational and Applied Mathematics B = Mathematics C = Statistics
2. Indicate your position: ___Graduate Student ___Undergraduate Student
3. Identify your research group(s) (PFUGs) by checking all those that apply
   __ Computational Algebraic Geometry __Computational Finance
   __ BioMath/Gene Networks __ BioMath/Mathematical Modeling
4. Rate the importance of communication skills to success in your major
   A = unimportant B = a little important C = somewhat important D = critically important
5. Rate the importance of communication skills to success in your future career
   A = unimportant B = a little important C = somewhat important D = critically important
6. Please describe why you decided to participate in this VIGRE program.
7. What education and/or experience do you have that is valuable to your research group?
8. What do you think you will get out of your participation in this VIGRE program?

II. PROBLEM SOLVING

9. What types of problems—related to your major field of study—do you like to solve?

10. Describe whether working with others helps you to solve these problems.
11. How skillful are you as a problem solver?
    A = very weak B = minimal C = competent D = professional

III. COMMUNICATION

Rate your own communication skills as they relate to your academic work.
12. Oral skills (both informal and formal interaction)
    A = very weak B = minimal C = competent D = professional
13. Written skills
    A = very weak B = minimal C = competent D = professional
14. Design skills (for example, designing documents, figures, graphs, etc.)
    A = very weak B = minimal C = competent D = professional
15. How does communication relate to the work that you will do in your group?

IV. GROUP WORK

Rate yourself as a collaborator by identifying your typical behavior in previous groups.
16. Attend meetings on time and prepared
    A = seldom B = occasionally C = often D = all the time
17. Do my fair share of the work
    A = seldom B = occasionally C = often D = all the time
18. Actively pose alternatives to reduce the likelihood of groupthink (for example, collaborators who engage in groupthink fail to consider all alternatives when making decisions and tend to desire harmony in their groups—often at the expense of effective decisions)
    A = seldom B = occasionally C = often D = all the time
19. Actively participate in group discussions
    A = seldom B = occasionally C = often D = all the time
20. Actively listen to other group members (that is, pay attention, clarify, and so on)
    A = seldom B = occasionally C = often D = all the time
21. Take careful notes
    A = seldom B = occasionally C = often D = all the time

To describe your interaction style in academic groups, circle true or false.
22. When I work in groups, I prefer that the work be split up so that each person does something different.
    A = True B = False
23. I prefer to work in an exploratory way. For example, when I'm given a task, I enjoy figuring out the best way to complete that task rather than being told what to do.
    A = True B = False

V. INTERDISCIPLINARY GROUPS

Questions 24 and 25 concern interdisciplinary groups in academic settings. We define a group as interdisciplinary when people from two or more disciplines work together to solve a problem.

24. What experience(s) do you have in interdisciplinary groups that will most affect how you plan to collaborate with your research group this semester?
25. How does working in an interdisciplinary group make the work you anticipate doing in your research group
    a)   more difficult?
    b)   more rewarding?

VI. CAREER PLANS

26. What is the most likely career for you in the long term?
27. List three communication skills that you believe may be required for this job.
28. How important do you believe participating in the VIGRE program will be fore preparing you for this career?
    A = unimportant B = a little important C = somewhat important
    D = critically important

Appendix 13.2: VIGRE Interview Questions (Program Directors and Faculty)

Name _____   PFUG(s) _____
Interview Date _____   Interview Duration _____

I. BACKGROUND and COLLABORATIVE RESEARCH EXPERIENCE*

1. So what surprised you the most about how things are going with your VIGRE group this semester?
2. What's changed about ways you've collaborated with other fields since you began?

II. EXPECTATIONS ABOUT PARTICIPATION

3. Would you characterize your role as a faculty member/program director in your PFUG(s)?
4. What do you personally hope to gain from this role(s)?
5. Who are the participating faculty in your PFUG(s)?
6. What expertise do you believe that you bring to your PFUG(s)?

7. What expertise do you see the other faculty bringing to your PFUG(s)?
8. What expertise do you see lacking in your PFUG(s)?
9. What challenges do you anticipate in the planning and/or implementation of your PFUG(s)?

III. PERCEPTIONS ABOUT VIGRE

10. VIGRE is characterized as a research and teaching approach that is multi-disciplinary and vertically integrated (that is, it involves people at all levels—faculty, postdocs, graduate students, undergrads). How does this VIGRE approach compare to a traditional research group, in your experience?
11. What do you see as the possible strengths of this VIGRE approach?
12. What are the possible weaknesses of this VIGRE approach?

IV. EXPECTATIONS FOR STUDENT LEARNING/ DEPARTMENT CURRICULUM

13. What are your objectives for the
    a) undergraduate students?
    b) graduate students?
14. What effect do you believe VIGRE will have on your department's curriculum (for example, increase in cross-listed courses)?
15. Do you have any comments or questions that these interview questions didn't allow you to make?

* Section I questions were added in interviews in the second round and thereafter through spring 2008.

Appendix 13.3: VIGRE Interview Questions (Postdocs)

Name _____ PFUG(s) _____
Interview Date _____ Interview Start Time _____
Stop Time _____

I. BACKGROUND and COLLABORATIVE RESEARCH EXPERIENCE

1. I'm interested to know why you decided to participate in the VIGRE program. Would you describe that for me?
2. What education and/or other experience do you have that you think is valuable to your PFUG?
3. What types of research is your PFUG conducting?

4. Have you been involved in research about a topic that's closely related to your PFUG's research?
   a) What were you studying?
   b) Were you the principal investigator?
   c) How many collaborators? What were their disciplines?
   d) Did the collaboration result in publication(s)?
   e) What was the most challenging aspect of collaborating across disciplines?
   f) What behaviors/expectations/assumptions are you bringing from that collaboration to VIGRE?

II. EXPECTATIONS ABOUT PARTICIPATION

5. Could you characterize your role as a postdoc in this PFUG?
6. What do you personally hope to gain from your role as a postdoc in this PFUG?
7. Who are the participating faculty in your PFUG?
8. What expertise do you see them bringing to your PFUG?
9. What expertise do you see lacking in your PFUG?
10. What challenges do you anticipate in the planning and/or implementation of your PFUG?

III. PERCEPTIONS ABOUT VIGRE

11. VIGRE is characterized as a research and teaching approach that is multi-disciplinary and vertically integrated (that is, it involves people at all levels—faculty, postdocs, graduate students, undergrads). How do you see this VIGRE approach as being different from a traditional research group?
12. What do you see as the possible strengths of this VIGRE approach?
13. What do you see as the possible weaknesses of this VIGRE approach?

IV. EXPECTATIONS FOR STUDENT LEARNING/ DEPARTMENT CURRICULUM

14. What are your objectives for the undergraduate students in your PFUG?
15. What effect do you believe VIGRE will have on your department's curriculum?
16. Do you have any comments or questions that these interview questions didn't allow you to make?

Appendix 13.4: Codes Used for Analyzing Interviews, Logs, and Observations (Processes of Self and Group Identification)

<NEGOTIATION OF MEANING>

- <ENGAGEMENT>
- <IMAGINATION>
- <ALIGNMENT>

<COMMUNICATION>

- Communication between departments
- Interaction between undergraduates and graduate students, undergraduates and faculty, etc.
- Ability to communicate information/events/etc. across the Rice campus

<BACKGROUND>

- The background of an undergraduate student or graduate student
- Explanations as to why an undergraduate student or graduate student signed up for the VIGRE program

<PRESENTATIONS>

- Descriptions of a student's development as a presenter and as a communicator (very similar to "communication" tag)
- References to the CAIN Project, and poster presentations
- Recall vs. thinking

<ADMINISTRATION>

- References to a lack of administration: extra strain, lack of recognition, lack of communication

<PFUG TOPIC>

- Talk that has to do with the process in which PFUG topics are chosen
- May also include talk that refers to processes in which individual projects are chosen for students (may also be defined as "volunteerism" and may include "hiring practices" as well)

<HIRING PRACTICES>

- Anything that describes the hiring practices of postdocs and other VIGRE faculty

<RETENTION/RECRUITMENT>

- Views about student retention and student recruitment

- Methods used to retain/recruit students
- The "graduation problem"

<EXPECTATION>
- Expectations of graduate students, faculty, and postdocs towards undergraduate students and VIGRE
- Expectations of undergraduate students to VIGRE, as well as research
    - Also includes disillusionment of undergraduate students

- Instances where students describe the style of learning they prefer
    - Ex: problem sets, vs. taking and running with something
- Instances where students describe different ways of thinking
- Explanations of VIGRE as an education in "teaching you how to think"

<BIG MOMENT>
- Moments a student or faculty member describes as having a big impact on a certain process

<LEARNING>
- Any instance where the student/professors describe learning a new skill

<TACIT>
- Any time a student/teacher displays a sense of awkwardness in describing a process that comes easily to them
- Any time a student demonstrates a lack of awareness of a process he/she has engaged in

<CULTURE>
- Any part in which the culture of one's own department is described

<INTERDISCIPLINARITY>
- Anything that has to do with specific differences between departments, or when a student compares his/her identity with different departments

<BEAUTY>
- When a student or teacher compares theoretical research problems to applied research problems
- When a student or professor distinguishes between what the faculty refers to as the difference between "beauty" and simply "throwing a computer at a problem"

<LEADERSHIP>

- Instances where students and graduate students take on more leadership and responsibilities in their PFUG (process)

<CONNECTIONS>

- Descriptions of other connections to research institutions outside of Rice
- Descriptions of additional connections within a department
- Descriptions of a lack of connections due to VIGRE
    - Ex: graduate student community feel has lessened due to VIGRE

<TEACHING/LEARNING STYLE>

- Any part in which a teaching style or perspective can be labeled or is described in the context
- Any part when someone describes how his/her teaching style has changed
- <VIGRE VS CLASS>
    - When a student describes the difference between
- <PLUGGING>
    - When a student or professor distinguishes (or fails to distinguish) between data entry and conceptual
    - Has to do very closely with the level of involvement a student has in research
- <MENTORING>
    - Any mentioning of mentorship of students
- <VOLUNTEERISM>
    - Evidence of personal willingness to meet an identified need

<RESEARCH TOPIC PREFERENCES>

- Values associated with research problem choices, including aim, type of evidence, method, interaction style
- Expectations of requisite knowledge, prior experience, or readiness for a topic
- Characteristics that make this topic accessible to people of different ranks

References

Allaire, Y., and M. E. Firsirotu. 1984. "Theories of Organizational Culture." *Organization Studies*. http://oss.sagepub.com/cgi/content/abstract/5/3/193.

Artemeva, N. 2005. "A Time to Speak, a Time to Act: A Rhetorical Genre Analysis of a

Novice Engineer's Calculated Risk Taking." *Journal of Business and Technical Communication* 19: 389–423.

———. 2008. "Toward a Unified Social Theory of Genre Learning." *Journal of Business and Technical Communication* 22: 160–85.

———, and A. Freedman, eds. 2006. *Rhetorical Genres Studies and Beyond*. Winnipeg, MB: Inkshed Publications.

Bazerman, C. 1994. "Systems of Genres and the Enactment of Social Intentions." In *Genre and the New Rhetoric*, ed. A. Freedman and P. Medway. London, UK: Taylor and Francis.

———. 2002. "Genre and Identity: Citizenship in the Age of the Internet and the Age of Global Capitalism." In *The Rhetoric and Ideology of Genre*, ed R. Coe, L. Lingard, and T. Teslenko. Cresskill, NJ: Hampton Press.

———. 2004. "Speech Acts, Genres, and Activity Systems: How Texts Organize Activity and People." In *What Writing Does and How it Does it: An Introduction to Analyzing Texts and Textual Practices*, ed. C. Bazerman and P. Prior. Mahwah, NJ: Lawrence Erlbaum.

———, and J. Paradis. 1991. *Textual Dynamics of the Professions: Historical and Contemporary Studies of Writing in Professional Communities*. Madison, WI: University of Wisconsin Press.

Bereiter, C. 2002. *Education and Mind in the Knowledge Age*. Mahwah, NJ: Lawrence Erlbaum.

Berkenkotter, C., and T. Huckin. 1995. *Genre Knowledge in Disciplinary Communication: Cognition/Culture/Power*. Hillsdale, NJ: Lawrence Erlbaum.

Blyler, N., and C. Thralls, eds. 1993. *Professional Communication: The Social Perspective*. Newbury Park, CA: Sage.

Bourdieu, P. 1987. *Distinction: A Social Critique of the Judgment of Taste*. Trans. R. Nice. Boston, MA: Harvard University Press.

Campbell, K. S. 1999. "Collecting Information: Qualitative Research Methods for Solving Workplace Problems." *Technical Communication* 46: 532–45.

Clark, D. 2007. "Rhetoric of Empowerment: Genre, Activity, and the Distribution of Capital." In *Communicative Practices in Workplaces and the Professions*, ed. M. Zachry and C. Thralls. Amityville, NY: Baywood Publishing.

Coe, R. M., L. Lingard, and T. Teslenko, eds. 2001. *The Rhetoric and Ideology of Genre: Strategies for Stability and Change*. Cresskill, NJ: Hampton Press.

Creswell, J. W., and D. L. Miller. 2000. "Determining Validity in Qualitative Inquiry." *Theory into Practice* 39: 124–30.

Denzin, N. K., and Y. S. Lincoln, eds. 1998. *Strategies of Qualitative Inquiry*. Thousand Oaks, CA: Sage Publications.

Driskill, L., and J. Zeleznik. 2003. "National Science Foundation Grant 0338507. A Project to Collect and Analyze Baseline Data for Studies of Communication Practices in Interdisciplinary, Vertically Integrated Research and Education Groups." Rice University, Houston, TX.

———, and ———. 2006. Final Report. "National Science Foundation Grant 0338507. A Project to Collect and Analyze Baseline Data for Studies of Communication Practices in Interdisciplinary, Vertically Integrated Research and Education Groups." Houston, TX: Rice University.

Engeström, Y. 1987. *Learning by Expanding: An Activity-Theoretical Approach to Developmental Research*. Helsinki, Finland: Orienta-Knosultit Oy.

Fisher, C. S. 1978. "Some Social Characteristics of Mathematicians and their Work." *American Journal of Sociology* 78: 1094–134.

Flynn, E. A., G. Savage, M. Penti, C. Brown, and S. Watke. 1991. "Gender and Models of Collaboration in a Chemical Engineering Design Course." *Journal of Business and Technical Communication* 5: 444–62.

Forman, R. 2003. National Science Foundation Proposal DMS 0240058. VIGRE Program for Rice University, Houston, TX.

Freedman, A., and C. Adam. 2000. "Write Where You Are: Situating Learning to Write in University and Workplace Settings." In *Transitions: Writing in Academic and Workplace Settings*, ed. P. Dias and A. Paré. Creskill, NJ: Hampton Press.

Gee, J. P. 1999. *An Introduction to Discourse Analysis*. London, Routledge.

——. 2004. "Discourse Analysis: What Makes It Critical?" In *An Introduction to Critical Discourse Analysis*, ed. R. Rogers. Mahwah, NJ: Lawrence Erlbaum Associates.

Halliday, M. 1978. *Language as Social Semiotic*. London: Edward Arnold.

——. 1994. *Introduction to Functional Grammar*. London: Edward Arnold.

Herrington, A. J. 1985. "Writing in Academic Settings: A Study of the Contexts for Writing in Two College Chemical Engineering Courses." *Research in the Teaching of English* 19: 331–61.

Huber, L. (1990). "Disciplinary Cultures and Social Reproduction." *European Journal of Education* 25 (3): 241–61.

Kent-Drury, R. 2000. "Bridging Boundaries, Negotiating Differences: The Nature of Leadership in Cross-Functional Proposal-Writing Groups." *IEEE Transactions on Professional Communication* 43/*Technical Communication* 47: 90–98.

Kleid, N. A. 2004. "Teaching and Practicing Teamwork in Industry and Academia." Society for Technical Communication Annual Conference *Proceedings*. http://www.stc.org/ConfProceed/2004/PDFs/0038.pdf.

Klein, J. T. 1990. *Interdisciplinarity: History, Theory, and Practice*. Detroit, MI: Wayne State University Press.

——. 1996. "Crossing Boundaries: Knowledge, Disciplinarities, and Interdisciplinarities." Charlottesville, VA: Umiversity of Virginia Press.

——, and A. Porter. 1990. "Preconditions for Interdisciplinary Research." In *International Research Management: Studies in Interdisciplinary Methods from Business, Government, and Academia*, ed. P. Birnbaum-More, F. Rossini, and D. Baldwin. New York, NY: Oxford University Press.

Koretz, D. M. 1992. Evaluating and Validating Indicators of Mathematics and Science Education. RAND report. N2900-NSF.

Lampert, M. 2000. "Knowing Teaching: The Intersection of Research on Teaching and Qualitative Research." *Harvard Educational Review* 70: 86-99.

Lauer, J. M., and Asher, J. W. 1988. *Composition Research: Empirical Designs*. New York, NY: Oxford University Press.

Lave, J. (1988). *Cognition in Practice*. New York, NY: Cambridge University Press.

——, and E. Wenger. 1991. *Situated Learning: Legitimate Peripheral Participation*. New York, NY: Cambridge University Press.

Leont'ev, A. N. 1978. *Activity, Consciousness, and Personality*. Englewood Cliffs, NJ: Prentice-Hall.

Lewis, D. 1997. Dear Colleague Letter Introducing the Report of the NSF DMS Special Emphasis Panel on VIGRE. http://www.nsf.gov/publications/pub_summ.jsp?ods_key=nsf97170.

Longo, B. 1998. "An Approach for Applying Cultural Study Theory to Technical Writing Research." *Technical Communication Quarterly* 7: 53–73.

Mackenzie, D. 2002. "NSF Moves with VIGRE to Force Changes in Academia." *Science* 296.5572 (May 24): 1389–90.

Marshall, C., and G. B. Rossman. 2006. *Designing Qualitative Research*, 4th edn. Thousand Oaks, CA: Sage Publications.

Martin, J. 1992. *Cultures in Organizations: Three Perspectives.* New York, NY: Oxford University Press.

Mehlenbacher, B.1992. "Rhetorical Moves in Scientific Proposal Writing: A Case Study from Biochemical Engineering." PhD dissertation, Carnegie Mellon University.

Miller, C. 1984. "Genre as Social Action." *Quarterly Journal of Speech* 70: 151–67.

National Science Foundation, Division of Science Resources Statistics. 2008. "Science and Engineering Degrees: 1966–2006. Detailed Statistical Tables NSF 08–321." http:///www.nsf.gov/statistics/nsf08321.

Orlikowski, W. J., and J. Yates. 1994. "Genre Repertoire: Examining the Structure of Communication Practices in Organizations." *Administrative Science Quarterly* 39: 541–74.

Paré, A. 2001. "Genre and Identity: Individuals, Institutions and Ideology." In *The Rhetoric and Ideology of Genre: Strategies for Stability and Change*, ed. R. M. Coe, L. Lingard, and T. Teslenko. Cresskill, NJ: Hampton Press.

——, and G. Smart. 1994. "Observing Genres in Action: Towards a Research Methodology." In *Genre and the New Rhetoric*, ed. A. Freedman and P. Medway. London: Taylor and Francis.

Rogers, R. 2004. "Setting an Agenda for Critical Discourse Analysis in Education." In *An Introduction to Critical Discourse Analysis*, ed. R. Rogers. Mahwah, NJ: Lawrence Erlbaum Associates.

Russell, D. 1997a. "Writing and Genre in Higher Education and Workplaces." *Mind, Culture, and Activity* 4: 224–37.

——. 1997b. "Rethinking Genre in School and Society: An Activity Theory Analysis." *Written Communication* 14: 504–54.

——, and C. Bazerman, eds. 1998. *The Activity of Writing/The Writing of Activity. A Special Issue of Mind, Culture, and Activity.* Mahwah, NJ: Lawrence Erlbaum Associates.

Rymer, J. 1989. "Scientific Composing Processes: How Eminent Scientists Write Journal Articles." In *Advances in Writing Research*, ed. D. Jolliffe, vol. 2. Norwood, NJ: Ablex.

Schryer, C. F. 1993. "Records as Genre." *Written Communication* 10: 200–34.

Schryer, C. F., L. Lingard, and M. Spafford. 2007. "Regularized Practices: Genres, Improvisation, and Identity Formation in Health-Care Professions." In *Communicative Practices in Workplaces and the Professions: Cultural Perspectives on the Regulation of Discourse and Organizations*, ed. M. Zachry and C. Thralls. Amityville, NY: Baywood.

Wenger, E. 1998. *Communities of Practice: Learning, Meaning, and Identity.* New York, NY: Cambridge University Press.

——, R. McDermott, and W. M. Snyder. 2002. *Cultivating Communities of Practice.* Cambridge, MA: Harvard Business School Press.

Winsor, D. A. 1996. *Writing like an Engineer: A Rhetorical Education.* Mahwah, NJ: Erlbaum.

——. 2001. "Learning to Do Knowledge Work in Systems of Distributed Cognition." *Journal of Business and Technical Communication* 15: 5–28.

Yates, J., and W. Orlikowski. 1992. "Genres of Organizational Communication: A Structurational Approach." *Academy of Management Review* 17: 299–326.

——, and ——. 2002. "Genre Systems: Structuring Interaction Through Communicative Norms." *Journal of Business Communication* 39: 13–35.

——, and ——. 2007. "The PowerPoint Presentation and Its Corollaries: How Genres Shape Communicative Action in Organizations." In *Communicative Practices in Workplaces and the Professions: Cultural Perspectives on the Regulation of Discourse and Organizations*, ed. M. Zachry and C. Thralls. Amityville, NY: Baywood Publishing.

Chapter 14

Team Learning in Usability Testing

Michael Hughes and Tom Reeves

Editors' Introduction

Hughes and Reeves have contributed a previously unpublished article to this collection that considers how processes of team learning enhance the effectiveness of usability tests. They acknowledge that previous research has established that team-based usability testing produces new knowledge that can be used to improve the design of a product, but they point out that researchers have thus far overlooked the ways that social learning processes on usability teams make these improvements possible.

These authors show us how much our conceptions of technical communication have changed over the past several years. In the 1990s, some still saw technical communication as a "bridge" over which knowledge travelled (Hart and Conklin 2006). Technical communicators were thus seen as passive conduits for the creative work of technical and scientific experts who created knowledge. In recent years, however, partly through the insights gleaned through qualitative research, technical communicators have posited an active constructivist alternative to the earlier passive conception: technical communicators are seen as active participants in a social process to construct new, usable knowledge. In this article, Hughes and Reeves add to this emerging view of the profession by suggesting that technical communicators might be seen as facilitators of team-based learning and knowledge creation.

Hughes and Reeves present us with a simple, elegant report on their case study. They begin with a clear and concise statement of their research purpose, and of the three questions that they sought to answer. Research is an inquiry that seeks to answer specific questions, and researchers must ensure that they have selected a design and methodology that are capable of producing answers to the research questions. It is thus always advisable to include the original research questions in the article that reports a study's findings. They also include a brief review of the relevant literature on team learning, and include tabular presentations of team learning modes and processes that could potentially be developed into diagnostic frames for assessing the learning capacity of a usability team.

They designed their inquiry as a case study, and gathered data through observations, videotaping, focus groups, and field notes. Data was coded and analyzed with a qualitative analysis program called NVivo. They provide a thorough description of how they used NVivo to perform top-down and bottom-up coding, which they say is similar to the constant comparative method of grounded theory. Qualitative research often results in the creation of enormous volumes of data, so the task of organizing and analyzing the data can be onerous. These authors describe the procedures they followed in detail, including those that led to analytical dead-ends, thus providing novice researchers with a realistic account of qualitative data analysis.

They also describe the steps they took to bring rigor to the inquiry. Aside from member checking, they based their study on an intact usability team that was engaged in a real usability test, and they held a peer review of their findings to benefit from additional perspectives.

Hughes and Reeves report their findings in relation to their three research questions. They make use of numerous quotations from their data to substantiate their interpretations. Their discussion includes some interesting observations about the relationship between power and expertise on usability testing teams, and they conclude with two specific, actionable recommendations to strengthen the performance of such teams.

Recalling the themes suggested by Giammona, Hughes and Reeves invite us to reflect on the current role and identity of technical communicators within a workplace environment increasingly characterized by cross-function teams. They suggest that technical communicators might become facilitators of learning and knowledge creation, thus making a significant contribution to the innovations produced in product development and enhancement processes. Perhaps most significantly, their paper indicates that technical communication does indeed continue to matter, though the profession's importance may be becoming less tied to the production of texts and more related to our ability to interact with others in productive ways that result in high levels of team performance and the enhanced quality and resilience of emerging technology.

Reference

Hart, H., and Conklin, J. 2006. "Towards a Meaningful Model for Technical Communication." *Technical Communication* 53: 395–415.

Introduction

Usability testing is an area in which technical communicators have become increasingly involved. Grice and Krull predicted in 2001 that technical writers would become usability testers, in their preface to a special issue of *Technical Communication* that focused on developments in the field of technical communication. More recently, Hayhoe (2007) reported that graduating seniors from an undergraduate program in technical communication were mostly interested in multimedia design or usability work. Not only do technical communicators find themselves participating in tests of their own information products (compare Postava-Davignon and colleagues 2004; Grayling 1998, 2002), they are also part of larger development teams involved in usability testing of the end product as well (Skelton 2002; Hughes 2002). Hughes and Hayhoe assert that "usability testing is a common form of technical communication research" (2008, 84).

Even though the use of teams in usability testing is an established practice (Barnum 2001), and even though the purpose of these tests is to acquire new knowledge that can inform the design of a product, little attention has been paid to the process of *team learning* within the context of usability testing. Team learning as defined by Kasl, Marsick, and Dechant is "a process through which a group creates knowledge for its members, for itself as a system, and for others" (1997, 229). Senge defines team learning as "the process of aligning and developing the capacity of a team to create the results its members truly desire. It builds on the discipline of developing shared vision" (1990, 234). The premise of this article is that technical communicators can be more valuable members of usability test teams if they understand and facilitate how those teams create knowledge that informs design decisions.

Research Purpose and Questions

The purpose of this study was to observe and describe a team-learning process in the context of usability testing using a cross-functional team. The specific research questions were:

1. What kinds of knowledge do team members learn during a usability test?
2. Does the team go through observable changes or phases during the process?
3. Is the interaction and influence within the team related to organizational status or technical expertise?

Review of the Literature on Team Learning

Although practitioners agree that usability testing can be especially effective when conducted by collaborative, cross-functional teams (Draper, Brown, Henderson, and McAteer 1996; Williams and Traynor 1994), no one has specifically studied its effect on team learning within this context. This review of the literature on team learning is organized according to the areas of: (a) types of learning; (b) team activities; and (c) power.

Types of Learning

Many of the studies of team learning categorize the types of learning that occur. Brooks (1994) speaks of *technical learning* to describe the facts and solutions learned that are aimed at the problem under study, and she uses *social learning* to describe learning about the process of being a team. Lynn (1998) differentiates between *within-team learning*, learning within the context of the team itself, and *cross-team learning*, which is experience gained by one team and transferred to another. He also identifies a third type, *market learning*, which is knowledge gained that is external to the firm; for example, from competitors, suppliers, and customers. Kogut and Zander (1992) speak of *organizational knowledge* as consisting of *information* (for example, who knows what) and *know-how* (for example, how to organize a research team).

Another set of categories for team learning is that of single-loop and double-loop learning:

> [*Single-loop* learning occurs when] failure to achieve the intended ends leads to a reexamination of means and a search for more effective means. [*Double-loop* learning occurs when] failure to achieve intended consequences may . . . lead to reflection on the original frame and the setting of a different problem.
> (Argyris and Schön 1978, 53)

Meyers and Wilemon (1989) describe single-loop learning as thermostat-like adjustments in response to error detection, and double-loop learning as error detection that modifies an organization's implicit norms and objectives. Watkins and Marsick note, "Single-loop learning is characterized by attempts to control or to protect oneself, while double-loop learning is characterized by shared control and inquiry" (1993, 79).

These various categories of team learning have a strong parallel in Mezirow's (1991) description of types of reflection in transformative adult learning: (a) reflection focused on the nature of the problem itself; (b) reflection focused on the process of problem solving; and (c) reflection focused on the premises or presuppositions that lie at the foundation of one's definition of a situation. Kasl and Elias (2000) used Mezirow's model as a theoretical framework for their investigation of small-group learning.

Team Activities

In addition to categorizing the types of learning that occur within teams, studies have also categorized how learning teams act. As detailed in Table 14.1, Kasl and colleagues (1997) and Watkins and Marsick (1993) delineate four modes of team learning: fragmented, pooled, synergistic, and continuous.

Watkins and Marsick (1993) identify five processes of team learning (framing, reframing, integrating perspectives, experimenting, and crossing boundaries). Table 14.2 presents a description of each of these processes.

Table 14.1 Team Learning Modes

Mode	Characteristics
Fragmented	Individuals learn separately. Members retain their separate views and are often not committed to working as a group.
Pooled	Individuals share information and perspectives in the interest of group efficiency and effectiveness, but the group as an entire unit does not learn. Knowledge is not created that is uniquely the group's own.
Synergistic	Members mutually create knowledge. Simple phrases or metaphors from the team's experience become code words for more elaborate meanings.
Continuous	A synergistic team where learning has become habitual. This is a posited mode, not one observed by Watkins and Marsick.

Table 14.2 Team Processes

Process	Description in usability-testing context
Framing	The group provides an initial perception of the problem based on past understanding and present input.
Reframing	The group transforms the original perceptions (frames) into a new understanding.
Integrating perspectives	Divergent views are synthesized and apparent conflicts resolved.
Experimenting	Members test new hypotheses.
Crossing boundaries	Members offer observations and recommendations outside of their own disciplines.

Power

Brooks (1994) is especially interested in the role that individual power plays in team learning. In her research she developed the following set of grounded propositions concerning power.

1. The collective production of new knowledge by teams requires active and reflective work.
2. Members with insufficient formal power have difficulty in carrying out active or reflective work.
3. Production of knowledge in the technical domain occurs only when power differences among team members are controlled.
4. Production of knowledge in the social domain occurs only when no power differences exist among team members. In a similar vein, Senge cites Bohm, "Hierarchy is antithetical to dialogue" (1990, 245).

Research Methodology

The specific method of inquiry we used was *case study*. Merriam (1998) points out that case study methods can be useful because they do the following:

- illustrate the complexities of a situation;
- explain the background of a situation, what happened, and why;
- explain why an innovation worked or failed to work.

Case Description

The case involved a team conducting a usability test of a Web-based training lesson for making car reservations and the course management software that gave the trainee access to the courses. The sponsoring company was a provider of a computer reservation system (CRS) used worldwide by leading travel service providers. Agent training for this company's computer reservation system had historically been conducted at the agency or at the company's training center. The training had been considered a standard feature of the company's CRS service and not a revenue-generating activity. As such, the travel cost of deploying the trainers or bringing in agents for training was a major cost for this company. For this reason, the Training Development Department made a commitment to move training to a web-delivered format. For this type of training to be accepted by the end user, both the courses and the course management software had to be as user-friendly as possible.

The cross-functional team who participated in the usability test comprised the following functions:

- manager of training development;
- training development supervisors (three);
- instructional designer;
- scripting programmer;
- technical writer;
- interactive courseware developers (two);
- product development manager;
- scripting programming supervisor.

The team was observed during the following usability test events:

- planning meeting;
- four user sessions and associated findings meetings after each user session;
- action meeting to decide what changes to make.

In addition to the usability test, the team participated in a post-event focus group to discuss their experience before, during, and after the usability test.

The study was conducted at a commercial usability-testing lab using its staff and standard testing protocols to facilitate the test. The planning meeting, the findings meetings after each user observation, and the action meeting were videotaped and observed by the primary researcher through a one-way mirror. During the user observation sessions the primary researcher sat with the team and took field observation notes. (Lighting and space limitations precluded videotaping the observation room.) The team focus group was conducted by the primary researcher and was also videotaped.

Data

The primary sources of data for this study (data that were coded and analyzed) consisted of the following:

- Videotapes and their transcripts from the team planning meeting, findings meetings, and the action meeting.
- Team focus group videotape and transcript.
- Researcher observation notes taken during the user observations.
- The log files maintained by a designated team member during each user observation.

The tapes covered over 16 hours of team interaction over a two-week period. Transcripts were analyzed using NVivo, a coding and model-building tool for qualitative research (Richards 1999).

Data Analysis

We analyzed the data using a combination of top-down and bottom-up coding. At the beginning of the analysis phase, we entered theoretical constructs from the literature review that we thought could be useful for organizing the observations. Specifically, we used *tree nodes* (NVivo terminology for codes and categories organized into a hierarchical structure) to create codes based on models taken from our review of the literature.

During the analysis we created *free nodes* (NVivo terminology for codes not yet associated with a category) and looked for opportunities to assign them into our theoretical models. Sometimes we would code data directly with one of the nodes from the initial theoretical models. We found early on, however, that this approach was not very useful. The constructs carried over from the literature were at too high a level to be useful for the initial organization of low-level codes.

During the coding, we began to organize some of the free nodes into what we felt were emerging themes. As these new categories or themes emerged, we would rearrange the existing nodes to fit the emerging models (by either creating new tree nodes and pulling in existing free nodes or moving nodes from an existing tree into the new tree). After analyzing each document, we would review all previously coded documents and apply any new codes or merge existing codes as appropriate. This routine recoding of previously coded transcripts resembles the constant comparative method employed in grounded theory study (Strauss and Corbin 1990).

As we progressed further into the analysis phase, we worked more and more within tree nodes. In some cases we created *parent nodes* (higher levels in a hierarchy) to document an emerging model. For example, we created a node called 'dependability' to collect the nodes that described the various criteria the team used to accept an individual user's actions as typical. This method represents a bottom-up approach, where the category is created based on existing low-level codes.

In other cases we created *children nodes* (lower levels in a hierarchy) to create a greater degree of granularity. For example, we had an early node called 'expert' which we soon had to expand into various children nodes as the data made us more sensitive to various ways a person could demonstrate expertise. In these cases we would conduct a computer search to locate all instances where data had been coded with the parent node to see if they should be more accurately recoded with the newly created child node. This represents a top-down approach, where low-level coding is being suggested by an existing category.

In summary, this combination of top-down and bottom-up coding during the analysis enabled us to challenge the constructs we brought into the study as well as those that emerged during the process, forcing us to support them with directly observable data, modify them, or dismiss them.

Rigor

To ensure rigor in the study, we specifically included the following elements in the design:

- Use of an intact organization and an actual product they were developing as the case study. This strengthened the credibility of the study over one that would be done using, for instance, ad hoc student teams working on class projects.
- Member checking of preliminary findings (having the participants review report drafts) to enhance credibility.
- Peer review of findings to enhance dependability.
- Use of a commercially practiced protocol for conducting the usability test. This makes the findings more readily transferable to real-world applications.
- Pilot study using the described test procedures to ensure that logistically we could run the test and that the process was worthy of study.

Findings

The following findings are organized around our three research questions.

What Was Learned

Our first research question was, "What kinds of knowledge do team members learn during a usability test?" Our coding and analysis of the data indicated that learning outcomes could be categorized as product knowledge and user knowledge primarily occurring in the context of single-loop learning, but with some double-loop learning episodes as well.

Product Knowledge

Part of what some team members learned by watching the users and in the findings meetings was how the educational software worked. For example, during a discussion about how scores were reported, one team member said, "This is user three, and I just now realized how that works." Often the learning required both collaborative observation and discussion. The following excerpt is from the focus group, when participants were asked about the team aspect of learning during the usability test.

> Maureen: And just hearing each other's comments in the room while we were watching.
>
> Alice: Yeah, because you had different levels of expertise on this. For Maureen and I, we had never seen it before, so without Cathy or somebody back there with us, there were things I was confused about, but at least I could ask . . .
>
> Maureen: But there were things that you and I noticed that Cathy might not have.

Another important type of product knowledge created involved more abstract observations that addressed design issues that could be applied across several products. For example, during one of the findings meetings, a team member made the following conclusion based on the team observation that users liked clicking on pop-ups but did not focus on the content of those pop-ups:

> They like to see the action but don't like to read it. I thought that was interesting to us as designers because we might not want to put very important things in something like that.

User Knowledge

Sometimes team members were surprised by what they learned about the users, as one member recalled during the focus group:

> I learned a lot from just watching the users. It was fascinating, seeing where they were confused, and sometimes in places where I never expected their confusion, especially when he was clicking 'refresh' to resume the course. That never in a million years would have occurred to me. Just a lot of things they did that were totally unexpected.

On the other hand, many comments related to how similar the user perceptions were to the members' own experiences with the product, as in the following excerpt where the team is discussing a user's reaction to a browser security message:

> Nicole: That security thing. Now the first time he cancelled out of that.
>
> Gretchen: He cancelled because he didn't know what to do with it.
>
> Bart: It's very intimidating.
>
> Gretchen: The first time I used it I did that. I did the exact same thing he did.
>
> Alice: I can see myself doing the same thing.
>
> Gretchen: I don't want to do anything. I'm scared I might load something on my hard-drive or on the network. Any time you see that, it's like you're going to download something and that just scares me to death.

There were several instances of this "Oh, I did the same thing [felt the same way, had the same problem, etc.]" reaction. It seems as if the team members had discounted their own individual problems with the product, blaming it on their own inadequacy. But they were open about sharing those instances when they saw actual users experience the same difficulties.

The team also learned how users applied the product in their own work context. During the discussion of logging in, the instructional designer learned that agents typically did not know their own login ID because it is programmed into a "ready-key" (a programmed key that serves as a shortcut for repetitive transactions). Yet they needed to know their login ID to get into the tutorial. During the focus group, the team members asserted that they would not get this kind of information from their current methods, including site visits and beta tests.

> Maureen: We do guessing but our current means of getting customer feedback is not the effective way.
>
> Gretchen: Not as effective as this, betas have never been effective for getting feedback on any training tool that we had developed.

Alice: Betas are not even good for the. . .

Gretchen: Product, exactly.

Alice: Exactly.

Maureen: And focus groups and by going out to their offices and meeting these people we're getting their feedback. [Pause] We're really not.

Gretchen: Mm-mm [shaking head no].

Single-Loop Learning

In general, the dominant learning outcomes consisted of single-loop learning. We categorized as single-loop learning any knowledge which was applied to fix specific aspects of the product. We classified as double-loop learning any knowledge that challenged the basic product or user definitions or the product objectives; that is, any knowledge that served to reframe those core definitions or objectives rather than merely "fix what didn't work."

Over 80 percent of the combined entries coded under "Learning Outcomes" occurred within the single-loop category. This is a conservative picture of what really happened, insofar as all of the double-loop learning entries involved only two incidents in the entire study. One of those incidents occurred during the focus group, which was not a part of the usability test itself, and the other incident was brushed aside by the facilitator. The client's motivation, as stated by the manager, was to ensure that the client's early development efforts, which would be the first exposure to the client's web-based training for many users, would be as user-friendly as possible. Given the problem-solving emphasis of the commercial test process and the objectives of the client, it is neither surprising nor negative that the learning was predominantly single loop.

Double-Loop Learning

The strongest evidence of a double-loop episode came during the focus group when the team was asked if anyone had learned anything that challenged his or her assumptions coming into the test. An instructional designer raised a concern that struck at the very core objective of the product:

> I had a belief that the tutorial really was going to be a stand-alone, successful tool, and I have to say that I now have my doubts. [Pause] I'm concerned about that. . . . What really nailed it for me was [user 3] saying it would still take an instructor. I don't know, certainly there could have been a little job security for her in there, because she is an instructor. But on the other hand, I think we've seen enough today and heard people say, "It's been ages since I did this lesson," that we have to put as much effort on driving them to the water as we do getting them to drink once they're there, or it's not going to work.

Team Phases

The second research question of this study was, "Does the team go through observable changes or phases during the process?" To answer this question, we examine our observations against theoretical constructs suggested by the literature.

The four modes of team learning discussed by Kasl and colleagues (1997) and Watkins and Marsick (1993)—fragmented, pooled, synergistic, and continuous—have a limited application to our findings. During the planning meeting the group acted in a predominantly pooled learning mode as individuals shared what they knew, but no new knowledge was created. The user sessions and the findings meetings were synergistic in that new knowledge was created from collective observation and processing of the team members' reflections on those observations. The clear mission and tight structure of the test process itself accounts for the lack of fragmentation that other types of teams might experience at start-up. The fact that it was an event tightly bounded by time and product made it unlikely that the continuous learning mode would have been reached. On the other hand, Watkins and Marsick's (1993) description of the processes of team learning identified in Table 14.3 was much richer in describing the behavior displayed in this study.

Status and Expertise

The third research question was, "Is the interaction and influence within the team related to organizational status or technical expertise?" Three roles in particular

Table 14.3 Team Processes

Process	How observed in this study
Framing	Initially, the team had a positivistic perspective of the product. "It is what it is and does what it does." When users did not interpret the user interface in the way the designers and developers designed it to be interpreted, it was described as *user error*.
Reframing	As the test progressed, the team developed a more pluralistic perspective of the product. They saw it as being contextually defined by the user and accepted that there could be multiple meanings to its terminology and affordances. Progressively during the user sessions, the team saw "user errors" as being reasonable, alternative interpretations caused by vagaries and ambiguities in the design.
Integrating perspectives	A new, collective vision emerged about the user and the product from the collective reflection and discussions held during the observation sessions and findings meetings.
Experimenting	Members proposed alternative designs and perspectives based on observing users. Scenarios were reworded to help clarify group theories about why something had happened and to test those interpretations.
Crossing boundaries	Design changes were recommended by all roles, not just the programmers or instructional designers.

seemed to carry higher degrees of power: *facilitator, technical expert,* and *resource controller.*

The facilitator controlled the agenda and could cut off discussion based on time constraints. He was also the official note taker and controlled if and how issues and solutions got recorded. For example, during a findings meeting he said, "You notice that I'm ignoring you guys [who are] saying it's not important, and I'm putting it up anyway—because I have the marker I can do that." He did this in an instance where he thought the team was ignoring data that he felt could be important. He also had power through his role as usability expert. By virtue of his "n of many" (having seen many subjects come through the lab) or his authoritative knowledge of usability heuristics, he could introduce arguments that were difficult to challenge. Statements by team members such as, "We saw users want to back up in the lesson," based on directly observable data, were discounted by the argument, "You're only seeing four users; I've seen five hundred." He did this in instances to counter data that did not support a conclusion he wanted the team to draw. Because of his dual role as usability expert and test administrator (time keeper/note taker), the facilitator had enormous power in tabling discussions. One team member's comment about the facilitator that was made in the focus group illustrates the weight of the facilitator's authority (and how she valued his expertise):

> The most valuable thing that I think he contributed in my opinion was hearing his experiential information "What I've gained with 15 years of seeing users." To me, that either reinforced what I was seeing to be true or say "Hey you're probably all wet 'cos this guy knows what he's talking about."

In the case study, the facilitator would make broad assertions based on general beliefs and models (schemata) not tied to precise memories of specific events or references, like "Users don't like to . . .". He had formed these beliefs based on his experiences, and although he felt that they were ultimately grounded in empirical observations, he could not always relate a belief to specific data. (This assertion is based on member-checking discussions with the facilitator where he expressly made this point.)

This reliance on general frameworks could be both enabling and constraining, in much the same way that Patton describes paradigms: "their strength [is] that it makes action possible, their weakness [is] that the very reason for action is hidden in the unquestioned assumptions of the paradigm" (1978, 203). This reliance on schemata rather than specific data would make it possible for the facilitator to be influenced by his or her own preferences as a user (his or her "n of one") while supporting it as being bound in empirical observations ("n of many"). The caveat is that when the facilitator said, "This is what users do," it could have been any one of the following:

- A reflection based on direct observation, although the ability to recall the specific events might not exist.

- A learned heuristic, accepted from a secondary, authoritative source.
- A personal schema projected or reconstructed as empirical data.

Another source of significant power was technical expertise. Unless the technical expert, such as the programmer, agreed that a solution could be done, it could not be accepted as a solution. In effect, this gave the technical expert veto power. Technical expertise could not be challenged effectively except by another technical expert.

Another source of power came from control of resources. In this case study, not only did one of the development supervisors exert power as a technical expert, she also had considerable control over resources and, therefore, had a major say in what priorities got assigned. The general theme seemed to be that everything that could get fixed technically would get fixed, and the overall prioritizing factor was how easily (technically) something could be fixed. Things easily fixed were put on the "fix right away" list and the more difficult items were classified as "fix later."

Conclusions

This study showed that cross-functional development teams can benefit from participating in usability tests by generating practical knowledge about the product being tested and its targeted users. Direct observations of the team and comments by the participants strongly indicated that the knowledge they collectively created exceeded what they had been able to do through individual efforts and user research methods that did not observe users in context, such as beta tests and focus groups. (Although beta tests involve users in context, they rely on self-reported data, not direct observations.) And because the usability test relied on directly observable data equally accessible to all team members, group conflict was largely avoided and the team could stay focused on solutions.

However, in spite of the democratic nature of the test, certain roles still maintained higher levels of influence over others. The two strongest influencers in this study were the technical expert and the facilitator/usability expert.

Given these conclusions, we make the following recommendations:

- Involve technical experts and resource managers as part of the test team. Since they are the ultimate gatekeepers of what will get done, they need to benefit from the transformative knowledge that the team participation creates.
- Facilitators and usability consultants need to be mindful that their opinions could be influenced by personal taste and biases and should make every effort to ground their observations and recommendations in the directly observed data of the test at hand. Not only does this serve as a check against personal bias, it also democratizes the process and improves collaborative reflection, an important element of team learning.

What this study did not get a chance to observe was whether the team learned anything about working as a team that got transferred back to the workplace. Future research that observes teams back on the job after conducting a collaborative usability test could be informative to see if team skills or synergies improve and get transferred to other work aspects.

References

Argyris, C., and D. Schön. 1978. *Organizational Learning: A Theory of Action Perspective.* San Francisco, CA: Jossey-Bass.

Barnum, C. 2001. *Usability Testing and Research.* Boston, MA: Allyn & Bacon/Longman Publishers.

Brooks, A. 1994. "Power and the Production of Knowledge: Collective Team Learning in Work Organizations." *Human Resource Development Quarterly* 5: 213–35.

Draper, S. W., M. I. Brown, F. P. Henderson, and E. McAteer. 1996. "Integrative Evaluation: An Emerging Role for Classroom Studies Of CAL." *Computers and Education* 26: 17–32.

Grayling, T. 1998. "Fear and Loathing of the Help Menu: A Usability Test of Online Help." *Technical Communication* 45: 168–79.

——. 2002. "If We Build It, Will They Come? A Usability Test of Two Browser-Based Embedded Help Systems." *Technical Communication* 49: 193–209.

Grice, R., and R. Krull. 2001. "2001, A Professional Odyssey." *Technical Communication* 48: 135–38.

Hart, H., and J. Conklin. 2006. "Toward a Meaningful Model for Technical Communication." *Technical Communication* 53: 395–415.

Hayhoe, G. 2007. "The Future of Technical Writing and Editing." *Technical Communication* 54: 281–82.

Hughes, M. 2002. "Moving from Information Transfer to Knowledge Creation: A New Value Proposition for Technical Communicators." *Technical Communication* 49: 275–85.

——, and G. Hayhoe. 2008. *A Research Primer for Technical Communication: Methods, Exemplars, and Analyses.* New York, NY: Lawrence Erlbaum Associates.

Kasl, E., and D. Elias. 2000. "Creating New Habits of Mind in Small Groups." In *Learning as Transformation: Critical Perspectives on a Theory in Progress,* ed. J. Mezirow and Associates. San Francisco, CA: Jossey-Bass.

——, V. Marsick, and K. Dechant. 1997. "Teams as Learners: A Research-Based Model of Team Learning." *Journal of Applied Behavioral Science* 33: 227–46.

Kogut, B., and U. Zander. 1992. "Knowledge of the Firm, Combinative Capabilities, and The Replication of Technology." *Organization Science* 3: 383–97.

Lynn, G. S. 1998. "New Product Team Learning: Developing and Profiting from your Knowledge Capital." *California Management Review* 40 (4): 74–81.

Merriam, S. B. 1998. *Qualitative Research and Case Study Applications in Education.* San Francisco, CA: Jossey-Bass.

Meyers, P. W., and D. Wilemon. 1989. "Learning in New Technology Development Teams." *The Journal of Product Innovation Management* 6 (2): 79–88.

Mezirow, J. 1991. *Transformative Dimensions of Adult Learning.* San Francisco, CA: Jossey-Bass.

Patton, M. Q. 1978. *Utilization-Focused Evaluation.* Beverly Hills, CA: Sage.

Postava-Davignon, C., C. Kamachi, C. Clarke, G. Kushmerek, M. Rhettger, P. Monchamp, and R. Ellis. 2004. "Incorporating Usability Testing into the Documentation Process." *Technical Communication* 51: 36–44.

Richards, L. 1999. *Using Nvivo in Qualitative Research*. Thousand Oaks, CA: Sage.

Senge, P. 1990. *The Fifth Discipline: The Art and Practice of the Learning Organization*. New York, NY: Doubleday.

Skelton, T. M. 2002. "Managing the Development of Information Products: An Experiential Learning Strategy for Product Developers." *Technical Communication* 49: 281–82.

Strauss, A., and J. Corbin. 1990. *Basics of Qualitative Research: Grounded Theory Procedures and Techniques*. Newbury Park, CA: Sage.

Watkins, K., and V. Marsick. 1993. *Sculpting the Learning Organization: Lessons in the Art and Science of Systemic Change*. San Francisco, CA: Jossey-Bass.

Williams, M. G., and C. Traynor. 1994. "Participatory Design of Educational Software." Paper presented at Recreating the Revolution. In *Proceedings* of the Annual National Educational Computing Conference.

Qualitative Research in Technical Communication

A Review of Articles Published from 2003 to 2007

Debbie Davy and Christina Valecillos

Editors' Introduction

As we were planning this collection, we wondered how common qualitative research techniques had become in the field today. Our sense was that they were used frequently but probably still represented a distinct minority approach to technical communication research. Since he was about to teach the capstone research course in Mercer University's master of science program in technical communication management during the summer of 2008, George Hayhoe decided to propose this question to one of the research teams in the course. "Qualitative Research in Technical Communication: A Review of Articles Published from 2003 to 2007" by Debbie Davy and Christina Valecillos is the result of that assignment.

What Davy and Valecillos found far exceeded our predictions. Between 2003 and 2007, the most recent complete five-year period at the time that they began their investigation, the five major journals in the field (*IEEE Transactions on Professional Communication*, *Journal of Technical and Business Writing*, *Journal of Technical Writing and Communication*, *Technical Communication*, and *Technical Communication Quarterly*) published 593 articles. Of these, 225 (37.9 percent) used qualitative research techniques. In addition, the authors surveyed the frequency with which article authors used various techniques, finding that content analysis alone, content analysis combined with interviews, and content analysis combined with both interviews and observation were the most commonly used methods, accounting for 172 of the 225 qualitative articles (76.4 percent).

Although some might argue that Davy and Valecillos did not use primarily qualitative methods, there is significant overlap between their techniques and those used in the studies they analyzed.

First, there can be no doubt that the Davy and Valecillos study is primarily based on content analysis. The two authors read the nearly 600 articles published in the five journals during the years they surveyed, and identified the research methods used in each one. Then, they coded each article—identifying the techniques described in it, not relying solely on the authors' methodology descriptions. Davy and Valecillos were then able to perform some simple quantitative analyses of the distribution of qualitative methods and the topics addressed by the articles of interest, which allowed them to identify gaps in the types of qualitative studies in the articles they surveyed.

Perhaps almost as surprising as the percentage of articles that used qualitative techniques during the years studied is the rarity with which the results of usability testing are reported in the literature during the period that Davy and Valecillos studied. Of the 593 articles analyzed, only 12 (about 2 percent) reported on usability tests. This phenomenon is easily explained since the vast majority of usability research is performed by or on behalf of companies interested in determining the usability of their own or their competitors' products. Publication of their findings could damage their own reputations or the reputations of their competitors. Still, it would be wonderful for practitioners and academic researchers alike to see exactly what companies learn as a result of these tests.

Davy and Valecillos complete their study by listing each of the 225 qualitative articles they identified. All told, this is a very useful article that offers a snapshot of the types of qualitative studies reported over a recent five-year period. And while it does not pretend to be comprehensive, their bibliography is an excellent starting point for anyone contemplating an all-inclusive list of qualitative research in technical communication.

Introduction

In performing qualitative research, researchers gather insight or knowledge about a topic in an attempt to understand perceptions, attitudes, and reasoning behind actions. Using qualitative methods, researchers are able to understand not only what is happening, but more importantly, why it is happening. Qualitative research is particularly useful for determining the opinions and attitudes of research participants, understanding how specific groups construct their sense of social reality, and discovering the reasons rather than the causes for these opinions.

As researchers, we were not just interested in historical accounts of qualitative research. While it is not possible to predict the future, it is possible to anticipate future trends. We believe that technical communicators are interested in qualitative research findings not only to learn about what is coming next but also to build on our profession's body of knowledge.

In this article, we review the findings from an examination of qualitative research articles published in five peer-reviewed technical communication journals and conclude with a snapshot of the state of qualitative research in the five-year period from 2003 to 2007. We began our research by reviewing the journals and selecting articles that demonstrated that the authors had performed quantitative and/or qualitative research. If the article used primarily quantitative methods, then the article was rejected as out of scope for our review. We focused on articles that used qualitative research methods as an integral part of the author's work, as defined in Table 15.1, and that used a sufficiently rigorous research design, data-gathering, and analytical approach.

Table 15.1 Types of Qualitative Research

Method	Description
Content analysis	The qualitative study and interpretation of written, oral, and visual material.
Interviews	Methods of gathering qualitative oral or written participant responses on facts, opinions, and attitudes that measure behavior.
Observation (and participant observation)	The qualitative observation and recording of behaviors, actions, or preferences within clearly defined criteria by the researcher or their representatives.
Focus group	Discussion by a group of people on specific ideas or topics, guided by a moderator.
Usability test	The qualitative observation of users completing a task with a product or document and collecting information from that study.

Through our review, we discovered five specific types of common qualitative research methods used in articles published during this period. The primary goal of our review was to identify these common types which created trends and areas of study in which qualitative research methods have been applied. Although surveys can be used to gather qualitative information, their use as a quantitative tool when calculating response rate, margin of error, and confidence intervals has excluded them from the list. Case studies can provide important insight into user behavior; however, they have not been included in our essay because they fall under the category of research design and are therefore out of scope for our holistic review. Table 15.1 lists the five types of qualitative research techniques found in the selected articles.

The Importance of Qualitative Research to Technical Communication

The qualitative research conducted in the field of technical communication has built a significant body of knowledge for professionals who are constantly being challenged by the demands of new technologies, an enhanced multimedia experience for users, and globalization. To truly understand the reasoning behind the actions of our users and readers, qualitative research is essential.

Research, the systematic collection and analysis of observations for the purpose of creating new knowledge that can inform actions and decisions, is what any profession needs in order to advance. Findings from qualitative research add to the domain of information for technical communicators on topics that are important to our field and demonstrate our mastery and skills as professionals. The field of technical communication must rely on research to inform best practices (Hughes and Hayhoe 2008, 4–12).

Quantitative analysis methods are typically associated with traditional research, which employs statistical analysis of data to allow researchers to discern trends and draw conclusions from the findings. Quantitative research is generally concerned with identifying causes, and with establishing correlations between variables. Qualitative research, on the other hand, relies largely on non-numerical data (such as observation journals and interview transcripts) that can be more difficult to analyze because of the volume of data that is generated during qualitative research, and because of the need for rigorous analytical procedures to minimize the problem of subjective interpretation (Hughes and Hayhoe 2008, 10–11).

Although qualitative research can be more challenging and time consuming, it allows researchers to detect trends in organization, techniques, audiences, deliverables, and other elements of technical communication. Through the analysis of their results, researchers can attribute influences and reasoning behind trends, which can help technical communicators become more adaptable and effective. For example, today's global marketplace provides many opportunities for qualitative research that can help improve our efforts in cross-cultural adaptation and localization. Some formatting and design decisions may be suitable for one country and ill-advised in another. Even within a single country, there may be diverse cultures, resulting in different reactions from individuals viewing the same document (St Germaine-Madison 2006).

Both quantitative and qualitative methods are systematic and involve the collection and analysis of data and creation of new knowledge. At the end, both types of research will ultimately articulate generalized truths from specific instances (Hughes and Hayhoe 2008, 5). It is qualitative research, however, that helps technical communicators better understand influences and answer the question "Why?"

Research Methodology

Overview

We reviewed qualitative research methods currently being used in the technical communication field. We identified 225 articles using qualitative methods from 2003 to 2007 in the five peer-reviewed print journals that publish the great majority of work in our field:

- *IEEE Transactions on Professional Communication*
- *Journal of Business and Technical Writing*
- *Journal of Technical Writing and Communication*
- *Technical Communication*
- *Technical Communication Quarterly*

The 225 articles that comprised our corpus were drawn from the total of 593 articles published in these journals during the years in question. We first reviewed the methodology section of each of these articles for techniques that indicated qualitative research. We reviewed the articles in detail again, and performed a cluster analysis based on the topics and types of qualitative research.

The 225 articles that we reviewed highlighted the issues and questions that are vexing technical communication researchers and academics. They shed light on the day-to-day working reality that technical communicators face and how practitioners are solving problems and identifying opportunities in the fast-changing work environment of the early 21st century. By looking at a range of articles published over several years, we begin to see the possibility of broader, longitudinal studies. Future studies could include mixed-methods designs that harness the strengths of both qualitative and quantitative research, perhaps bringing significant new insights to our understanding of the current and potential contribution of technical communication to contemporary organizations. The findings reported here suggest some trends and gaps related to qualitative research in technical communication and might be helpful for researchers who are in the process of developing interesting and relevant research questions to guide their future work.

Why 2003 to 2007?

The years 2003 to 2007 represent the most recent complete five-year period prior to our research during the summer of 2008. By 2003, globalization and its effect on technical communication experts (and their own interest in this phenomenon) had been firmly established. Cultural and localization factors played an important role in articles during this period, indicating that these components of our modern world influenced the type of research conducted. For example, in "The Triumph of Users: Achieving Cultural Usability Goals with User Localization," the author discusses text messaging and its adaptation in cultural localization (Sun 2006).

Further to this point, the researchers themselves have become more diversified, representing a shift away from exclusively North American and European-dominated research and toward Asian and other global markets. For example, in "Chinese and American Technical Communication: A Cross-Cultural Comparison of Differences," the authors examine Chinese government proposals and feasibility studies, as well as examples of Chinese student work (Barnum and Huilin 2006).

We hypothesized that five years of research reported in the field's major peer-reviewed journals would provide a good overview of the most recent work in the field. Technical communicators are looking for current information that can be used to advance their knowledge of their discipline. For example, "Seeing Technical Communication from a Career Perspective: The Implications of Career Theory for Technical Communication Theory, Practice, and Curriculum Design" discusses four broad but interrelated strands of inquiry that technical communication researchers might pursue based on research in career studies (Jablonsky 2005).

It is reasonable to expect that technical communicators would look at qualitative research in the field as yielding first-hand accounts of how others are using emerging technologies, evolving their areas of expertise, and adding business value. Articles prior to 2003 could be perceived as "old news," especially considering how the Internet, real-time communications, mobility, and Web 2.0 technologies have changed current business practices and knowledge sharing.

Why These Five Journals?

The five journals reviewed in our research for this essay represent the major sources of qualitative research in technical communication. The journals cover a wide range of topics within the field. More importantly, articles in these journals are peer reviewed, lending them credibility and authenticity. Our study excluded books. While some books published in the field during this period used qualitative research methods, journals offered us more diverse, numerous, and easily accessible examples.

Summary of Trends in Qualitative Research, 2003 to 2007

Our approach to secondary research synthesized features of qualitative and quantitative research in the 593 articles we surveyed. Most similar to the coding of patterns in data that some qualitative research methods use, our approach looked for patterns rather than the more common "summary and evaluative comments" approach. Our approach was not strictly qualitative research, despite the use of textual analysis and the search for patterns. Nor was it quantitative research, although we have provided counts and percentages.

An initial cluster analysis enabled us to group articles and find common themes to identify at a high level the most significant trends in qualitative research during the five-year period that we examined.

Since the articles were collected from the five peer-reviewed journals, the authors are generally university professors who are performing content analyses or reporting on research conducted within the classroom or industry. Although *Technical Communication* and *IEEE Transactions on Professional Communication* contain articles that are more practitioner-focused, the audience for most of the articles consists primarily of academics.

Most Frequently Used Qualitative Research Methods

In our review, we identified the top qualitative research approaches used by authors. The top methods comprise content analysis (the study of an information product most often developed by another technical communicator), interviews with study participants, and observation of study participants.

In content analysis, documents are often coded to facilitate statistical analysis. Our own research involved high-level content analysis, which helped us to classify the articles.

Most Frequent Qualitative Research Methods

1. Content Analysis Only (114 articles)
2. Content Analysis with Interviews (38 articles)
3. Content Analysis with Interviews and Observation (20 articles)

Our review indicated that content analysis is the most frequently used qualitative research method, though this method was often augmented with other qualitative methods. In addition to these approaches, the authors of the remaining 53 articles used 19 other combinations of qualitative research methods with content analysis. These other methods included usability tests, focus groups, participant observation (in which the author is also a study participant), and pilot studies. However, none of these remaining articles contained research combinations used frequently enough to indicate a favored approach.

As shown in Table 15.2 and Figure 15.1, authors tend to use multiple methodologies. The percentages found from one method to another show that there is a significant decrease in the use of singular methods outside of performing content analysis.

Within the five qualitative research types identified (content analysis, interviews, observation and participant observation, focus group, and usability test), it is important at this stage to establish the hierarchy of the methods and

Table 15.2 Remaining Research Method Combinations

Remaining research method combinations	Number of articles
Content analysis with observation and a usability test	8
Content analysis with a usability test	7
Content analysis with observation	6
Content analysis with interviews and observation and a usability test	4
Content analysis with focus groups	3
Content analysis with interviews and focus groups	3
Content analysis with interviews and participant observation	3
Content analysis with interviews and focus groups and participant observation	2
Content analysis with interviews and observation and focus groups	2
Content analysis with interviews and observation and participant observation and focus groups	2
Content analysis with interviews and observation and pilot study	2
Content analysis with observation and focus groups	2
Content analysis with observation and participant observation	2
Content analysis with participant observation	2
Content analysis with interviews and observation and participant observation	1
Content analysis with interviews and a pilot study	1
Content analysis with interviews and a usability test	1
Content analysis with observation and a pilot study	1
Content analysis with usability test and a pilot study	1

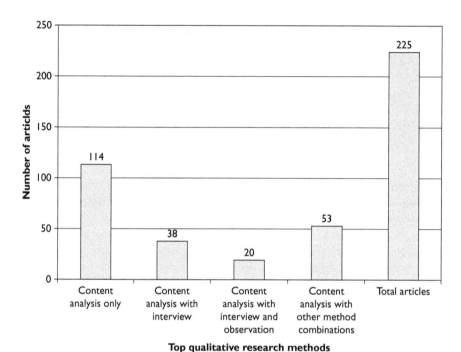

Figure 15.1 Top Three Most Frequently Used Qualitative Research Methods in 225 Articles

explain the significance of these findings. Out of 225 articles selected for this review, 114 (50 percent) use content analysis as the sole qualitative research method. Through content analysis, researchers review examples of technical communication and focus on a specific question by analyzing textual or graphical artifacts.

To understand the reason why so many qualitative research articles include content analysis only, consider the evolution and current state of the technical communication courses and programs in United States universities, where the vast majority are located. Most technical communication programs are housed in English departments, and most of the articles that we studied were written by professors in those departments. For faculty seeking promotion and tenure, content analysis can be performed on a limited budget and does not require the time and resources needed for other research methods involving access to users or test participants. It is also a more accessible method for colleagues in English studies, who are often not familiar with other qualitative research methods.

Although surveys produce mostly quantitative data, they were used in the research for 67 of our 225 articles. Interviews and observation were two other common qualitative research methods. Both methods allow for the collection of extensive details. Although interviews may be time consuming for both the researcher and the participant, they occur frequently in the journal articles, reflecting the value that researchers find in the direct feedback and conversations that occur during interviews. Observation is a method that may be considered invasive, but understanding readers' and users' motivation is important to our understanding of audience. From the beginning, we are taught to know our audience, and to do so we must also be knowledgeable about their environments and behaviors.

The frequent use of multiple techniques indicates the necessity to include more than one source of information gathering. This trend of using multiple methods is a form of triangulation in qualitative studies and enhances the dependability of research results.

Topics

We reviewed all articles to ascertain the subject matter and identified 15 different topic categories, ranked in order from the most to the fewest articles represented (see Table 15.3). In the wide range of topics in the articles, the top five were: rhetoric, followed by pedagogy, usability, cross-cultural studies, and the evolving technical communication profession, reflecting the diversity of research in the technical communication profession.

In Table 15.3, we highlight the topics appearing most frequently in the top five qualitative research topics (152 articles) and accounting for 67 percent of the topics covered in our review.

Rhetoric appears most frequently as the topic in the articles in our corpus (51 articles). The authors, often technical communication instructors, are more likely to be skilled in rhetoric and rhetorical analysis, and interested in examining the power of rhetoric and the technical communicator's influence on audiences. From current to historical examples of rhetoric dealing with technology, science, and society, authors examined the use of ethos and pathos, and approaches to solving ethical dilemmas.

Pedagogy or instructional methodology (36 articles) is another topic reflective of the journals' contributors and audience. This topic covers practical applications of theory, tools, technologies, and processes. The articles address teaching technical communication, simulation activities, and problem-based learning.

The third most prevalent topic was usability (28 articles), including focuses on Web site design, documents, processes, and usability testing. This is a topic of great interest to technical communicators, as good usability reflects our business value and purpose. These articles provide tactical and easily applied methodologies for improving usability and conducting usability tests on the content we author as technical communicators.

Cross-cultural studies (20 articles) constitute another significant topic in the articles we analyzed. Technical communicators often work globally and in distributed teams that require an awareness of cross-cultural communication.

Table 15.3 Qualitative Research Topics

Rank	Topics	Number of articles
1	Rhetoric	51
2	Pedagogy	36
3	Usability	28
4	Cross-cultural studies	20
5	Evolving technical communication profession	17
6	Virtual teams	14
7	Business communication (organizational/corporate announcements, policy statements, press releases)	12
8	Visual communication	12
9	Electronic communication (e-mails, instant messaging, discussion boards)	8
10	Gender studies	6
11	Instructional design	6
12	Relationship between academia and industry	5
13	Ethics	4
14	Research methods	4
15	Conducting reviews and critiques	2
Total		225

Following cross-cultural studies as a favored topic is a trend in articles that showcase the evolving technical communicator's role in the profession (17 articles). These articles cover career research, needed skills, job performance, and definitions of the technical communication profession. Articles about the technical communication profession are vital for new technical communicators or those who are working to establish credibility within a company.

And beyond these topics are 10 others that, although important in their own right, represent the 73 articles in the bottom third of the topics (32 percent).

Gaps in Qualitative Research

By examining the types of qualitative research methods being used and the areas of study, we have discovered some gaps in existing qualitative research in our field.

In terms of qualitative research methods, focus groups and usability tests were not very commonly represented in the articles we examined. Focus groups and usability tests are becoming more frequently used methods for technical communication practitioners who must design effective interfaces. Although 28 of the articles discussed usability, usability tests and evaluations were not frequently used as a qualitative research method. Despite the fact that usability tests and focus groups can be time consuming to plan, conduct, and analyze, the effort and feedback obtained will further validate the researcher's findings and provide solid reference materials for readers.

A review of the topics also revealed that substantive qualitative research on our business value and the ways that technical communicators provide a return on investment for companies is significantly absent. Current real-world applications and experiences supported by qualitative research are also needed to strengthen the confidence of technical communicators in making decisions and establishing credibility for the technical communication field. Technical communicators also need research that will help them to understand how critical skills can best be applied, no matter how the technology landscape changes. Exploratory qualitative research is often needed before more precise quantitative research can be undertaken to identify specific correlations or causal relationships. To understand its value, it might be necessary for us to understand significant qualitative explorations of how technical communicators bring about valued results; quantitative researchers can then attempt to develop more precise hypotheses based on the results of the explorations.

Qualitative research derived from large-scale global projects involving multiple countries is another category that is missing. As our work becomes increasingly virtual and global, articles are beginning to provide insight into cross-cultural studies and the ways that technology is helping to bridge the geographical distance. Although a number of articles include case studies involving China, technical communicators need best practices and lessons learned from diverse types of international experiences.

Discussion of Findings

Qualitative research, as demonstrated by the number of articles found in the five journals for the period reviewed, is a critical component in the technical communication field. The research can be grouped into six main categories based on research methods used: content analysis, interviews, observation and participant observation, focus groups, and usability tests. We sorted the 225 articles we reviewed by method. Not surprisingly, most articles fell into more than one category. Table 15.1 provides a description of the methods; the discussion below provides an overview of the state of qualitative research in the field in each.

Content Analysis

Content analysis, the study and interpretation of written, oral, and visual material, was the most widely used type of qualitative research in technical communication within the five-year period we reviewed. Content analysis tries to answer questions about research and content:

- Research questions include: Is the research authentic? Is the research credible? Is the research representative?
- Content questions include: What is the significance of the content? For whom is the content significant?

In the 225 articles we examined, all used some form of content analysis and 114 (51 percent) used content analysis as the primary research technique.

In "The Two Shuttle Accident Reports: Context and Culture in Technical Communication" (Dombrowski 2006), the author compares the official space shuttle accident reports for the final tragic flights of *Columbia* (2003) and *Challenger* (1986). As Dombrowski examines both reports rhetorically, he also examines the positive impact that technical communicators have had on industry.

In content analysis, documents are often coded to facilitate statistical analysis. At a high level, our own research on articles using qualitative research methods involved content analysis that allowed us to classify articles based on topics and qualitative research methods. But much finer coding of document content was frequently used by the researchers whose work we studied.

For example, in "A Corpus Study of Canned Letters: Mining the Latent Rhetorical Proficiencies Marketed to Writers-in-a-Hurry and Non-Writers" (Kaufer and Ishizaki 2006), the authors used text-analysis software to examine words and sentence structure in 728 sample documents. Although this study also involved a quantitative research method, the software assisted the authors with further analysis of the common rhetorical features used in the letters.

In addition to the reasons mentioned earlier, another reason why content analysis was the most popular research method may be the fact that there is a great body of documents available in print media that lend themselves to analysis, providing insight into specific fields of study not covered directly by the articles or secondary research. The advantages and preponderance of content analysis may lead academics and practitioners to adopt this method even more often in the future.

Interviews

Interviews were used in 67 articles (30 percent) to gather information from participants and were either conducted face to face or during a scheduled telephone call. With interviews, the researcher has the ability to ask follow-up questions to gain a better understanding of the participant's opinions and understanding of the subject.

Interviews can be conducted at the beginning of the study to help the researcher focus on specific areas of interest before initiating quantitative research methods. In "Behavioral Adaptation within Cross-Cultural Virtual Teams" (Anawati and Craig 2006), the researchers used in-depth interviews to help establish the questions and structure for an anonymous online questionnaire sent to members of an international information technology (IT) firm. Of the 67 articles that used interviews, 31 also used a survey.

Interviews can also be helpful during and toward the end of a study to help the researcher better understand trends or unexpected findings. In "What We Teach and What They Use: Teaching and Learning in Scientific and Technical Communication Programs and Beyond" (Brady 2007), interviews were used throughout a six-year study of eight graduate students, generating over 700 pages of tape transcripts. The interviews allowed the researcher to see how opinions and practices shifted as the students progressed in their professional careers.

In "What is 'Good' Technical Communication? A Comparison of the Standards of Writing and Engineering Instructors" (Smith 2003), the researcher analyzed the think-aloud sessions of four technical writing instructors and four engineering instructors as they each reviewed the same student papers. Following the think-aloud sessions, interviews captured additional information and feedback explaining either the comments the instructors made on the papers or comments not uttered during the think-aloud sessions. Interview sessions helped to determine whether or not engineering and writing instructors hold students to different standards for writing quality.

Conducting interviews is a very common qualitative research method in the articles we reviewed. The data collected from one-on-one feedback from participants increase the credibility of the research by answering questions and in some cases providing qualitative insights along with statistics derived from quantitative methods to strengthen research findings.

Observation and Participant Observation

In our analysis, all articles had some type of observation and combinations of observation with other methods, which accounted for 51 of the 225 articles (23 percent). We found that the use of observation as a research tool is often combined with another type of research methodology. The challenge for observational research would be the resources and availability of the researcher to conduct the study. Observational research, as with most other types, requires time with subjects and permission to conduct the research.

Observation can be seen in many articles where observational analyses are used in conjunction with other types of research. For example, Sun (2006) studied participants in the United States and China, where she reviewed 2,370 text messages and conducted interviews in two stages with the 19 Americans and 22 Chinese participants. Sun looked at the basic questions of the five Ws (Who, What, Where, When, Why) and the cultural aspects of messaging that indicated the rhetoric used by the participants. In doing so, she showed the value of the observational type of research, although her research required significant contact with participants from two continents. The interviews lasted from 45 to 75 minutes each, and in the second stage of the study several participants were shadowed and observed for one to two days each.

Participant observation requires the researcher to take part in the study along with the other participants. In one study where participant observation was used, James Conklin created a worksheet for himself and participants to fill out when they attended or observed team meetings (Conklin 2007). In cases when Conklin could not be present, the participant observers could document their observations about the team dynamics. This type of preparation resulted in qualitative research data from the direct observations of his participants. For Conklin, this method provided access to information that would otherwise have been unobtainable, and provided opportunities to gather information from a larger study.

This combination of observation and participant observation is not as commonly found in the journals as one might expect. For example, in *Technical Communication Quarterly*, there is only one instance in the articles that we reviewed where the two techniques are combined. The most common technique is the use of researchers' own observation of subjects. It is possible that another reason why participant observation is infrequently used is that it is difficult for the researcher to maintain control in such cases.

Observations are often conducted initially by researchers as silent participants, as in Conklin's research or Ann Brady's research on students' writing styles and their development as they move from academia to the workplace (Brady 2007). Brady conducted personal observations that she later used to formulate a survey for her subjects to complete online. This use of personal observation provided valuable data to Brady's research, and was then reused in a significant part of another research project.

Focus Group

Focus group research has specific orientations that are conducive and beneficial to research in technical communication. For example, in "Ten Engineers Reading: Disjunctions between Preference and Practice in Civil Engineering Faculty Responses" (Taylor and Patton 2006), a focus group consisting of professors provides the researcher with significant feedback from instructors who not only had experience in teaching but in most cases also had industry work experience. In this article, the researchers use a focus group of 10 instructors to determine how instructors evaluate and comment on students' technical communication skills. Taylor and Patton provided the focus group with one student report that yielded consistent results. Taylor and Patton's use of the focus group methodology is an excellent example of producing results from a minimal sample.

Focus group research plays an important role in the practitioners' world in analyzing users. "Comparative User-Focused Evaluation of User Guides: A Case Study" (Ganier 2007) demonstrates the practical application of having experts in the field of customer service, quality control, marketing, and R&D test a model user guide. Unlike Taylor and Patton's research, Ganier concentrated on a single company, with focus groups consisting of various experts. These varied interpretations gave researchers the opportunity to troubleshoot issues for clients in an environment that must adapt to varied users, whereas in Taylor and Patton's research, the evaluation showed how to adapt a standard across a narrower client base.

Practical applications are also critical examinations completed by focus group research. Nicole Amare and Charlotte Brammer (2005) conducted an examination of the response of 31 engineers, 15 instructors, and 45 students to three carefully selected engineering memos. The practical application of this academic institutional research is to prepare a student's transition from academia to the working world. This study showed differences across the different sectors of the focus groups. The practitioners and academics displayed divergent preferences in a fundamental area that technical communication faces on a daily basis.

However, there seems to be a great limitation to the use of focus groups based on their minimal usage as demonstrated by the results of the literature review. Focus groups were used in only 12 of the 225 articles (5 percent), so the big question is, "Why?" There appear to be some answers based on the observations made so far; that is, focus groups have a very narrow scope of application and require a trained facilitator, a skill that many researchers may lack. However, focus groups are extremely practical in gathering information for a narrow study.

Focus groups are generally small—for example, a "group of 10 engineers"— therefore, the sampling is insufficient to make generalizations that might benefit the field as a whole. The other problem with focus groups (and, of course, also with interviews and observation) is the source of participants. As with the engineers' sample, researchers need to recruit a group from industry, academe, or other sources that contains members who meet a set of predetermined qualifications.

Usability Test

Usability testing is a type of usability study that involves observing users performing tasks. Usability testing appears to be the least-used type of qualitative research. Given the importance of usability testing in industry, it is surprising that it was not used often in the corpus of articles we studied. Of the 28 articles focusing on usability (12 percent), only 13 were reports of usability testing. Of the articles that relied on usability testing, "Constructive Interaction: An Analysis of Verbal Interaction in a Usability Setting" (van den Haak, de Jong, and Schellens 2006) is an excellent example of the benefits inherent in this type of research. The authors outline how they incorporated visual and audio recordings of 20 pairs of students from the University of Twente. The research evaluated what comments were made, who made the comments, and when comments were uttered during tasks involving use of an online library catalog. This article reveals how useful this type of research can be, and explores avenues other than the responses found in typical usability testing.

Technical Communication and the *IEEE Transactions on Professional Communication* lead the other journals in articles containing reports of usability testing, perhaps reflecting these journals' more practical research focus. This conclusion is reinforced by the fact that most of these articles that report usability research focus on the customer service industry, whether in IT or other fields, and therefore the importance of this type of research to today's growing service economy cannot be understated. Evidence supporting this conclusion is found in "Using Customer Contact Center Technicians to Measure the Effectiveness of

Online Help Systems" (Downing 2007), where the author analyzes information collected by a company team created to review technicians' online help systems. The team reviewed 600 technicians' reports that included three major usability issues. The consequence of this usability testing resulted in an increase of efficiency for the company.

The infrequent use of usability testing in published research in our field may result from budgetary and resource constraints. Essentially, usability testing can require a large commitment of researcher time and resources, or work with data collected and shared by companies, and that approach may not be practical for researchers to access.

Conclusion

The articles we reviewed represent a snapshot of recent studies that have contributed to the field of technical communication. In the five peer-reviewed journals we surveyed, we identified 225 articles from the total of 593 articles that included qualitative research methods (content analysis, case studies, interviews, observation, focus groups, and usability tests) during the five-year period from 2003 to 2007. The number of articles that used qualitative research is a significant portion (almost 38 percent) of the articles published in that period. This volume of work supports the importance of qualitative research in technical communication.

Our analysis of trends showed that content analysis was the method most frequently used. Our review also indicated areas where more research is needed, as well as an important finding that focus groups and usability tests seem to be infrequently used and reported in research studies.

Research in technical communication provides a significant body of knowledge for technical communication professionals, who are constantly being challenged to keep up with emerging trends. Because all of these peer-reviewed journals are available online, technical communicators can quickly search for and access reports of qualitative research (Luzón 2007).

Every day, technical communicators are making strides in their professional careers, discovering innovative approaches, and applying the best practices garnered from published research. Because qualitative research does not require familiarity with inferential statistical analysis, the results are easier for practitioners to access. In other words, technical communicators are not only able to understand what happened, but why. This knowledge enables technical communicators to make informed decisions and to articulate confidently to others the reasoning behind the techniques they adopt.

Technical communicators need to understand the reasoning behind the actions of their users to deliver quality work, and it is qualitative research that most effectively enables practitioners to achieve this understanding. Qualitative research is not only important to growing the body of knowledge of technical communicators but is also critical to establishing respect for and understanding of our field within other disciplines and professions Most importantly, using the results of qualitative research, technical communicators can better answer the question, "Why?," and help their users.

Reference

Hughes, M. A., and G. F. Hayhoe. 2008. *A Research Primer for Technical Communication: Methods, Exemplars, and Analyses*. New York, NY: Lawrence Erlbaum Associates, Taylor and Francis Group.

Articles Reviewed

Abbott, C., and P. Eubanks. 2005. "How Academics and Practitioners Evaluate Technical Texts: A Focus Group Study." *Journal of Business and Technical Communication* 19: 171–218.

Adkins, K. 2003. "Serpents and Sheep: The Harriman Expedition, Alaska, and the Metaphoric Reconstruction of American Wilderness." *Technical Communication Quarterly* 12: 423–37.

Alley, M., and A. N. Katheryn. 2005. "Rethinking the Design of Presentation Slides: A Case for Sentence Headline and Visual Evidence." *Technical Communication* 52: 417–26.

Alley, M., M. Schreiber, K. Ramsdell, and J. Muffo. 2006. "How the Design of Headlines in Presentation Slides Affect Audience Retention." *Technical Communication* 53: 225–34.

Amare, N. 2006. "To Slideware or Not to Slideware: Students' Experiences with PowerPoint vs. Lecture." *Journal of Technical Writing and Communication* 36: 297–308.

Amare, N., and C. Brammer. 2005. "Perceptions of Memo Quality: A Case Study of Engineering Practitioners, Professors, and Students." *Journal of Technical Writing and Communication* 35:179–90.

Anawati, D., and A. Craig. 2006. "Behavioral Adaptation within Cross-Cultural Virtual Teams." *IEEE Transactions on Professional Communication* 49: 44–56.

Anderson, D. L. 2004. "The Textualizing Functions of Writing for Organizational Change." *Journal of Business and Technical Communication* 18: 141–64.

Anthony, L. 2006. "Developing a Freeware, Multiplatform Corpus Analysis Toolkit for the Technical Writing Classroom." *IEEE Transactions on Professional Communication* 49: 275–86.

Artemeva, N. 2005. "A Time to Speak, a Time to Act: A Rhetorical Genre Analysis of a Novice Engineer's Calculated Risk Taking." *Journal of Business and Technical Communication* 19: 389–421.

Baake, K. 2003. "Archaeology Reports: When Context Becomes an Active Agent in the Rhetorical Process." *Technical Communication Quarterly* 12: 389–403.

Barker, T., and N. Matveeva. 2006. "Teaching Intercultural Communication in a Technical Writing Service Course: Real Instructors' Practices and Suggestions for Textbook Selection." *Technical Communication Quarterly* 15: 191–214.

Barnum, C. M., and L. Huilin. 2006. "Chinese and American Technical Communication: A Cross-Cultural Comparison of Differences." *Technical Communication* 53: 143–66.

Barr, J., P. Jack, and S. Rosebaum. 2003. "Documentation and Training Productivity Benchmarks." *Technical Communication* 50: 471–84.

Baskerville, R., and J. Nandhakumar. 2007. "Activating and Perpetuating Virtual Teams: Now That We're Mobile, Where Do We Go?" *IEEE Transactions on Professional Communication* 50: 17–34.

Battalio, J. T. 2006. "Teaching a Distance Education Version of the Technical Communication Service Course: Timesaving Strategies." *Journal of Technical Writing and Communication* 36: 273–96.

Beamer, L. 2003. "Directness in Chinese Business Correspondence of the Nineteenth Century." *Journal of Business and Technical Communication* 17: 201–37.

Benbunan-Fich, R., and S. Altschuller. 2005. "Web Presence Transformations in the 1990s: An Analysis of Press Releases." *IEEE Transactions on Professional Communication* 48: 131–46.

Blakeslee, A. M., and R. Spilka. 2004. "The State of Research in Technical Communication." *Technical Communication Quarterly* 13: 73–92.

Bowdon, M. 2004. "Technical Communication and the Role of the Public Intellectual: A Community HIV-Prevention Case Study." *Technical Communication Quarterly* 13: 325–40.

Bradner, E., G. Mark, and T. D. Hertel. 2005. "Team Size and Technology Fit: Participation, Awareness, and Rapport in Distributed Teams." *IEEE Transactions on Professional Communication* 48: 68–77.

Brady, A. 2007. "What We Teach and What They Use: Teaching and Learning in Scientific and Technical Communication Programs and Beyond." *Journal of Business and Technical Communication* 21: 37–61.

Bragge, J., H. Merisalo-Rantanen, and P. Hallikainen. 2005. "Gathering Innovative End-User Feedback for Continuous Development of Information Systems: A Repeatable and Transferable E-Collaboration Process." *IEEE Transactions on Professional Communication* 48: 55–67.

Brumberger, E. R. 2003. "The Rhetoric of Typography: The Persona of Typeface and Text." *Technical Communication* 50: 206–23.

——. 2004. "The Rhetoric of Typography: Effects on Reading Time, Reading Comprehension, and Perceptions of Ethos." *Technical Communication* 51: 13–24.

——. 2007. "Visual Communication in the Workplace: A Survey of Practice." *Technical Communication Quarterly* 16: 369–95.

Bryson, M. A. 2003. "Nature, Narrative, and the Scientist-Writer: Rachel Carson's and Loren Eiseley's Critique of Science." *Technical Communication Quarterly* 12: 369–87.

Byrne, J. 2005. "Evaluating the Effect of Iconic Linkage on the Usability of Software User Guides." *Journal of Technical Writing and Communication* 35: 155–78.

Campbell, J. A., and R. K. Clark. 2005. "Revisioning the Origin: Tracing Inventional Agency through Genetic Inquiry." *Technical Communication Quarterly* 14: 287–93.

Carliner, S. 2003. "Modeling Information for Three-Dimensional Space: Lessons Learned from Museum Exhibit Design." *Technical Communication* 50: 554–70.

——. 2004. "What do we Manage? A Survey of the Management Portfolios of Larger Technical Communications Groups." *Technical Communication* 51: 45–67.

Carter, M., M. Ferzli, and E. N. Wiebe. 2007. "Writing to Learn by Learning to Write in the Disciplines." *Journal of Business and Technical Communication* 21: 282–302.

Casper, C. F. 2007. "In Praise of Xarbon, in Praise of Science: The Epideictic Rhetoric of the 1996 Nobel Lectures in Chemistry." *Journal of Business and Technical Communication* 21: 303–23.

Clark, D. 2004. "Is Professional Writing Relevant? A Model for Action Research." *Technical Communication Quarterly* 13: 307–23.

Codone, S. 2004. "Reducing the Distance: A Study of Course Websites as a Means to Create a Total Learning Space in Traditional Courses." *IEEE Transactions on Professional Communication* 47: 190–99.

Conklin, J. 2007. "From the Structure of Text to the Dynamic of Teams: The Changing Nature of Technical Communication Practice." *Technical Communication* 54: 210–31.

Cook, K. C., C. Thralls, and M. Zachry. 2003. "Doctoral-Level Graduates in Professional, Technical and Scientific Communication 1995–2000: A Profile." *Technical Communication* 50: 160–73.

——. 2007. "Immersion in a Digital Pool: Training Prospective Online Instructors in Online Environments." *Technical Communication Quarterly* 16: 55–82.

Cooke, L. 2003. "Information Acceleration and Visual Trends in Print, Television, and Web News Sources." *Technical Communication Quarterly* 12: 155–81.

Coppola, N. W., and N. Elliot. 2007. "A Technology Transfer Model for Program Assessment in Technical Communication." *Technical Communication* 54: 459–74.

Coppola, N. W., S. R. Hiltz, and N. G. Rotter. 2004. "Building Trust in Virtual Teams." *IEEE Transactions on Professional Communication* 47: 95–104.

Corman, Steven R. 2006. "On Being Less Theoretical and More Technological in Organizational Communication." *Journal of Business and Technical Communication* 20: 325–38.

Cozijn, R., A. Maes, D. Schackman, and N. Ummelen. 2007. "Structuring Job Related Information on the Intranet: An Experimental Comparison of Task vs. an Organization-Based Approach." *Journal of Technical Writing and Communication* 37: 203–16.

Dannels, D. P. 2003. "Teaching and Learning Design Presentations in Engineering: Contradictions Between Academic and Workplace Activity Systems." *Journal of Business and Technical Communication* 17: 139–69.

Dautermann, J. 2005. "Teaching Business and Technical Writing in China: Confronting Assumptions and Practices at Home and Abroad." *Technical Communication Quarterly* 14: 141–59.

Dawley, D. D., and W. P. Anthony. 2003. "User Perceptions of E-Mail at Work." *Journal of Business and Technical Communication* 17: 170–200.

Dayton, D. 2003. "Electronic Editing in Technical Communication: A Survey of Practices and Attitudes." *Technical Communication* 50: 192–205.

——. 2004. "Electronic Editing in Technical Communication: The Compelling Logics of Local Contexts." *Technical Communication* 51: 86–101.

——. 2006. "A Hybrid Analytical Framework to Guide Studies of Innovative IT Adoption by Work Groups." *Technical Communication Quarterly* 15: 355–82.

Dayton, D., and M. McShane Vaughn. 2007. "Developing a Quality Assurance Process to Guide the Design and Assessment of Online Courses." *Technical Communication* 54: 475–85.

de Groot, E. B., H. Korzilius, C. Nickerson, and M. Gerritsen. 2006. "A Corpus Analysis of Text Themes and Photographic Themes in Managerial Forewords of Dutch-English and British Annual General Reports." *IEEE Transactions on Professional Communication* 49: 217–25.

de Jong, M., and D. Rijnks. 2006. "Dynamics of Iterative Reader Feedback: An Analysis of Two Successive Plus-Minus Evaluation Studies." *Journal of Business and Technical Communication* 20: 159–76.

Ding, D. D. 2004. "Context-Driven: How is Traditional Chinese Medicine Labeling Developed?" *Journal of Technical Writing and Communication* 34: 173–88.

——. 2006. "An Indirect Style in Business Communication." *Journal of Business and Technical Communication* 20: 87–100.

Dinolfo, J., B. Heifferon, and L. A. Temesvari. 2007. "Seeing Cells: Teaching the Visual/Verbal Rhetoric of Biology." *Journal of Technical Writing and Communication* 37: 395–417.

Dombrowski, P. 2003. "Ernst Haeckel's Controversial Visual Rhetoric." *Technical Communication Quarterly* 12: 303–19.

——. 2006. "The Two Shuttle Accident Reports: Context and Culture in Technical Communication." *Journal of Technical Writing and Communication* 36: 231–52.

Donehy-Farina, S., P. W. Callas, M. A. Ricci, M. P. Caputo, J. L. Amour, and F. B. Rogers. 2003. "Technical Communication and Clinical Health Care: Improving Rural Emergency Trauma Care through Synchronous Videoconferencing." *Journal of Technical Writing and Communication* 33: 111–23.

Donnell, J. 2005. "Illustration and Language in Technical Communication." *Journal of Technical Writing and Communication* 35: 239–71.

Dossena, M. 2006. "Doing Business in Nineteenth-Century Scotland: Expressing Authority, Conveying Stance." *IEEE Transactions on Professional Communication* 49: 246–53.

Downing, J. 2007. "Using Customer Contact Center Technicians to Measure the Effectiveness of Online Help Systems." *Technical Communication* 54: 201–09.

Dragga, S., and D. Voss. 2003. "Hiding Humanity: Verbal and Visual Ethics in Accident Reports." *Technical Communication* 50: 61–82.

Duan, P., and W. Gu. 2005. "Technical Communication and English for Specific Purposes: The Development of Technical Communication in China's Universities." *Technical Communication* 52: 434–48.

Durack, K. T. 2003. "Observations on Entrepreneurship, Instructional Texts, and Personal Interaction." *Journal of Technical Writing and Communication* 33: 87–109.

Eubanks, P., and C. Abbott. 2003. "Using Focus Groups to Supplement the Assessment of Technical Communication Texts, Programs, and Courses." *Technical Communication Quarterly* 12: 25–45.

Evia, C. 2004. "Quality Over Quantity: A Two-Step Model for Reinforcing User Feedback in Transnational Web-Based Systems Through Participatory Design." *IEEE Transactions on Professional Communication* 47: 71–74.

Farkas, D. K. 2005. "Explicit Structure in Print and On-Screen Documents." *Technical Communication Quarterly* 14: 9–30.

Fisher, D., T. Bowers, A.Ellerton, T. J. Brumm, and S. K. Mickelson. 2003. "As the Case May Be: The Potential of Electronic Cases for Interdisciplinary Communication Instruction." *IEEE Transactions on Professional Communication* 46: 313–19.

Ford, J. D. 2004. "Knowledge Transfer across Disciplines: Tracking Rhetorical Strategies from a Technical Communication Classroom to an Engineering Classroom." *IEEE Transactions on Professional Communication* 47: 301–15.

Freiermuth, M. R. 2003. "Case-Based Simulations in the EST Classroom." *IEEE Transactions on Professional Communication* 46: 221–30.

Gallivan, M. J., and R. Benunan-Fich. 2005. "A Framework for Analyzing Levels of Analysis Issues in Studies of E-Collaboration." *IEEE Transactions on Professional Communication* 48: 87–104.

Ganier, F. 2007. "Comparative User-Focused Evaluation of User Guides: A Case Study." *Journal of Technical Writing and Communication* 37: 305–22.

Garner, J. T. 2006. "It's Not What You Know: A Transactive Memory Analysis of Knowledge Networks at NASA." *Journal of Technical Writing and Communication* 36: 329–51.

Gerritsen, M., and E. Wannet. 2005. "Cultural Differences in the Appreciation of Introductions of Presentations." *Technical Communication* 52: 194–208.

Glasbeek, H. 2004. "Solving Problems on your Own: How Do Exercises in Tutorials Interact with Software Learners' Level of Goal-Orientedness?" *IEEE Transactions on Professional Communication* 47: 44–53.

Graham, M. B., and N. Lindeman. 2005. "The Rhetoric and Politics of Science in the Case of the Missouri River System." *Journal of Business and Technical Communication* 19: 422–48.

Griggs, K. 2007. "Non-Rule Environmental Policy: A Case Study of a Foundry Sand Land Disposal NPD." *Journal of Technical Writing and Communication* 37: 17–36.

Gross, A. G. 2004. "Finding Funding; Writing Winning Proposals for Research Funds." *Technical Communication* 51: 25–35.

———. 2007. "Medical Tables, Graphics, and Photographs: How They Work." *Journal of Technical Writing and Communication* 37: 419–33.

Gurak, L. J. 2003. "Towards Consistency in Visual Information: Standardized Icons Based in Task." *Technical Communication* 50: 492–96.

Hall, M., M. de Jong, and M. Steehouder. 2004. "Cultural Differences and Usability Evaluation: Individualistic and Collectivistic Participants Compared." *Technical Communication* 51: 489–503.

Hargie, O., D. Dickson, and S. Nelson. 2003. "Working Together in a Divided Society: A Study of Intergroup Communication in the Northern Ireland Workplace." *Journal of Business and Technical Communication* 17: 285–318.

Harris, R. A. 2005. "Reception Studies in the Rhetoric of Science." *Technical Communication Quarterly* 14: 249–55.

Hart, H., and J. Conklin. 2006. "Towards a Meaningful Model for Technical Communication." *Technical Communication* 53: 395–415.

Hartley, J. 2003. "Using New Technology to Assess the Academic Writing Styles of Male and Female Pairs and Individuals." *Journal of Technical Writing and Communication* 33: 243–61.

Hass, B., and M. Kleine. 2003. "The Rhetoric of Junk Science." *Technical Communication Quarterly* 12: 267–84.

Henze, B., R. 2004. "Emergent Genres in Young Disciplines: The Case of Ethnological Science." *Technical Communication Quarterly* 13: 393–421.

Herndl, C. G., and G. Wilson. 2007. "Reflections on Field Research and Professional Practice." *Journal of Business and Technical Communication* 21: 216–26.

Hirst, R. 2004. "Herbert Spencer's Philosophy of Style: Conserving Mental Energy." *Journal of Technical Writing and Communication* 34: 265–90.

Humphreys, L. 2005. "Social Topography in a Wireless Era: The Negotiation of Public and Private Space." *Journal of Technical Writing and Communication* 35: 367–84.

Hunt, K. 2003. "Establishing a Presence on the World Wide Web: A Rhetorical Approach." *Technical Communication* 50: 519–28.

Isakson, C. S., and J. H. Spyridakis. 2003. "The Influence of Semantics and Syntax on What Readers Remember." *Technical Communication* 50: 538–53.

Ishii, K. 2005. "The Human Side of the Digital Divide: Media Experience as the Border of Communication Satisfaction with Email." *Journal of Technical Writing and Communication* 35: 385–402.

Jablonsky, J. 2005. "Seeing Technical Communication from a Career Perspective: The Implications of Career Theory for Technical Communication Theory, Practice, and Curriculum Design." *Journal of Business and Technical Communication* 19: 5–41.

Johnson, C. S. 2006. "A Decade of Research: Assessing Change in the Technical Communication Classroom using Online Portfolios." *Journal of Technical Writing and Communication* 36: 413–31.

———. 2006. "Prediscursive Technical Communication in the Early American Iron Industry." *Technical Communication Quarterly* 15: 171–89.

———. 2007. "The Steel Bible: A Case Study of 20th Century Technical Communication." *Journal of Technical Writing and Communication* 37: 281–303.

Jones, A. A., and T. E. Freeman. 2003. "Imitation, Copying, and the Use of Models: Report Writing in an Introductory Physics Course." *IEEE Transactions on Professional Communication* 46: 168–84.

Jones, S. L. 2005. "From Writers to Information Coordinators: Technology and the Changing Face of Collaboration." *Journal of Business and Technical Communication* 19: 449–67.

Kain, D. J. 2005. "Constructing Genre: A Threefold Typology." *Technical Communication Quarterly* 14: 375–409.

Kain, D., and E. Wardle. 2005. "Building Context: Using Activity Theory to Teach About Genre in Multi-Major Professional Communication Courses." *Technical Communication Quarterly* 14: 113–39.

Kaufer, D., and S. Ishizaki. 2006. "A Corpus Study of Canned Letters: Mining the Latent Rhetorical Proficiencies Marketed to Writers-In-A-Hurry and Non-Writers." *IEEE Transactions on Professional Communication* 49: 254–66.

Khalifa, M., and V. Liu. 2006. "Semantic Network Discussion Representation: Applicability and Some Potential Benefits." *IEEE Transactions on Professional Communication* 49: 69–81.

Khatri, V., I. Vessey, S. Ram, and V. Ramesh. 2006. "Cognitive Fit Between Conceptual Schemas and Internal Problem Representations: The Case of Geospatio-Temporal Conceptual Schema Comprehension." *IEEE Transactions on Professional Communication* 49: 109–27.

Killoran, J. B. 2006. "Self-Published Web Résumés: Their Purposes and their Genre Systems." *Journal of Business and Technical Communication* 20: 425–59.

Kim, L. 2005. "Tracing Visual Narratives: User-Testing Methodology for Developing a Multimedia Museum Show." *Technical Communication* 52: 121–37.

Kim, L., and Albers, M. J. 2003. "Presenting Information on the Small-Screen Interface: Effects of Table Formatting." *IEEE Transactions on Professional Communication* 46: 94–104.

Kimball, M. A. 2006. "Cars, Culture, and Tactical Technical Communication." *Technical Communication Quarterly* 15: 67–86.

Kock, N. 2003. "Action Research: Lessons Learned from a Multi-Iteration Study of Computer-Mediated Communication in Groups." *IEEE Transactions on Professional Communication* 46: 105–28.

——. 2003. "Communication-Focused Business Process Redesign: Assessing a Communication Flow Optimization Model through an Action Research Study at a Defense Contractor." *IEEE Transactions on Professional Communication* 46: 35–54.

Koerber, A. 2005. "You Just Don't See Enough Normal: Critical Perspectives on Infant-Feeding Discourse and Practice." *Journal of Business and Technical Communication* 19: 304–27.

——. 2006. "Rhetorical Agency, Resistance, and the Disciplinary Rhetorics of Breastfeeding." *Technical Communication Quarterly* 15: 87–101.

Krahmer, E., and N. Ummelen. 2004. "Thinking about Thinking Aloud: A Comparison of Two Verbal Protocols for Usability Testing." *IEEE Transactions on Professional Communication* 47: 105–17.

Lavid, J., and M.Taboada. 2004. "Stylistic Differences in Multilingual Administrative Forms: A Cross-Linguistic Characterization." *Journal of Technical Writing and Communication* 34: 43–65.

Laviosa, S. 2006. "Data-Driven Learning for Translating Anglicisms in Business Communication." *IEEE Transactions on Professional Communication* 49: 267–74.

Lay, M. M. 2004. "Reflections on *Technical Communication Quarterly*, 1991–2003: The Manuscript Review Process." *Technical Communication Quarterly* 13: 109–19.

Lee, M. J., M. Tedder, and X. Gangxin. 2006. "Effective Computer Text Design to Enhance Readers' Recall: Text Formats, Individual Working Memory Capacity, and Content Type." *Journal of Technical Writing and Communication* 36: 57–73.

Levis, J. M., and G. M. Levis. 2003. "A Project-Based Approach to Teaching Research Writing to Non-Native Writers." *IEEE Transactions on Professional Communication* 46: 210–20.

Lin, C. 2007. "Organizational Website Design as a Rhetorical Situation." *IEEE Transactions on Professional Communication* 50: 35–44.

Lindeman, N. 2007. "Creating Knowledge for Advocacy: The Discourse of Research at a Conservation Organization." *Technical Communication Quarterly* 16: 431–51.

Lippincott, G. 2003. "Moving Technical Communication into the Post-Industrial Age: Advice from 1910." *Technical Communication Quarterly* 12: 321–42.

——. 2004. "Something in Motion and Something to Eat Attract the Crowd: Cooking with Science at the 1893 World's Fair." *Journal of Technical Writing and Communication* 33: 141–64.

Longo, B. 2004. "Toward an Informed Citizenry: Readability Formulas as Cultural Artifacts." *Journal of Technical Writing and Communication* 34: 165–72.

——, C. Weinert, and K. T. Fountain. 2007. "Implementation of Medical Research Findings through Insulin Protocols: Initial Findings from an Ongoing Study of Document Design and Visual Display." *Journal of Technical Writing and Communication* 37: 435–52.

Loorbach, N., M. Steehouder, and E. Taal. 2006. "The Effects of Motivational Elements in User Instructions." *Journal of Business and Technical Communication* 20: 177–99.

Lowry, P. B., J. F. Nunamaker Jr, Q. E. Booker, A. Curtis, and M. R. Lowry. 2004. "Creating Hybrid Distributed Learning Environments by Implementing Distributed Collaborative Writing in Traditional Educational Settings." *IEEE Transactions on Professional Communication* 47: 171–89.

Lutz, J., and M. Fuller. 2007. "Exploring Authority: A Case Study of a Composition and a Professional Writing Classroom." *Technical Communication Quarterly* 16: 201–32.

Luzón, M. J. 2007. "The Added Value Features of Online Scholarly Journals." *Journal of Technical Writing and Communication* 37: 59–73.

Mackiewicz, J. 2003. "Which Rules for Online Writing are Worth Following? A Study of Eight Rules in Eleven Handbooks." *IEEE Transactions on Professional Communication* 46: 129–37.

———. 2004. "The Effects of Tutor Expertise in Engineering Writing: A Linguistic Analysis of Writing Tutors' Comments." *IEEE Transactions on Professional Communication* 47: 316–28.

———. 2005. "How to Use Five Letterforms to Gauge a Typeface's Personality: A Research-Driven Method." *Journal of Technical Writing and Communication* 35: 291–315.

———. 2007. "Compliments and Criticisms in Book Reviews about Business Communication." *Journal of Business and Technical Communication* 21: 188–215.

Maier, C. C., C. Kampf, and P. Kastberg. 2007. "Multimodal Analysis: An Integrative Approach for Scientific Visualizing on the Web." *Journal of Technical Writing and Communication* 37: 453–78.

Manning, A., and N. Amare. 2006. "Visual-Rhetoric Ethic: Beyond Accuracy and Injury." *Technical Communication* 53: 195–211.

Markel, M. 2005. "The Rhetoric of Misdirection in Corporate Privacy-Policy Statements." *Technical Communication Quarterly* 14: 197–214.

Matveeva, N. 2007. "The Intercultural Component in Textbooks for Teaching a Service Technical Writing Course." *Journal of Technical Writing and Communication* 37: 151–66.

McGill, K. 2003. "Field Study and the Rhetoric Curriculum." *Technical Communication Quarterly* 12: 285–302.

McGovern, H. 2007. "Training Teachers and Serving Students: Applying Usability Testing in Writing Programs." *Journal of Technical Writing and Communication* 37: 323–46.

Melenhorst, M., T. van Der Geest, and M. Steehouder. 2005. "Noteworthy Observations about Note-Taking by Professionals." *Journal of Technical Writing and Communication* 35: 317–29.

Mirel, B., and N. Johnson. 2006. "Social Determinants of Preparing a Cyber-Infrastructure Innovation for Diffusion." *Technical Communication Quarterly* 15: 329–53.

Moore, P. 2004. "Questioning the Motives of Technical Communication and Rhetoric: Steven Katz's 'Ethic Of Expediency.'" *Journal of Technical Writing and Communication* 34: 5–29.

Moran, M. G. 2003. "Ralph Lane's 1586 'Discourse on the First Colony': The Renaissance Commercial Report as Apologia." *Technical Communication Quarterly* 12: 125–54.

———. 2005. "Figures of Speech as Persuasive Strategies in Early Commercial Communication: The Use of Dominant Figures in the Raleigh Reports about Virginia in the 1580s." *Technical Communication Quarterly* 14: 183–96.

Mudraya, O.V. 2004. "Need for Data-Driven Instruction of Engineering English." *IEEE Transactions on Professional Communication* 47: 65–70.

Munkvold, B. E. 2005. "Experiences from Global E-Collaboration: Contextual Influences on Technology Adoption and Use." *IEEE Transactions on Professional Communication* 48: 78–86.

Myers, M. 2004. "The Million Dollar Letter: Some Hints on How to Write One." *Journal of Technical Writing and Communication* 34: 133–43.

——. 2007. "The Use of Pathos in Charity Letters: Some Notes toward a Theory and Analysis." *Journal of Technical Writing and Communication* 37: 3–16.

Narita, M., K. Kurokawa, and T. Utsuro. 2003. "Case Study on the Development of a Computer-Based Support Tool for Assisting Japanese Software Engineers with their English Writing Needs." *IEEE Transactions on Professional Communication* 46: 194–209.

Nelson, S. 2003. "Engineering and Technology Student Perceptions of Collaborative Writing Practices." *IEEE Transactions on Professional Communication* 46: 265–76.

Ocker, R. 2005. "Influences on Creativity in Asynchronous Virtual Teams: A Qualitative Analysis of Experimental Teams." *IEEE Transactions on Professional Communication* 48: 22–39.

O'Hara, K. 2004. "'Curb Cuts' on the Information Highway: Older Adults and the Internet." *Technical Communication Quarterly* 13: 423–45.

Ortiz, L. A., and J. Dyke Ford. 2007. "Choose Sunwest: One Airline's Organizational Communication Strategies in a Campaign Against the Teamsters Union." *Journal of Technical Writing and Communication* 37: 215–47.

Panteli, N., and R.M. Davison. 2005. "The Role of Subgroups in the Communication Patterns of Global Virtual Teams." *IEEE Transactions on Professional Communication* 48: 191–200.

——. 2003. "Managing Nature/Empowering Decision-Makers: A Case Study of Forest Management Plans." *Technical Communication Quarterly* 12: 439–59.

Paretti, M. C. 2006. "Audience Awareness: Leveraging Problem-Based Learning to Teach Workplace Communication Practices." *IEEE Transactions on Professional Communication* 49: 189–98.

Paretti, M. C., L. D. McNair, and L. Holloway-Attaway. 2007. "Teaching Technical Communication in an Era of Distributed Work: A Case Study of Collaboration Between U.S. And Swedish Students." *Technical Communication Quarterly* 16: 327–52.

Pearce, G. C., I.W. Johnson, and R. T. Barker. 2003. "Assessment of the Listening Styles Inventory." *Journal of Business and Technical Communication* 17: 84–113.

Philbin, A. I., and M. D. Hawthorne. 2007. "Applying Assessment in a Self-Standing Program." *Technical Communication* 54: 490–502.

Propen, A. 2007. "Visual Communication and the Map: How Maps as Visual Objects Convey Meaning in Specific Contexts." *Technical Communication Quarterly* 16: 233–54.

Ran, B., and P. R. Duimering. 2007. "Imaging the Organization: Language Use in Organizational Identity Claims." *Journal of Business and Technical Communication* 21: 155–87.

Rehling, L. 2005. "Teaching in a High-Tech Conference Room: Academic Adaptations and Workplace Simulations." *Journal of Business and Technical Communication* 19: 98–113.

Reinsch Jr, L. N., and J. W. Turner. 2006. "Ari, R U There? Reorienting Business Communication for a Technological Era." *Journal of Business and Technical Communication* 20: 339–56.

Richards, A. R. 2003. "Argument and Authority in the Visual Representations of Science." *Technical Communication Quarterly* 12: 183–206.

Richards, A. R., and C. David. 2005. "Decorative Color as a Rhetorical Enhancement on the World Wide Web." *Technical Communication Quarterly* 14: 31–48.

Riley, K., and J. Mackiewicz. 2003. "Resolving the Directness Dilemma in Document Review Sessions with Non-Native Speakers." *IEEE Transactions on Professional Communication* 46: 1–16.

Robert, L. P., and A. R. Dennis. 2005. "Paradox of Richness: A Cognitive Model of Media Choice." *IEEE Transactions on Professional Communication* 48: 10–21.

Roberts, T. L., P. B. Lowry, and P. D. Sweeney. 2006. "An Evaluation of the Impact of Social Presence through Group Size and the Use of Collaborative Software on Group Member 'Voice' in Face-to-Face and Computer-Mediated Task Groups." *IEEE Transactions on Professional Communication* 49: 28–43.

Rogers, P. S., and S. M. Lee-Wong. 2003. "Reconceptualizing Politeness to Accommodate Dynamic Tensions in Subordinate-to-Superior Reporting." *Journal of Business and Technical Communication* 17: 379–412.

Roy, D., and R. Grice. 2004. "Helping Readers Connect Text and Visuals in Sequential Procedural Instruction: Developing Reader Comprehension." *Technical Communication* 51: 517–25.

Royal, C. 2005. "A Meta-Analysis of Journal Articles Intersecting Issues of Internet and Gender." *Journal of Technical Writing and Communication* 35: 403–29.

Rude, C. D. 2004. "Toward an Expanded Concept of Rhetorical Delivery: The Uses of Reports in Public Policy Debates." *Technical Communication Quarterly* 13: 271–88.

Rude, C., and K. C. Cook. 2004. "The Academic Job Market in Technical Communication, 2002–2003." *Technical Communication Quarterly* 13: 49–71.

Ryan, C. 2005. "Struggling to Survive: A Study of Editorial Decision-Making Strategies At MAMM *Magazine*." *Journal of Business and Technical Communication* 19: 353–76.

Salvo, M., M. W. Zoethewey, and K. Agena. 2007. "A Case of Exhaustive Documentation: Re-Centering System Oriented Organizations Around User Need." *Technical Communication* 54: 46–57.

Samson, C. 2006. ". . . Is Different From . . .: A Corpus-Based Study of Evaluative Adjectives in Economics Discourse." *IEEE Transactions on Professional Communication* 49: 236–45.

Sapp, D. A. 2006. "The Lone Ranger as Technical Writing Program Administrator." *Journal of Business and Technical Communication* 20: 200–19.

Sarker, S., D. B. Nicholson, and K. D. Joshi. 2005. "Knowledge Transfer in Virtual Systems Development Teams: An Exploratory Study of Four Key Enablers." *IEEE Transactions on Professional Communication* 48: 201–18.

Schneider, S. 2005. "Usable Pedagogies: Usability, Rhetoric, and Sociocultural Pedagogy in the Technical Writing Classroom." *Technical Communication Quarterly* 14: 447–67.

Schryer, C. F., and P. Spoel. 2005. "Genre Theory, Health-Care Discourse, and Professional Identity Formation." *Journal of Business and Technical Communication* 19: 249–78.

Schultz, L. D., and J. H. Spyridakis. 2004. "The Effect of Heading Frequency on Comprehension of Online Information: A Study of Two Populations." *Technical Communication* 51: 504–16.

Schwender, C., and C. Kohler. 2006. "Introducing Seniors to New Media Technology: New Ways of Thinking for a New Target Group." *Technical Communication* 53: 464–70.

Shaver, L. 2007. "Eliminating the Shell Game: Using Writing-Assignment Names to Integrate Disciplinary Learning." *Journal of Business and Technical Communication* 21: 74–90.

Shelby, A. N., and L. N. Reisch Jr. 2003. "Writing in Non-Interpersonal Settings." *Journal of Business and Technical Communication* 17: 50–83.

Sivunen, A., and M. Valo. 2006. "Team Leaders' Technology Choice in Virtual Teams." *IEEE Transactions on Professional Communication* 49: 57–68.

Slattery, S. 2007. "Undistributing Work Through Writing: How Technical Writers Manage Texts in Complex Information Environments." *Technical Communication Quarterly* 16: 311–25.

Smith, S. 2003. "What is 'Good' Technical Communication? A Comparison of the Standards of Writing and Engineering Instructors." *Technical Communication Quarterly* 12: 7–24.

Spafford, M. M., C. F. Schryer, M. Mian, and L. Lingard. 2006. "Look Who's Talking: Teaching and Learning Using the Genre of Medical Case Presentations." *Journal of Business and Technical Communication* 20: 121–58.

Spinuzzi, C. 2005. "Lost in the Translation: Shifting Claims in the Migration of a Research Technique." *Technical Communication Quarterly* 14: 411–46.

St Germaine-Madison, N. 2006. "Instructions, Visuals, and the English-Speaking Bias of Technical Communication." *Technical Communication* 53: 184–94.

Starke-Meyerring, D., A. Hill Duin, and T. Palyetzian. 2007. "Global Partnerships: Positioning Technical Communication Programs in the Context of Globalization." *Technical Communication Quarterly* 16: 139–74.

Still, B. 2006. "Talking to Students: Embedded Voice Commenting as a Tool for Critiquing Student Writing." *Journal of Business and Technical Communication* 20: 460–75.

Sun, H. 2006. "The Triumph of Users: Achieving Cultural Usability Goals with User Localization." *Technical Communication Quarterly* 15: 457–81.

Swarts, J. 2004. "Technological Mediation of Document Review: The Use of Textual Replay in Two Organizations." *Journal of Business and Technical Communication* 18: 328–60.

———. 2005. "PDAs in Medical Settings: The Importance of Organization in PDA Text Design." *IEEE Transactions on Professional Communication* 48: 161–76.

———. 2007. "Mobility and Composition: The Architecture of Coherence in Non-Places." *Technical Communication Quarterly* 16: 279–309.

Taylor, S. S. 2006. "Assessment in Client-Based Technical Writing Classes: Evolution of Teacher and Client Standards." *Technical Communication Quarterly* 15: 111–39.

———, and M. Patton. 2006. "Ten Engineers Reading: Disjunctions Between Preference and Practice in Civil Engineering Faculty Responses." *Journal of Technical Writing and Communication* 36: 253–71.

Thatcher, B. 2006. "Intercultural Rhetoric, Technology Transfer, and Writing In U.S.-Mexico Border Maquilas." *Technical Communication Quarterly* 15: 383–405.

Thayer, A. 2004. "Material Culture Analysis and Technical Communication: The Artifact Approach to Evaluating Documentation." *IEEE Transactions on Professional Communication* 47: 144–47.

———, M, Evans, A. McBride, M. Queen, and J. Spyridakis. 2007. "Content Analysis as a Best Practice in Technical Communication Research." *Journal of Technical Writing and Communication* 37: 267–79.

Thomas, G. F., and C. L. King. 2006. "Reconceptualizing E-Mail Overload." *Journal of Business and Technical Communication* 20: 252–87.

Tillery, D. 2003. "Radioactive Waste and Technical Doubts: Genre and Environmental Opposition to Nuclear Waste Sites." *Technical Communication Quarterly* 12: 405–21.

——. 2005. "The Plain Style in the Seventeenth Century: Gender and the History of Scientific Discourse." *Journal of Technical Writing and Communication* 35: 273–89.

Turns, J., and J. Ramey. 2006. "Active and Collaborative Learning in the Practice of Research: Credit-Based Directed Research Groups." *Technical Communication* 53: 296–307.

——, and T. S. Wagner. 2004. "Characterizing Audience for Informational Web Site Design." *Technical Communication* 51: 68–85.

van den Haak, M. J., M. D. T. de Jong, and P. Jan Schellens. 2006. "Constructive Interaction: An Analysis of Verbal Interaction in a Usability Setting." *IEEE Transactions on Professional Communication* 49: 311–24.

van der Meij, H., and M. Gellevij. 2004. "The Four Components of a Procedure." *IEEE Transactions on Professional Communication* 47: 5–14.

Wahl, S. 2003. "Learning at Work: The Role of Technical Communication in Organizational Learning." *Technical Communication* 50: 247–58.

Wareham, J., V. Mahnke, S. Peters, and N. Bjorn-Andersen. 2007. "Communication Metaphors-in-Use: Technical Communication and Offshore Systems Development." *IEEE Transactions on Professional Communication* 50: 93–108.

Warnick, B. 2005. "Looking to the Future: Electronic Texts and the Deepening Interface." *Technical Communication Quarterly* 14: 327–33.

Wegner, D. 2004. "The Collaborative Construction of a Management Report in a Municipal Community of Practice: Text and Context, Genre and Learning." *Journal of Business and Technical Communication* 18: 411–51.

Welch, K. E. 2006. "Technical Communication and Physical Location: Topoi and Architecture in Computer Classrooms." *Technical Communication Quarterly* 14: 335–44.

White, J. V. 2003. "Color: The Newest Tool for Technical Communicators." *Technical Communication* 50: 485–91.

Whithaus, C., and J. Magnotto Neff. 2006. "Contact and Interactivity: Social Constructionist Pedagogy in a Video-Based, Management Writing Course." *Technical Communication Quarterly* 15: 431–56.

Wiles, D. 2003. "Single Sourcing and Chinese Culture: A Perspective on Skills Development within Western Organizations and the People's Republic of China." *Technical Communication* 50: 371–84.

Willerton, R. 2005. "Visual Metonymy and Synecdoche: Rhetoric for Stage-Setting Images." *Journal of Technical Writing and Communication* 35: 3–31.

Williams, M. F. 2006. "Tracing W. E. B. Dubois' 'Color Line' in Government Regulations." *Journal of Technical Writing and Communication* 36: 141–65.

Wilson, G., and C. G. Herndl. 2007. "Boundary Objects As Rhetorical Exigence: Knowledge Mapping And Interdisciplinary Cooperation at the Los Alamos National Laboratory." *Journal of Business and Technical Communication* 21: 129–54.

Winsor, D. 2006. "Using Writing to Structure Agency: An Examination of Engineers' Practice." *Technical Communication Quarterly* 15: 411–30.

Wolfe, J. 2006. "Meeting Minutes as a Rhetorical Genre: Discrepancies between Professional Writing Textbooks and Workplace Practice Tutorial." *IEEE Transactions on Professional Communication* 49: 354–64.

Yohon, T., and D. Zimmerman. 2006. "An Exploratory Study of Adoption of Software and Hardware by Faculty in the Liberal Arts and Sciences." *Journal of Technical Writing and Communication* 36: 9–27.

Zappen, J. P. 2005. "Digital Rhetoric: Toward an Integrated Theory." *Technical Communication Quarterly* 14: 319–25.

Zdenek, S. 2007. "Just Roll Your Mouse over Me: Designing Virtual Women for Customer Service on the Web." *Technical Communication Quarterly* 16: 397–430.

Zhou, L. 2005. "An Empirical Investigation of Deception Behaviour in Instant Messaging." *IEEE Transactions on Professional Communication* 48: 147–60.

Zhu, P., and K. St.Amant. 2007. "Taking Traditional Chinese Medicine International and Online: An Examination of the Cultural Rhetorical Factors Affecting American Perceptions of Chinese-Created Web Sites." *Technical Communication* 54: 171–86.

Zimmerman, B. B., and D. Paul. 2007. "Technical Communication Teachers as Mentors in the Classroom: Extending an Invitation to Students." *Technical Communication Quarterly* 16: 175–200.

Chapter 16

Conclusion

George F. Hayhoe and James Conklin

Predicting the future is always a crap shoot, a matter of luck rather than skill. But by studying patterns of research in the last 20 years as well as more recent trends, we will surely be luckier in forecasting the paths that technical communication researchers are likely to choose in the immediate future than the diviners and augurs of ancient times.

We are convinced that qualitative studies, which have grown in significance and gained in respect from the research community during the past 25 years, will become increasingly important in the coming decade. We base this forecast on the following premises.

- Qualitative techniques are particularly well suited to the discipline of technical communication.
- A significant percentage of the articles published from 2003 through 2007, as well as frequently before that time, utilized qualitative methods.
- Practitioners in our field can understand and apply the results of qualitative studies much more readily than the results of quantitative studies.
- Qualitative methods are especially appropriate for workplace research conducted by practitioners.
- Qualitative techniques are well suited to explore themes that are central to current practice, such as those addressed by Giammona (2004), and emerging themes that are beginning to become apparent. The same is true of many other topics of research in the field.

Therefore, we are comfortable in predicting that qualitative methods will continue to play a major role and probably an increasingly important one in technical communication research. We will explore the truth of these premises and the validity of this conclusion in the following pages.

The Fit Between Qualitative Methods and the Field

As Faigley noted in 1985, "Because qualitative research offers the potential for describing the complex social situation that any act of writing involves, empirical researchers are likely to use qualitative approaches with increasing frequency" (1985, 243). Qualitative research attempts to make meaning by interpreting social phenomena, and there can be no doubt that writing and speaking are among the richest forms of social interaction as well as of meaning making.

Although some elements of technical communication can be explored using quantitative techniques, some of the most crucial cannot really be measured or counted. For example, the author's awareness of purpose and audience is usually cited as the most critical factor in successful technical communication, but is not susceptible to quantification in any significant way. Purpose and audience are, however, especially easy to explore by asking writers or presenters to tell the story behind the experience of producing a piece of discourse, or by asking the audience about the meaning and effectiveness of what they have heard or read. What was the writer or speaker trying to accomplish through the act of communication? Who were the intended readers or listeners? How closely did the writer's or speaker's purpose correspond to that of the audience? Was the audience successful in achieving their goals?

The central role of storytelling that we mentioned in the Preface to this volume really gets to several ultimate truths about technical communication.

- Technical communication is perhaps the most complicated form of communication because it adds complexity of subject matter to the already complex equation of human communication. Let us say that Marie must inform Jean about the date, time, and subject of a meeting, as well as list the names of others who will attend. Date and time are fairly easy to specify with precision, but explaining why the group is meeting and why Bob will not attend adds several degrees of complexity. Now let us say we are using this example to explain to a novice how to include this meeting information in an event invitation using his online calendar, and we add the complication of technology to the message.

- Stories can make complex subject matter easier to understand. For example, a story about two couples' situations as they approach retirement can make it easier to understand complicated eligibility criteria for various government benefits or the advantages of one alternative over another.

- Even when writers or speakers have not adequately analyzed purpose and audience, we can ask them to tell us in retrospect what purposes the discourse addresses and what audiences it reaches successfully. For example, a writer who didn't do any conscious audience analysis prior to preparing an instruction set can be asked to read what she has written, describe the characteristics of the people who would find that particular set of instructions helpful, and then speculate on the needs of other potential audiences.

- Audiences can tell us their reactions to pieces of discourse—most especially, how effective the authors have been and how successful they think that they themselves would be in accomplishing what they perceive as the goals of the communication. For example, a person with decreased visual acuity could examine a Web page and tell us how easy it is to read the words and interpret the photographs on the page.

- Storytelling is powerful precisely because it gets not only to the who and the what, but also to the why and the how. And after all, these are probably the most common questions we ask of those who attempt to communicate with us, from early childhood onward.
- Storytelling can also encompass (and even integrate) the reason and logic that underlie science and technology, and the values and ethical considerations that often underlie negotiations between groups of people. Technical communicators often work on teams that are seeking to introduce new technologies, products, and approaches into workplaces and communities. The decision to accept and use the innovations offered by science and technology is often not merely a matter of logic, but also involves complicated choices that are resolved through a balancing of values and points of view. Does this new computer system mean that some of my friends will lose their jobs? How can we convince people that this new environmental technology is safe and will make our communities more sustainable? If we install this new equipment in our company, will my contributions no longer matter as much? As Fisher (1978, 1984) showed in his work on narrative rationality, making decisions that involve cherished beliefs in social contexts is always intersubjective, and involves a consideration of facts, values, the self, and the social. Narrative approaches, such as those associated with qualitative research, can unleash a meaning-making process that allow groups to gain insights into each other's perspectives, and to collaborate on outcomes that satisfy everyone.

The fit between qualitative methods and technical communication that Faigley pointed to 25 years ago has certainly not decreased over time, nor does it promise to do so in the next decade or more.

The Frequency of Qualitative Methods in Articles Published 2003 to 2007

As Davy and Valecillos note in Chapter 15, 225 of the 593 articles (37.9 percent) published in the five major journals in the field during the five years between 2003 and 2007 utilized qualitative techniques. Although some of these articles mixed qualitative and quantitative methods, we believe that the fact that nearly 38 percent of the articles reported on research that used qualitative methods is very significant.

We cannot really generalize about whether a trend toward use of qualitative methodology has been developing without looking at a much longer period than the five years examined by Davy and Valecilllos. Nevertheless, in Table 16.1 we can see a relatively high degree of consistency in the frequency of qualitative methods for the period on which they report.

Table 16.1   Qualitative Research in Technical Communication, 2003–2007

Year	Number of qualitative articles	Percentage of articles published that year
2003	47	39
2004	36	30
2005	47	39
2006	52	43
2007	43	36
Mean	45	37

The mean number of articles using qualitative methods during this five-year period is 45. It is interesting to note that although the curve is not linear, the number is consistently 43 or higher with the exception of one year. We can also observe that the percentage of articles using qualitative techniques published during this five-year period never falls below 30 percent, with a mean percentage of 37 percent.

Again, we cannot discern a trend based on only five years of data. However, these numbers do reveal that there has been a lot of recent interest in qualitative research methods, and the Davy and Valecillos bibliography shows that these articles cross many topics of interest within the field. We see no reason why this interest should not at least continue at the current level, if not increase, especially as the number of researchers who have been trained in these techniques continues to grow each year.

The Suitability of Qualitative Research for the Practitioner Audience

One of us spent 12 years as editor of *Technical Communication*, the journal in our field with the largest circulation and the largest percentage (> 90%) of non-academic subscribers of the five major technical and professional communication journals. Because of the composition of that subscriber base, which consists overwhelmingly of people who do not have degrees in technical communication, and for whom the profession is a second or subsequent career, a common complaint heard from subscribers during Hayhoe's editorship was that articles reporting the results of quantitative research were difficult to understand because of the attention that the authors paid to statistical significance. This audience, usually trained in the humanities or social sciences, has no background in inferential statistics and doesn't appreciate the need for or the meaning of the statistical tests used to determine the reliability of the inferences made in such studies. In other words, they don't understand why they should care about such things.

On the other hand, many of the techniques used by qualitative researchers are essentially the same as some that technical communication practitioners employ in their own work. For example, in preparing the print and online user documentation for the latest release of a software product, practitioners might observe users working with the current release, conduct a focus group to identify problems with the current documentation, or analyze the current documents and compare them to the documents produced by competitors. The familiarity of these methods makes it easy for practitioners to understand the research design of a qualitative study, and the lack of inferential statistical analysis makes the qualitative research report less intimidating and more accessible.

It is impossible to know whether this familiarity and accessibility translate into greater appreciation for qualitative research in technical communication by most practitioners today. Nor do we want to suggest that quantitative studies should not be pursued. Research results of any kind are of relatively minor interest to practitioners who must attend to increasing workloads with decreasing resources. When you are doing the work that used to be done by several people, as is often the case for practitioners in today's economy, you don't have a lot of time to peruse journals to learn the latest research results. Furthermore, there are many contexts in which quantitative studies are extremely helpful. For example, they can help us compare the readability of information products or the retention of information presented in different ways.

All other things being equal, however, we can be certain that the familiarity and accessibility of qualitative research methods make studies that use those methods seem less foreign and easier to understand by technical communication practitioners who choose to read them. We can also be assured that as more practitioners enter the profession each year from academic programs in which they have been introduced to examples of qualitative research, and perhaps even to the methodologies that they use, these practitioners will likely be even more receptive to qualitative studies.

The Suitability of Qualitative Methods for Workplace Research

Although some qualitative research techniques are familiar to most technical communication practitioners because they use them to do preliminary investigations of audiences, processes, and products, very few practitioners pursue this "research" in a systematic way over an extended period, nor do they typically report their findings in any formal way, even internally to their organizations. Still, there is great potential for practitioners to undertake research projects that address workplace problems using qualitative techniques.

One of us taught a summer graduate course recently in methods of workplace research. Because the summer session was only 10 weeks long and the students were required to design, conduct, and report on a research project that addressed a concern within their organizations during that period, it wasn't surprising that all of the technical communication students in the class undertook a qualitative or mixed-methods research project. More notable, however, was the fact that two of the three engineers in the class did likewise, despite their suspicions about the use of social scientific methods rather than quantitative techniques.

Qualitative methods are highly suitable for workplace research for several reasons.

- A qualitative study doesn't require large numbers of participants to yield reliable results. Therefore, participant recruiting is easier and less time consuming.
- For people accustomed to working with texts and narratives (as most technical communicators are), a simple qualitative design can usually be executed more quickly—and its results can be analyzed more quickly—than a similarly simple quantitative design, simply because of the smaller number of participants required.
- Technical communication practitioners can easily partner on such studies with colleagues in marketing and tech support, for example, who may also be untrained in quantitative methods.
- Such studies can often be piggybacked onto related qualitative tasks such as audience or document analysis that are already being performed.
- Existing staff can generally be trained to perform and analyze the results of qualitative studies more quickly and easily than training them to perform statistical analysis, which may require resources external to the organization, and the purchase of expensive and complicated statistical computer programs.
- Practitioners can easily learn to communicate the results of this internal research in brief memo-style reports that follow the IMRAD model (introduction including literature review, methodology, results, and discussion).

Qualitative workplace studies can prove extremely valuable to an organization. Such a project could, for example, explore why one type of information product is preferred by users over another, or why it generates fewer calls to tech support. It could also reveal why one agile documentation method is more effective than another. As technical communicators are increasingly adding value to their organizations in new ways, the recommendations resulting from workplace research projects using qualitative methods offer a great way to make significant contributions to the company's bottom line.

The Use of Qualitative Methods to Explore Topics of Interest

We devoted a brief section of the introductions to chapters 4 through 14 in this volume to considering how the research in those chapters dealt with the major themes that Giammona uncovered in her article (Chapter 3):

- What is a technical communicator today?
- What forces are affecting the field?
- What is our future role in organizations?

- What should managers of technical communicators be concerned with?
- How do we contribute to innovation?
- How should we be educating future practitioners?
- What technologies are impacting on us?
- Where do we go from here?
- Does technical communication matter?

These are central questions that we must tackle to understand the current state of the profession as well as its future direction, and they are thus vital topics that future research must address. What is particularly notable is that each of these questions must be addressed at least in part using qualitative research methods. Indeed, although only two of these nine questions begin with "how" and none begin with "why," *why* and *how* are both questions underlying all of them. For example, we can't understand the forces affecting the field once they've been identified without knowing why they are important. Understanding the *why* and even identifying the *what* will require interviewing or conducting focus groups with those who are knowledgeable about those external forces, including experienced technical communicators, their managers, technology and marketing experts, corporate executives, and those who utilize their products and services.

Beyond the themes that Giammona uncovered in her surveys and interviews, there are many other areas of concern to technical communicators that can best be addressed using qualitative techniques, either alone or in combination with quantitative methods.

- How do the rhetorical conventions of languages and cultures differ?
- Why do some page designs work better than others?
- How do the visual rhetorics of different cultures affect the process of localization?
- Why do agile software developers sometimes resist developing documentation at the same time as the code?
- How can technical communicators whose native language is English write most effectively for translation?

All of these issues and many more as well need to be addressed so that our discipline becomes even more responsive to those who rely on us for clear and precise information products.

Conclusion

Will qualitative research studies overtake quantitative research in technical communication at some time in the future? We cannot make that prediction with any confidence, nor can we foresee whether it will ever increase beyond the approximately 38 percent of articles in our major journals that it accounted for in the period 2003 to 2007. What we can say, though, is that qualitative methods have proven extremely useful for exploring important issues that confront us as a profession, as well as very helpful for revealing how people produce discourse in workplace settings. And all of the premises that we have probed in the course of this chapter suggest that researchers in our field are not likely to abandon qualitative methods and may well use them more frequently in the future.

We hope that this anthology will assist those in our field to understand qualitative research in the same way that the work of Odell and Goswami (1985), Spilka (1993), and Redish and Ramey (1995) did, and particularly that it will assist students and practitioners in appreciating how qualitative studies can enrich not only their practice but also their professional lives.

References

Faigley, L. 1985. "Nonacademic Writing: The Social Perspective." In *Writing in Nonacademic Settings*, ed. L. Odell and D. Goswami. New York, NY: The Guilford Press.

Fisher, W. R. 1978. "Toward a Logic of Good Reasons." *The Quarterly Journal of Speech* 64: 376–84.

———. 1984. "Narration as a Human Communication Paradigm: The Case of Public Moral Argument." *Communication Monographs* 31: 1–22.

Giammona, B. 2004. "The Future of Technical Communication: How Innovation, Technology, Information Management, and Other Forces are Shaping the Future of the Profession." *Technical Communication* 51: 349–66. (Reprinted here as Chapter 3.)

Odell, L., and D. Goswami, eds. 1985. *Writing in Nonacademic Settings*. New York, NY: The Guilford Press.

Ramey, J. 1995. "What Technical Communicators Think About Measuring Value Added: Report on a Questionnaire." *Technical Communication* 42: 40–51.

Redish, J. 1995. "Adding Value as a Professional Technical Communicator." *Technical Communication* 42: 26–39.

Spilka, R. 1993. "Preface." In *Writing the Workplace: New Research Perspectives*, ed. Rachel Spilka. Carbondale, IL: South Illinois University Press.

Contributor Biographies

Ann M. Blakeslee earned her PhD in rhetoric in 1992 from Carnegie Mellon University. She is the author of *Interacting with Audiences: Social Influences on the Production of Scientific Writing* (Lawrence Erlbaum Associates, 2001), based on her dissertation research, and is co-author (with Cathy Fleischer) of *Becoming a Writing Researcher* (Lawrence Erlbaum, 2007). She has also published numerous book chapters and journal articles addressing disciplinary genres, writing for audiences, ethics in qualitative research, the state of research in technical communication, and the acquisition of disciplinary writing skills. She is professor for the Department of English Language and Literature at Eastern Michigan University, and she directs the Writing Across the Curriculum program there. She is also active in the Association of Teachers of Technical Writing and the Society for Technical Communication.

Lee-Ann Kastman Breuch is an associate professor in the Department of Writing Studies at the University of Minnesota, where she teaches courses in first-year writing, technical communication, computer pedagogy, teacher training, and usability testing. Her research addresses writing theory and pedagogy in technical disciplines, instructional technology, and evaluation of online environments.

Saul Carliner is an associate professor of educational technology at Concordia University in Montreal. His research focuses on the design and management of workplace learning and communications. He is a former president of the Society for Technical Communication.

Caroline M. Cole earned her PhD in rhetoric and composition, and is currently a lecturer at the University of California, Berkeley, where she teaches courses in business communication for the College Writing Programs and the Haas School of Business. She is also a writing consultant for businesses and industry professionals.

Theresa Conefrey completed her PhD in the Institute of Communications Research at the University of Illinois. Currently, she teaches graduate courses in technical communication in the Engineering Management program at Santa

Clara University. Her research addresses professionalization in STEM fields. She enjoys occasional consulting opportunities with Bay Area technology companies.

James Conklin is an assistant professor of applied human sciences at Concordia University in Montreal. His research focuses on the role of knowledge transfer and exchange in efforts to bring about change to social systems, and for five years he has led the formative and developmental evaluation of the Seniors Health Research Transfer Network in Ontario, Canada. He also has more than 25 years of experience as a consultant and manager in the fields of technical communication, organization development, and knowledge transfer. He was elected a Fellow of the Society for Technical Communication in 2002. He has published articles and conference papers on topics in technical communication, social learning, and organization effectiveness. He holds a PhD from Concordia University.

Debbie Davy is an IEEE senior member, a peer reviewer for the IEEE Transactions on Professional Communication, a graduate of Mercer University's master of science program in technical communication management, and a student in the PhD program at Texas Tech University. She is also a professional technical communicator and practice leader of Mastertechwriter and DK Consultants, niche consulting companies based in Toronto, Canada. Her professional practice provides compliance, technical communication, and business analysis services to organizations in the private, public, and broader public sectors.

David Dayton taught technical communication from 1994 to 2010 at three different universities, most recently at Towson University in Maryland. He was the principal investigator on a multimodal research project to study the use and impacts of single sourcing and content management in technical communication. The first-hand reports Web site was conceived as part of that project, which received grant funding from the Society for Technical Communication. He now works as a communications analyst for the Government Accountability Office in Washington, DC.

Menno de Jong is professor of communication studies at the University of Twente, Enschede, the Netherlands, and the editor of *Technical Communication*. His main research interest concerns the methodology of applied communication research.

Linda Phillips Driskill is a professor of English and management communications at Rice University. Her research areas include engineering and professional communication, academic writing, international communication, and writing in the disciplines.

Barbara A. Giammona has managed technical communicators for over 20 years in a variety of industries, including medical products, business software, financial services, and manufacturing. She holds a BA in English from the

University of California, Irvine, and an MS in the management of technology from the Polytechnic Institute of NYU.

Hillary Hart is distinguished senior lecturer in the Department of Civil Engineering at the University of Texas at Austin. A Fellow of the Society for Technical Communication, she is the 2010–2011 vice-president of STC and will be president in 2011–2012. Her funded research focuses on engineering ethics, the social construction of technology, and environmental communication.

George F. Hayhoe spent 18 years as a technical communicator for Du Pont and Westinghouse, as well as for several smaller companies. He has taught in the master's and doctoral programs in technical communication at Utah State University and East Carolina University, and is currently professor of technical communication and director of the master of science program in technical communication management at Mercer University. A fellow of the Society for Technical Communication since 1997, he was editor of its journal, *Technical Communication*, from 1996 to 2008. He is also a senior member of the Institute of Electrical and Electronics Engineers and the IEEE Professional Communication Society, and is a past president of that society. He is the co-author of *A Research Primer for Technical Communication* and the co-editor of *Connecting People with Technology*. He has also published book chapters, articles, and conference papers on a variety of topics in technical and professional communication. He holds a PhD from the University of South Carolina.

Michael Hughes has a PhD in instructional technology from the University of Georgia and a master's in technical and professional communication from Southern Polytechnic State University. He is co-author of *A Research Primer for Technical Communication* and works for IBM Internet Security Systems as a senior user experience architect. His professional and research focus is designing user interfaces that accommodate the user as learner.

Leo Lentz is an associate professor of communication studies at Utrecht University in the Netherlands. Web site usability and document design are the main focuses of his research.

Scott A. Mogull is an assistant professor at Clemson University and the development editor of *Technical Communication Quarterly*. His primary research interest is technical communication in the pharmaceutical/biotechnology industry. He has a PhD in Technical Communication and Rhetoric from Texas Tech University and master's degrees in technical communication from the University of Washington and in Microbiology from the University of Texas at Austin. A former scientist, he has published in *Infection & Immunity*, the most-cited journal in infectious diseases and the third most-cited journal in immunology.

Tiffany Craft Portewig is currently a consultant for a performance improvement firm and serves as the book review editor for IEEE Transactions on Professional

Communication. She has published in *Technical Communication* and the *Journal of Technical Writing and Communication*, and has taught at Auburn University and Texas Tech University. She received her PhD in technical communication and rhetoric from Texas Tech University.

Tom Reeves is professor emeritus of learning, design, and technology at the University of Georgia. Since earning his PhD at Syracuse University, he has developed and evaluated numerous interactive learning programs for education and training. He is a former Fulbright lecturer and former editor of the *Journal of Interactive Learning Research*. His research interests include evaluation of educational technology, design-based research, authentic learning environments, and educational technology in developing countries. In 2003, he became the first Fellow of the Association for the Advancement of Computing in Education.

Marika Seigel is an assistant professor of rhetoric and technical communication at Michigan Technological University, where she teaches undergraduate and graduate courses in technical communication to students from a variety of disciplines. In addition to technical communication, her research interests include usability, gender studies, and rhetorics of science and technology. She received her bachelor's degree in English from the University of Michigan, and her master's and PhD in English (with a focus in rhetoric and composition) from Penn State University.

Rachel Spilka is associate professor at the University of Wisconsin-Milwaukee. She is editor of *Writing in the Workplace: New Research Perspectives* (SIUP, 1993) and *Digital Literacy for Technical Communication: 21st Century Theory and Practice* (Routledge, 2010), and co-editor (with Barbara Mirel) of *Reshaping Technical Communication: New Directions and Challenges for the 21st Century* (Erlbaum, 2002). She has also managed STC's Research Grants Committee and Ken Rainey Excellence in Research Award Committee.

Patricia Sullivan is a professor of English and director of the graduate program in rhetoric and composition at Purdue University. She is currently working on a pedagogical history of engineering writing.

Christopher Thacker holds both a master of arts in professional studies and a master of science in professional writing from Towson University. He is the managing partner of Insubordination Records of Columbia, Maryland, and maintains the company's content-managed Web site. He is also a full-time instructor in the Business Excellence program of the College of Business and Economics at Towson University.

Christina Valecillos is a senior communications representative for Lockheed Martin in Orlando, Florida. She is a graduate of the Mercer University master of science program in technical communication management.

Thomas Vosecky is a PhD candidate in rhetoric and technical communication at Michigan Technological University, where he is involved in the development of a writing center for the MBA program. His research interests include the ancient Greek concept of techné (the capacity to make), research methods, and the case study as a means of simulating practical experience in the classroom. He received a bachelor's degree in psychology from the University of Minnesota, Minneapolis, a master's degree in rhetoric and technical communication from Michigan Tech, and Automotive Service Excellence automobile technician certification.

Charles Wallace is an associate professor of computer science at Michigan Technological University. He has been involved in the undergraduate Software Engineering degree program at Michigan Tech since its inception in 2004. His research and teaching interests lie in software requirements, documentation, verification, and usability. He holds a bachelor's degree in linguistics from the University of Pennsylvania, a master's degree in linguistics from the University of California, Santa Cruz, and a doctorate in computer science and engineering from the University of Michigan.

Julie Zeleznik Watts is an associate professor in the English and philosophy department at the University of Wisconsin-Stout. She is program director for the online MS in technical and professional communication and teaches courses in composition, document design, and technical writing. Her research addresses program assessment in technical communication and student learning in learning communities.

Marieke Welle Donker-Kuijer is a PhD candidate at the University of Twente, Enschede, the Netherlands. Her PhD research focuses on the methodology of evaluating informative Web sites; more specifically on the merits and drawbacks of heuristic expert evaluation.

Russell Willerton is an associate professor at Boise State University, where he teaches undergraduate and graduate courses in technical communication. His case study about work at a health information company appeared in a special issue of *Technical Communication Quarterly*. He is a senior member and an officer in STC's Snake River chapter.

Index

eBooks – at www.eBookstore.tandf.co.uk

A library at your fingertips!

eBooks are electronic versions of printed books. You can store them on your PC/laptop or browse them online.

They have advantages for anyone needing rapid access to a wide variety of published, copyright information.

eBooks can help your research by enabling you to bookmark chapters, annotate text and use instant searches to find specific words or phrases. Several eBook files would fit on even a small laptop or PDA.

NEW: Save money by eSubscribing: cheap, online access to any eBook for as long as you need it.

Annual subscription packages

We now offer special low-cost bulk subscriptions to packages of eBooks in certain subject areas. These are available to libraries or to individuals.

For more information please contact webmaster.ebooks@tandf.co.uk

We're continually developing the eBook concept, so keep up to date by visiting the website.

www.eBookstore.tandf.co.uk

Milton Keynes UK
Ingram Content Group UK Ltd.
UKHW031348071024
449327UK00033B/3049